Modern EMC Analysis Techniques

Volume I: Time-Domain Computational Schemes

Modern EMC Analysis Techniques. Volume I: Time-Domain Computational Schemes
Nikolaos V. Kantartzis and Theodoros D. Tsiboukis

ISBN: 978-3-031-00577-0   paperback
ISBN: 978-3-031-01705-6   ebook

DOI: 10.1007/978-3-031-01705-6

A Publication in the Springer series

SYNTHESIS LECTURES ON COMPUTATIONAL ELECTROMAGNETICS #21

Lecture #21

Series Editor: Constantine A. Balanis, Arizona State University

Series ISSN
ISSN 1932-1252   print
ISSN 1932-1716   electronic

Modern EMC Analysis Techniques

Volume I: Time–Domain Computational Schemes

Nikolaos V. Kantartzis and Theodoros D. Tsiboukis
Aristotle University of Thessaloniki

SYNTHESIS LECTURES ON COMPUTATIONAL ELECTROMAGNETICS #21

ABSTRACT

The objective of this two-volume book is the systematic and comprehensive description of the most competitive time-domain computational methods for the efficient modeling and accurate solution of contemporary real-world EMC problems. Intended to be self-contained, it performs a detailed presentation of all well-known algorithms, elucidating on their merits or weaknesses, and accompanies the theoretical content with a variety of applications. Outlining the present volume, the analysis covers the theory of the finite-difference time-domain, the transmission-line matrix/modeling, and the finite integration technique. Moreover, alternative schemes, such as the finite-element, the finite-volume, the multiresolution time-domain methods and many others, are presented, while particular attention is drawn to hybrid approaches. To this aim, the general aspects for the correct implementation of the previous algorithms are also exemplified. At the end of every section, an elaborate reference on the prominent pros and possible cons, always in the light of EMC modeling, assists the reader to retrieve the gist of each formulation and decide on his/her best possible selection according to the problem under investigation.

KEYWORDS

electromagnetic compatibility (EMC), time-domain methods, computational electromagnetics

Preface

In writing a book on contemporary electromagnetic compatibility (EMC) analysis techniques, the authors are well aware that they are delving into a field of challenging innovations pertinent to a large readership ranging from students and academics to engineers and seasoned professionals. Nowadays, EMC technology is more pervasive than ever in many educational, social, industrial, and commercial sectors. Essentially, the widely recognized significance of EMC problems and applications has turned the interest of the scientific community toward their in-depth investigation, with an emphasis on the simulation of the relevant electromagnetic phenomena via highly advanced time-domain methodologies. This is exactly the aim of the two-volume book, which basically reflects the outgrowth of a systematic research performed by the authors for almost a 10-year period in the area of electromagnetic compatibility/interference measurement and modeling. Intended to be self-contained from the computational perspective, the book performs a detailed presentation of most time-domain algorithms, elucidating on their merits or weaknesses, and accompanies the theoretical content with a variety of real-world applications. Thus, having acquired the necessary evidence for every numerical approach, the reader is then free to decide on the best possible scheme for his/her requirements.

Outlining the first volume, devoted to the theoretical establishment of time-domain methods in terms of EMC analysis, Chapter 1 highlights the key issues for their selection as the basic toolkit and gives a brief description along with their most important categories. Thus, apart from the three well-known fundamental formulations, a record of several alternative approaches is also provided. After this preparatory work, Chapter 2 covers the theory of the former concepts, namely, the finite-difference time-domain, the transmission-line matrix/modeling, and the finite integration techniques. Assuming only some mathematics background, each algorithm is examined from different viewpoints concerning its discretization strategy, conditional stability, and convergence rate. Moreover, at the end of every section, an elaborate reference on the prominent pros and possible cons, always in the light of EMC modeling, helps the reader to get the gist of the specific method. On the other hand, all new schemes are illustrated in Chapter 3. Among them, one can discern the finite-element, the finite-volume, the multiresolution, the alternating-direction implicit finite-difference, or the pseudospectral time-domain methods, and particular attention is drawn to higher-order realizations and nonstandard spatial/temporal operators. The last section of the chapter is devoted to hybrid approaches, as a flexible means with combined computational power and adjustable features. Furthermore, the general aspects for the correct implementation of the previous algorithms are included in Chapter 4, where the most significant programming details and modeling hints receive the proper discussion. In this context, a lot of meaningful abstractions related to the excitation procedure, the lattice reflection-error suppression, the absorbing boundary conditions, the curvilinear tessellations, the frequency-dependence of constitutive parameters, and the surface impedance boundary conditions, can be pursued therein.

To bridge the gap between a cursory undergraduate and a formidable specialist's handbook, this book contains a variety of simple EMC examples, so that the core of each method is appreciated without being overwhelmed by cumbersome problems. It is stressed that the interested reader can take avail of the extensive reference list at the end of each chapter to obtain the appropriate information.

Finally, the authors thank Dr. T. T. Zygiridis for his thorough proofreading and valuable suggestions during the preparation of the manuscript. Above all, they do anticipate that the theoretical formulations and numerical results provided in both volumes will inspire the reader to expand its material beyond the stated purpose of developing robust EMC models in the time-domain and help him/her to acquire a better knowledge on the constant evolutions in this pivotal scientific area.

Thessaloniki   **N. V. Kantartzis**
December 2007   **T. D. Tsiboukis**

Contents

CHAPTER 1

Introduction

1.1 ADVANCES AND REQUIREMENTS IN MODERN EMC PROBLEMS

A significant aspect of contemporary electronic equipment is the customarily unpredicted production of electromagnetic energy, which has the potential of imposing redundant interference to other electrical appliances. Because such devices constitute an indispensable part of everyday life—ranging from communications and transportation to life-support medical systems and entertainment—any effort toward the classification of their radiated emissions has turned to a growing prerequisite. As a matter of fact, this deterioration in a device's performance induced by an electromagnetic disturbance is generally designated as *electromagnetic interference (EMI)* and refers, principally (although not exclusively), to radio-frequency signals in the nature of wideband noise or change in the propagation medium itself. In essence, EMI is intimately related to the systematic control of *electromagnetic compatibility (EMC)*, i.e., the competence of an electrical apparatus to operate in its prescribed electromagnetic environment within balanced thresholds of safety and efficiency, without suffering from or triggering intolerable degradation as an upshot of unwanted disruptions. Apparently, the proper investigation of a structure's EMC attitude substantiates its profitable function, eliminates serious reliability hazards, and contributes to the fulfillment of key fabrication standards.

The assortment of EMC/EMI interactions may be divided into two distinct types according to the traveling path followed by the transmitted energy. Particularly, the first kind occurs because of the degeneration of electromagnetic waves along the propagation direction and contains cases of spurious coupling among the circuits of an integrated board as well as field coupling between interior components. On the other hand, the second type is separated into two smaller categories: emissions and immunity. The former are attributed to the periodicity of several wave fronts that should be carefully confined to isolate the remaining parts of the appliance, whereas the latter concentrates on propagating fields that are transferred in the structure of intercst.

The fact that EMC/EMI processes are acknowledged as a problem of critical concern in the design and manufacture of highly skilled and state-of-the-art devices is justified by the serious theoretical and experimental research performed nowadays [1–7]. Actually, the rapid evolution of electrical and electronics engineering has led to major advances in the area of information technology, digital computing, telecommunications, and instrumentation. Furthermore, the ever-augmenting demands for additional services or capacities in larger frequency spectra continue to pose novel and more challenging requirements with sizeable testing and fabrication costs. Some indicative applications that entail meticulous movements during their initial perception and construction include high-speed printed circuit boards (PCBs), smart antennas, general-purpose radio-frequency microwave devices, biomedical appliances, monolithic circuitry and electronic packaging interconnects, avionics, vehicle-based communication systems, wideband measurement/test facilities, wireless components, mobile telephony, shielding equipment, chiral media together with the more recent double-negative metamaterials, nanoelectromechanical systems (NEMSs), and optical configurations. Nonetheless, as sophisticated as the previous structures might be, they are *still* susceptible to electromagnetic noise and the most important: they are vulnerable to external sources of interference. Hence, their elaborate study seems more mandatory than ever because of the enhanced accuracy, reliability, and compliance with emission/immunity protection stipulations.

Taking into consideration these deductions, it becomes obvious that EMI mitigation approaches that satisfy the acceptable specifications have departed from the traditional trial-and-error practices because EMC has developed into a very interdisciplinary subject. In fact, it is hardly affordable for almost any manufacturer to increase their overall budget or deteriorate their market reputation through hasty attempts, without first determining the principal design directions. The ideal period to estimate all these aspects of EMC is during the preliminary sketch cycle, long before the first item is brought to production. This is exactly where the field of *computational electromagnetics (CEM)* can offer its invaluable qualifications. Having matured quite a lot, CEM involves a number of powerful methodologies [8–12] capable of solving realistic engineering problems. Its prevalent recognition is supported by the huge strides in computing systems both in CPU speed and memory sizes. Moreover, most CEM algorithms keep on improving via intelligently established concepts and robust mathematical models, through which they try to suppress inherent discretization errors or ameliorate stability and convergence weaknesses. In this manner, scientific research is enriched with very significant results, before any costly fabrication procedure, that enable the correct tuning of all parameters, accomplish instructive physical interpretations, and offer attractive opportunities for innovations.

1.2   NECESSITY OF ANALYZING EMC APPLICATIONS IN THE TIME-DOMAIN

Depending on the nature of the application under study, the kernel of a CEM solver can be formulated either in the frequency or in the time domain. Frequency-domain methods for the harmonic steady-state Maxwell's laws have been the initial representatives of numerical modeling in the

EMC field [8, 9]. Based on asymptotic and integral-like notions, they provide acceptable outcomes for diverse situations, especially for the characterization of scattering properties, the treatment of eigenvalue problems, and the handling of relatively involved geometries or media inhomogeneities. However, owing to their implicit profile, they require the solution of a system of equations, whose order becomes prohibitively large in the case of modern EMC structures, a fact that may lead to cumbersome or even unfeasible implementations. Not to mention their single-frequency orientation that comes into opposition to the broadband needs of recent design standards.

Conversely, time-domain techniques deal with the direct solution of Maxwell's time-dependent curl equations without engaging variational functionals, potentials, or weighted residuals [10–12]. For this aim, they sample electric and magnetic field vectors according to a volumetric pattern over a temporal frame, paying attention to the suitable choice of sampling parameters to avoid nonphysical oscillations via the proper stability criteria. Of notable interest is their update mechanism that lies on a fully explicit marching-on-in-time integration for the calculation of each unknown component. Therefore, no system of equations is generated or needs to be solved, and the resulting simulations are very easy to configure; merely a set of stencils linking adjacent nodal values. Moreover, spatial discretization in time-domain schemes allocates field quantities on a structured uncollocated tessellation with a subwavelength resolution tailored to pertinently model the frequency content of the excitation scheme without violating the consistency of the arrangement. To the preceding advantages, one must add the realization simplicity of most methods and the straightforward handling of different media compositions, such as frequency-dependent or anisotropic ones. Also, it is imperative to stress their truly wideband behavior attained by the impressive variety of excitation forms that can be supported. Regarding the termination of open-boundary applications, time-domain techniques can be combined with advanced absorbing boundary conditions, thus restricting the total domain to minimum dimensions. In contrast, some nonnegligible defects, such as the somewhat inflexible meshing or the appearance of artificial discretization errors, tend to be surpassed in a viable level thanks to sound theoretical works on effective hybridizations with other approaches or nonorthogonal variants of the main algorithms.

From the above assets, the demanding stipulations of contemporary problems, and the promising future of EMC technology, it turns out that time-domain numerical modeling can grant significant aid toward the accurate manipulation of many real-world cases in an extended frequency range. Indeed, the principal characteristics of explicit solvers seem to be ideal for the large electrical sizes, the complex geometries, the fine details, the multifrequency environments, the irregular boundaries, the tangential/radial singularities, or the prominent constitutive parameter variations encountered in EMC devices. Essentially, given the interleaved integration profile and the advanced stencil management aptitudes of time-domain methods, Maxwell's curl equations can now be rigorously transformed to algebraic links between field vectors, from which the corresponding differential operators are readily extracted [13, 14]. As a natural consequence, lattice self-consistency is guaranteed and the well-posed representation of arbitrary discontinuities is adequately facilitated.

1.3   BACKGROUND AND MAIN CATEGORIES OF TIME-DOMAIN METHODS

Albeit the era of time-domain modeling began in the decade of 1960s, most of the methods belonging to this category remained in relative obscurity for several years. This can be attributed to algorithmic implementation reasons (e.g., termination of infinite domains, treatment of inhomogeneous interfaces) and some practical issues with an emphasis on the development of computer systems. Because most of the time-domain techniques entail high-speed processing power and considerable amounts of memory, it becomes evident why they were not recognized by the research community as soon as they appeared. However, the tremendous growth of computing machines, in conjunction with the presentation of diverse efficient boundary termination conditions, gave a real thrust to their acknowledgment, which has led to the construction of commercial software packages for the general use of time-domain CEM concepts in applied EMC problems.

1.3.1   Principal Techniques

Based on the requirements of EMC analysis as observed in the analogous literature, the most popular representatives of the category are: the *finite-difference time-domain (FDTD)* method [15–17], the *finite integration technique (FIT)* [18], and the *transmission-line matrix/modeling (TLM)* algorithm [19–21]. Specifically, the FDTD approach creates a robust connection of field components on a mesh staggered in space and time. So for spatial sampling, electric quantities, placed on cell edges, are enclosed by four circulating magnetic quantities, located on cell faces and vice versa. A similar idea holds for the time update as well, where the so-called leapfrog scheme computes the unknown variables via data from the opposite field at an intermediate time instant. Despite its simplicity, this setup is meant to be crucial because it complies with the rules for optimal discretization and offers a natural contour mapping of Ampère's and Faraday's integral equations. On the contrary, FIT delves into the integral form of Maxwell's laws, rather than the differential one, and involves six components of electric and magnetic voltages/fluxes on a dual grid. Particularly, voltages are associated to contours, whereas fluxes to surfaces, hence allowing the generation of a matrix system, suitable for arbitrary curvatures and very fine geometrical details. In this way, numerical simulations, founded on integral balances, are proven to be stable and energy conserving, without inducing vector artifacts. Adopting a different space-time rationale, the TLM technique approximates electromagnetic waves by scattered and propagating pulses, placed at prefixed nodes of the elementary cells. This realization, being similar to the multiport network gist, defines field quantities by means of reference planes, with the wave front amplitudes perpendicular to them. On each plane, isolating two cells, a sampling node for the tangential electric and magnetic component is defined, whereas one individual network port is associated to every polarization for supplementary efficiency at media interfaces.

1.3.2 Enhanced Algorithms

As already depicted, EMC is a very active scientific area with many emerging applications and, therefore, prone to constant changes in terms of complexity and performance. Although the three aforementioned methodologies have proven their worth in the treatment of numerous cases,[1] the latest progress has elevated the existing thresholds of accuracy and reliability. To follow these stimulating advances, novel time-domain schemes appeared either in the form of proficient combinations or hybrid approaches. Among them, one can possibly distinguish the *finite-element time-domain (FETD)* algorithm [22–26] for the handling of highly irregular geometries through the use of adaptive meshes and the *finite-volume time-domain (FVTD)* technique [27] for generalized curvilinear arrangements. In addition, abrupt field fluctuations can be more drastically confronted by the *multiresolution time-domain (MRTD)* method [28], whereas the *pseudospectral time-domain (PSTD)* algorithm [29] provides considerably reduced lattice resolutions for periodic structures. On the other hand, important reductions of the total simulation time are performed by the *alternating-direction implicit (ADI) FDTD* approach [30, 31]. Also, diminution of the serious, and always misleading, rudimentary dispersion-error mechanisms is achieved by the *nonstandard FDTD* method or other *higher-order* counterparts [32]. Finally, it is noteworthy to indicate that the above formulations can be, under the appropriate modeling circumstances, combined with the three main time-domain techniques to yield one compact tool with problem-adjustable competences.

REFERENCES

1. C. Christopoulos, *Principles and Techniques of Electromagnetic Compatibility*. Boca Raton, FL: CRC Press, 1995.
2. P. Kodali, *Engineering Electromagnetic Compatibility*, 2nd ed. New York: IEEE Press, 2001.
3. C. R. Paul, *Introduction to Electromagnetic Compatibility*, 2nd ed. New York: Wiley, 2006.
4. F. M. Tesche, M. V. Ianoz, and T. Karlsson, *EMC Analysis Methods and Computational Models*. New York: Wiley Interscience, 1997.
5. M. I. Montrose and E. D. Nakauchi, *Testing for EMC Compliance—Approaches and Techniques*. New York: IEEE Press and Wiley Interscience, 2004.

---

[1] For an elaborate investigation of all computational methods, described in this book, via many realistic EMC configurations, extensive numerical results, and comparisons with reference/measurement data, the reader is referred to its second volume, *Modern EMC Analysis Techniques: Models and Applications*, of the same authors and publisher.

6. T. Williams, *EMC for Product Designers*, 4th ed. Oxford, UK: Newnes, 2007.

7. C. A. Balanis, *Antenna Theory: Analysis and Design*, 3rd ed. New York: IEEE Press and Wiley Interscience, 2005.

8. A. F. Peterson, S. L. Ray, and R. Mittra, *Computational Methods for Electromagnetics*. Piscataway, NJ: IEEE Press and Oxford University Press, 1998.

9. W. C. Chew, J.-M. Jin, E. Michielssen, and J. Song, *Fast and Efficient Algorithms in Computational Electromagnetics*. Boston, MA: Artech House, 2001.

10. F. L. Teixeira, Ed., *Geometric Methods for Computational Electromagnetics (Progress in Electromagnetics Research Series—PIER 32)*. Cambridge, MA: EMW Publishing, 2001.

11. D. B. Davidson, *Computational Electromagnetics for RF and Microwave Engineering*. Cambridge, UK: Cambridge University Press, 2005.

12. H.-D. Brüns, C. Schuster, and H. Singer, "Numerical electromagnetic field analysis for EMC problems," *IEEE Trans. Electromagn. Compat.*, vol. 49, no. 2, pp. 253–262, May 2007.

13. B. Gustafsson, H.-O. Kreiss, and J. Oliger, *Time Dependent Problems and Difference Methods*. New York: Wiley, 1995.

14. J. G. Van Bladel, *Electromagnetic Fields*. New York: IEEE Press and Wiley Interscience, 2007.

15. K. S. Yee, "Numerical solution of initial boundary value problems involving Maxwell's equations in isotropic media," *IEEE Trans. Antennas Propagat.*, vol. AP-14, no. 3, pp. 302–307, May 1966.

16. K. S. Kunz and R. J. Luebbers, *The Finite Difference Time Domain Method for Electromagnetics*. Boca Raton, FL: CRC Press, 1993.

17. A. Taflove and S. C. Hagness, *Computational Electrodynamics: The Finite-Difference Time-Domain Method*, 3rd ed. Norwood, MA: Artech House, 2005.

18. T. Weiland, "A discretization method for the solution of Maxwell's equations for six-component fields," *Electron. Commun. (AEÜ)*, vol. 31, no. 3, pp. 116–120, 1977.

19. W. J. R. Hoefer, "The transmission line matrix (TLM) method," in *Numerical Techniques for Microwave and Millimeter Wave Passive Structures*, T. Itoh, Ed. New York: Wiley, 1989, pp. 496–451.

20. M. Krumpholz and P. Russer, "A field theoretical derivation of TLM," *IEEE Trans. Microwave Theory Tech.*, vol. 42, no. 9, pp. 1660–1668, Sept. 1994, doi:10.1109/22.310559.

21. C. Christopoulos, *The Transmission-Line Modeling Method: TLM*. New York: IEEE Press, 1995.

22. J.-F. Lee, R. Lee, and A. C. Cangellaris, "Time-domain finite-element methods," *IEEE Trans. Antennas Propagat.*, vol. 45, no. 3, pp. 430–442, Mar. 1997.

23. J.-M. Jin, *The Finite Element Method in Electromagnetics*, 2nd ed. New York: Wiley, 2002.

24. J. Volakis, A. Chatterjee, and L. Kempel, *Finite Element Method for Electromagnetics: Antennas, Microwave Circuits, and Scattering Applications*. Piscataway, NJ: IEEE Press and Oxford University Press, 1998.

25. Y. Zhu and A. C. Cangellaris, *Multigrid Finite Element Methods for Electromagnetic Field Modeling*. New York: IEEE Press and Wiley Interscience, 2006.

26. A. C. Polycarpou, *Introduction to the Finite Element Method in Electromagnetics*. San Rafael, CA: Morgan & Claypool Publishers, 2006.

27. N. K. Madsen and R. W. Ziolkowski, "A three-dimensional modified finite volume technique for Maxwell's equations," *Electromagnetics*, vol. 10, pp. 147–161, 1990.

28. N. Bushyager and M. Tentzeris, *MRTD (Multi Resolution Time Domain) Method in Electromagnetics*. San Rafael, CA: Morgan & Claypool Publishers, 2005, doi:10.2200/S00009ED1V01Y200508 CEM002.

29. Q. H. Liu, "The PSTD algorithm: A time-domain method requiring only two cells per wavelength," *Microwave Opt. Technol. Lett.*, vol. 15, no. 3, pp. 158–165, 1997, doi:10.1002/(SICI)1098-2760(19970620)15:3<158::AID-MOP11>3.0.CO;2-3.

30. T. Namiki, "A new FDTD algorithm based on ADI method," *IEEE Trans. Microwave Theory Tech.*, vol. 47, no. 10, pp. 2003–2007, Oct. 1999.

31. F. Zheng, Z. Chen, and J. Zhang, "A finite-difference time-domain method without the Courant stability conditions," *IEEE Microwave Guided Wave Lett.*, vol. 9, no. 11, pp. 441–443, Nov. 1999.

32. N. V. Kantartzis and T. D. Tsiboukis, *Higher-Order FDTD Schemes for Waveguide and Antenna Structures*. San Rafael, CA: Morgan & Claypool Publishers, 2006, doi:10.2200/S00018ED1V01Y20060 4CEM003.

· · · ·

CHAPTER 2

Fundamental Time-Domain Methodologies for EMC Analysis

2.1 INTRODUCTION

The widespread adoption of time-domain numerical methods to complement the ordinary means of analysis in modern EMC/EMI problems is deeply related to the remarkable evolution of computing and storage systems as well as to the persistent need for precise and affordable models, before any fabrication process. Combining the accomplishment of reliable full-wave solutions with the extraction of insightful interpretations, these advanced tools have grown to the point where they comprise an indispensable part of engineering practice and design. Hence, in this chapter, the focus falls on the three principal representatives of the class, namely, the *finite-difference time-domain (FDTD)* algorithm [1–5], the *transmission line matrix/modeling (TLM)* method [6–9], and the *finite integration technique (FIT)* [10, 11]. Apart from their different topological and realization perspectives, all of them share the key concept of advancing a set of unknown electromagnetic properties along a purely explicit way.

In particular, the first algorithm approximates the differential operators in Maxwell's curl equations on a staggered mesh in space and establishes an interleaved temporal marching with field quantities being offset by half a spatial increment to each other. The overall approach uses second-order finite-difference approximations to achieve second-order accuracy, whereas its stability is guaranteed by certain mathematical conditions. On the other hand, the second method constructs an equivalent array of short transmission lines, instead of directly discretizing Maxwell's equations, by means of a systematic boundary mapping between the wave amplitudes and the electric/magnetic constituents in a properly generated node-based grid. Finally, the third technique performs a consistent discretization of Ampère's and Faraday's laws in their integral version. The resulting matrix equations exhibit conservation competences with regard to energy and charge of the discrete formulation, thus allowing a one-to-one involvement of the underlying measurable quantities.

In this context, the basic aim of the chapter is to provide the fundamental background of the aforesaid methodologies and shed light to the derivation of their most important renditions with an emphasis on the treatment of realistic EMC applications. The discrete models, so constructed, are thoroughly examined, and several critical issues concerning prominent simulation capabilities or inherent drawbacks are exemplified. To this end, the description of each algorithm is accompanied by a brief, yet careful, review of its practical aspects, computational overhead, and potential extensions that offer instructive guidelines for research.

2.2   THE FDTD METHOD

The FDTD method was introduced in 1966 [1] and since then sprang to prominence as the most powerful and trustworthy numerical tool for the treatment of time-dependent Maxwell's [2–5, 12–16]. Taking avail of the constant computer developments, its concepts have experienced a prevalent acceptance in EMC modeling and exhibited many competences for various categories of problems [17–33]. The origins and the strength of the FDTD method lie on the extraction of a mathematically simple and elegant process for the explicit treatment of wave propagation in a 3-D lossless medium. Using field theory and employing no potentials or variational functionals, the algorithm establishes a well-posed field assignment on a grid staggered in time and space. The properties of the resulting scheme are then subjected to analysis to establish the well-posedness of the quantitative setup. The latter issue implied, nonetheless, an additional step after the modeling stage, namely, the reformulation of the problem in discrete terms as a finite set of algebraic expressions that are more suitable to the philosophy of modern computing machines. This is exactly where the essential abstraction of FDTD technique has been launched, where second-order accurate central-difference approximants are directly applied to the differential operators of Ampère's and Faraday's laws. Consequently, the method attains a significant sampled-data decrease of continuous fields in a discretized volume over a period. In this way, electric and magnetic components are spatially interleaved to permit a natural satisfaction of tangential conditions at media interfaces, whereas the explicit character of the method evades huge systems and enables the accurate investigation of almost every realistic EMC/EMI problem.

2.2.1   Grid-Staggering and Leapfrog Time-Stepping

Let us consider a source-free region of space where all field vectors are continuous and single-valued functions of position and time. Then, the time domain—differential, and integral—equations of Ampère's and Faraday's laws (curl equations) receive the following form

$$\nabla \times \mathbf{H} = \mathbf{J}_c + \frac{\partial \mathbf{D}}{\partial t} \xrightarrow{\text{Integral form}} \oint_C \mathbf{H} \cdot d\mathbf{l} = \iint_S \mathbf{J}_c \cdot d\mathbf{s} + \frac{\partial}{\partial t} \iint_S \mathbf{D} \cdot d\mathbf{s} \,, \qquad (2.1)$$

$$\nabla \times \mathbf{E} = -\mathbf{M}_c - \frac{\partial \mathbf{B}}{\partial t} \xrightarrow{\text{Integral form}} \oint_C \mathbf{E} \cdot \mathbf{dl} = -\iint_S \mathbf{M}_c \cdot \mathbf{ds} - \frac{\partial}{\partial t} \iint_S \mathbf{B} \cdot \mathbf{ds}, \qquad (2.2)$$

where

$\mathbf{E}$	electric field intensity	$\mathbf{B}$	magnetic flux density
$\mathbf{D}$	electric flux density	$\mathbf{H}$	magnetic field intensity
$\mathbf{J}_c$	conduction electric current density	$\mathbf{M}_c$	conduction magnetic current density

Moreover, S is an arbitrary surface that has the closed contour C as its boundary. Likewise, electric/magnetic Gauss' laws become

$$\nabla \cdot \mathbf{D} = 0 \xrightarrow{\text{Integral form}} \oiint_S \mathbf{D} \cdot \mathbf{ds} = 0, \qquad (2.3)$$

$$\nabla \cdot \mathbf{B} = 0 \xrightarrow{\text{Integral form}} \oiint_S \mathbf{B} \cdot \mathbf{ds} = 0 \qquad (2.4)$$

For the aims of the analysis, assume that the media are linear, isotropic, time-invariant, and non-dispersive, so simplifying the constitutive relations between fluxes ($\mathbf{D}$ and $\mathbf{B}$) and field intensities ($\mathbf{E}$ and $\mathbf{H}$), as

$$\mathbf{D} = \varepsilon \mathbf{E} \text{ and } \mathbf{B} = \mu \mathbf{H} \qquad (2.5)$$

To complete the set of Maxwell's equations, electric and magnetic conduction current densities, which are responsible for the corresponding loss mechanisms in the media via the analogous currents, are given by

$$\mathbf{J}_c = \sigma \mathbf{E} \text{ and } \mathbf{M}_c = \rho' \mathbf{H} \qquad (2.6)$$

Taking into account the properties of (2.5) and (2.6) and substituting them into (2.1) and (2.2), one can easily derive a set of six coupled 3-D partial differential equations, each involving the appropriate field components along the axes of a Cartesian coordinate system (x,y,z). So,

$$\frac{1}{\varepsilon} \left(\frac{\partial H_z}{\partial y} - \frac{\partial H_y}{\partial z} - \sigma E_x \right) = \frac{\partial E_x}{\partial t}, \qquad (2.7a)$$

$$\frac{1}{\varepsilon} \left(\frac{\partial H_x}{\partial z} - \frac{\partial H_z}{\partial x} - \sigma E_y \right) = \frac{\partial E_y}{\partial t}, \qquad (2.7b)$$

$$\frac{1}{\varepsilon}\left(\frac{\partial H_y}{\partial x} - \frac{\partial H_x}{\partial y} - \sigma E_z\right) = \frac{\partial E_z}{\partial t}, \qquad (2.7c)$$

$$\frac{1}{\mu}\left(\frac{\partial E_y}{\partial z} - \frac{\partial E_z}{\partial y} - \rho' H_x\right) = \frac{\partial H_x}{\partial t}, \qquad (2.8a)$$

$$\frac{1}{\mu}\left(\frac{\partial E_z}{\partial x} - \frac{\partial E_x}{\partial z} - \rho' H_y\right) = \frac{\partial H_y}{\partial t}, \qquad (2.8b)$$

$$\frac{1}{\mu}\left(\frac{\partial E_x}{\partial y} - \frac{\partial E_y}{\partial x} - \rho' H_z\right) = \frac{\partial H_z}{\partial t} \qquad (2.8c)$$

Actually, these equations constitute the *key origin* for the founding of the FDTD technique, which in its basic rendition performs a straightforward space-time discretization to track the time-varying fields throughout the computational domain. In this manner, numerical simulations lend themselves well to scientific visualization packages that, subsequently, furnish remarkable deductions or sound interpretations on the behavior of complex electromagnetic interactions and transient phenomena.

As already mentioned, the FDTD algorithm applies to the solution of initial boundary value electromagnetic problems. To accomplish spatial discretization, the grid elements receive the structure of Figure 2.1. Hence, electric variables, located in the middle of cell edges, are surrounded by four circulating magnetic variables. Similarly, magnetic quantities, placed at the centers of cell faces, are surrounded by four circulating electric quantities. This arrangement is extremely important because it follows the general topological rules for the optimal discretization and gives a straightforward contour allocation process for Ampère's and Faraday's integral laws (via dually interlinked current flux loops) at the elements of the artificial domain. In the occasion of smoothly curved media interfaces or arbitrary 3-D objects embedded in the computational region, the scheme maps constitutive parameters directly to the corresponding field components of the lattice. This setup yields a "staircase" policy—with a resolution depending on the cell dimensions—and maintains the tangential continuity of state variables at the interfaces without any special match conditions.

The same concept is used to advance fields in time. Hence, **E** and **H** vectors are temporally staggered by one half of a time step. In particular, these quantities are supposed to be the average of the components during a temporal step. This procedure, known as *leapfrog integration* and depicted in Figure 2.2, updates the variables by using information from the opposite field type at an intermediate time instant. For example, every electric component is advanced in terms of past magnetic values and vice versa. By leading to conditionally stable simulations, time-stepping continues until the algorithm reaches the desired steady state. The most striking trait of the leapfrog scheme is its

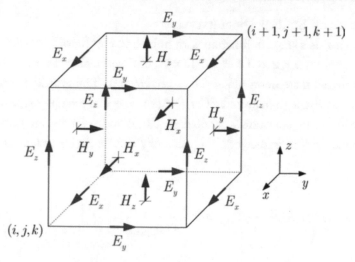

FIGURE 2.1: The Yee unit cell as the elementary spatial module of an FDTD lattice.

fully explicit character—calculations depend only on earlier results—which avoids the solution of any system of equations and laborious matrix inversions. Thus, the overall burden is kept relatively low, depending only on the number of field unknowns. It is stressed that the accuracy of the resulting expressions is of second order, whereas their central profile achieves the gentle propagation of every wave in the computational domain devoid of serious nonphysical amplitude reductions.

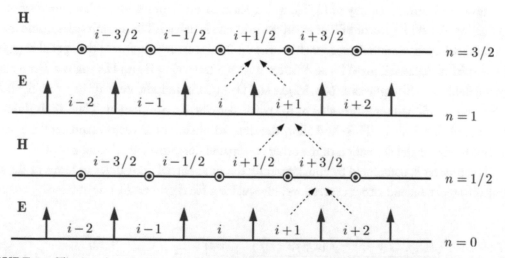

FIGURE 2.2: The leapfrog time integration procedure (1-D case) according to which fields are updated in time via information coming from the opposite field type at an intermediate time point.

2.2.2   Temporal and Spatial Discretization

Assume that $f(x,y,z,t)$ is a smooth function with finite and continuous derivatives with respect to the independent variables x,y,z and t in a 3-D region of space Ω, which is subdivided into cells of $\Delta x, \Delta y$, and Δz spatial increments along the respective axes. The typical representation of every mesh node in this domain, is then given by $(i,j,k) \equiv (x = x_0 + i\Delta x, y = y_0 + j\Delta y, z = z_0 + k\Delta z)$, where (x_0, y_0, z_0) is the grid origin, commonly denoted at $(0,0,0)$ and i, j, k are integers indexing the discretization directions. In this framework, the value of f and its derivatives at any lattice node, are expressed as

$$f(x,y,z,t) = f(i\Delta x, j\Delta y, k\Delta z, n\Delta t) = f|_{i,j,k}^{n} \quad \text{and} \quad \left.\frac{\partial f}{\partial z}\right|_{(x,y,z,t)} = \left.\frac{\partial f}{\partial z}\right|_{(i\Delta x, j\Delta y, k\Delta z, n\Delta t)} = \left.\frac{\partial f}{\partial z}\right|_{i,j,k}^{n}, \quad (2.9)$$

in which Δt is the analogous time increment accountable for the discrete temporal division and n is an integer. For an adequately small Δz, $\partial f/\partial z$ can be approximated by a simple twofold central finite-difference scheme. Thus, for a fixed time instant $t = n\Delta t$, the spatial derivative becomes

$$\left.\frac{\partial f}{\partial z}\right|_{(x,y,z,t)} = \frac{f\left(x,y,z + \frac{\Delta z}{2}, t\right) - f\left(x,y,z - \frac{\Delta z}{2}, t\right)}{\Delta z} \Rightarrow \left.\frac{\partial f}{\partial z}\right|_{i,j,k}^{n} = \frac{f|_{i,j,k+1/2}^{n} - f|_{i,j,k-1/2}^{n}}{\Delta z} + O\left[(\Delta z)^2\right]$$

$$(2.10)$$

Note the second-order accuracy of (2.10), which leads to the suppression of lattice reflection and discretization errors by a factor of four when the grid size is halved. This permits reasonable modeling at minimal computational expense. Also, the $\pm 1/2$ increment guarantees the spatial staggering of the algorithm because it provides an efficient way of interleaving **E** and **H** variables. For instance, magnetic field uses the arguments $(k + 1/2)\Delta z$ and $(k - 1/2)\Delta z$ to indicate that the two neighboring **H** values are positioned on either side of the particular electric component. Thus, their difference could be used to calculate $\partial \mathbf{H}/\partial z$ and allow the time advancing of **E** vector. Apparently, a parallel treatment holds for the derivation of the other two spatial derivatives $\partial f/\partial x$ and $\partial f/\partial y$.

The process is completed by approximating the time partial derivatives in terms of the same central-difference second-order expressions. Hence, for a fixed grid node, it is obtained

$$\left.\frac{\partial f}{\partial t}\right|_{(x,y,z,t)} = \frac{f\left(x,y,z,t + \frac{\Delta t}{2}\right) - f\left(x,y,z,t - \frac{\Delta t}{2}\right)}{\Delta t} \Rightarrow \left.\frac{\partial f}{\partial t}\right|_{i,j,k}^{n} = \frac{f|_{i,j,k}^{n+1/2} - f|_{i,j,k}^{n-1/2}}{\Delta t} + O\left[(\Delta t)^2\right],$$

$$(2.11)$$

where superscripts $n \pm 1/2$ display that field values are temporally adjacent to time step n (occurring somewhat after and before). Such intervals of $\pm 1/2\Delta t$ are ideal for realizing the leapfrog marching.

Bearing in mind the terminology, so described, let us recall the medium of Section 2.2.1, which is selected to be lossy and isotropic, both from an electric and magnetic viewpoint. In addition, it is presumed that Ampère's law is applied at time steps $n + 1/2$, namely, the curl operator of $\mathbf{H}$ vector, the conduction electric density, and the time partial derivative of $\mathbf{E}$ field, in (2.7), are associated with instants $n + 1/2$. For the leapfrog temporal integration to be developed, Faraday's law will receive the suitable staggering by one half of a time step, which similarly means that all terms at both sides of (2.8) will be evaluated at intervals of n. As a consequence, these equations, in a 3-D compact vector notation, become

$$\frac{1}{\varepsilon}\nabla \times \mathbf{H}^{n+1/2} = \frac{\sigma}{\varepsilon}\mathbf{E}^{n+1/2} + \frac{\partial \mathbf{E}}{\partial t}^{n+1/2} \rightarrow \varepsilon\left(\frac{\mathbf{E}^{n+1} - \mathbf{E}^n}{\Delta t}\right) = \nabla \times \mathbf{H}^{n+1/2} - \sigma\mathbf{E}^{n+1/2}, \quad (2.12)$$

$$\frac{1}{\mu}\nabla \times \mathbf{E}^n = -\frac{\rho'}{\mu}\mathbf{H}^n - \frac{\partial \mathbf{H}}{\partial t}^n \rightarrow \mu\left(\frac{\mathbf{H}^{n+1/2} - \mathbf{H}^{n-1/2}}{\Delta t}\right) = -\nabla \times \mathbf{E}^n - \rho'\mathbf{H}^n \quad (2.13)$$

Observing the right-hand side of (2.12), it is deduced that electric field $\mathbf{E}$ must be computed at time step $n + 1/2$. However, such an evaluation cannot be directly performed because $\mathbf{E}$ is stored in the computer only at the previous $n - 1$ time steps. To circumvent this oddity, an efficient approximation is used, which simply substitutes the problematic $\mathbf{E}$ quantities by the central average of their already stored past values at time-step $n - 1$ and the unknown values at time step $n + 1$ [3]. Thus, such a manipulation results in

$$\mathbf{E}^{n+1/2} = \frac{\mathbf{E}^{n+1} + \mathbf{E}^n}{2} \quad \text{and} \quad \mathbf{H}^n = \frac{\mathbf{H}^{n+1/2} + \mathbf{H}^{n-1/2}}{2}, \quad (2.14)$$

where the equivalent treatment for the magnetic components of $\mathbf{H}$ in (2.13), due to the presence of ρ', is also supplied. Although not unique, this approach attains the most satisfactory and stable results without generating any fictitious inconsistencies, while the extra introduced terms do not require the solution of any simultaneous equations. Upon these rearrangements, (2.12)–(2.14) give

$$\mathbf{E}^{n+1} - \mathbf{E}^n = \frac{\Delta t}{\varepsilon}\left[\nabla \times \mathbf{H}^{n+1/2} - \sigma\left(\frac{\mathbf{E}^{n+1} + \mathbf{E}^n}{2}\right)\right], \quad (2.15)$$

$$\mathbf{H}^{n+1/2} - \mathbf{H}^{n-1/2} = -\frac{\Delta t}{\mu}\left[\nabla \times \mathbf{E}^n + \rho'\left(\frac{\mathbf{H}^{n+1/2} - \mathbf{H}^{n-1/2}}{2}\right)\right], \quad (2.16)$$

which, through the separation of the unknown $\mathbf{E}$ and $\mathbf{H}$ terms on their left-hand side, yield

$$\mathbf{E}^{n+1} = \left(\frac{2\varepsilon - \sigma\Delta t}{2\varepsilon + \sigma\Delta t}\right)\mathbf{E}^n + \left(\frac{2\Delta t}{2\varepsilon + \sigma\Delta t}\right)\nabla \times \mathbf{H}^{n+1/2}, \tag{2.17}$$

$$\mathbf{H}^{n+1/2} = \left(\frac{2\mu - \rho'\Delta t}{2\mu + \rho'\Delta t}\right)\mathbf{H}^{n-1/2} - \left(\frac{2\Delta t}{2\mu + \rho'\Delta t}\right)\nabla \times \mathbf{E}^n \tag{2.18}$$

Apparently, these update relations confirm the robust nature of leapfrog integration for the time advancing of every field constituent with a second-order accuracy, as deduced from the truncation term of central difference and average approximations. Besides, the above formulation grants a functional means of handling *perfect electric conductor (PEC)* conditions. When $\sigma \gg 1$, (2.17) reduces to $\mathbf{E}^{n+1} \simeq -\mathbf{E}^n$. Therefore, if the components of $\mathbf{E}$, tangential to such a boundary, are initially set to zero, they will basically retain this value throughout the entire simulation. Actually, this way of enforcing PEC walls is very useful in general models, as it necessitates only the ordinary FDTD [2]. Comparable remarks hold also for the imposition of magnetic boundaries in the more rare case of $\rho' \gg 1$ appearing in (2.18).

On the other hand, the key idea for the correct discretization of spatial derivatives in the medium stems from the fact that the FDTD lattice is designed to comply with the rotational nature of the curl operator acting upon electric and magnetic vectors. This elegant regime enables the uncollocated determination of $\mathbf{E}$ and $\mathbf{H}$ components in a spatially proportional manner. Another considerable aspect is the location of the reference origin in the elementary cell. Among the possible choices, two seem to be the most widespread. In accordance with the first and more complicated notation, a specific lattice point (i,j,k) plays the role of the origin in the cell—defining usually one of its corners—as presented in Figure 2.1. Then, all field quantities are expressed with respect to (i,j,k) via the appropriate $\pm 1/2$ or ± 1 stencil offsets. So, by applying the central-difference scheme of (2.10) to the last terms of (2.17) and (2.18), it is derived that

$$\hat{\mathbf{x}} \cdot \nabla \times \mathbf{H}|_{i+1/2,j,k}^{n+1/2} = \frac{H_z|_{i+1/2,j+1/2,k}^{n+1/2} - H_z|_{i+1/2,j-1/2,k}^{n+1/2}}{\Delta y} - \frac{H_y|_{i+1/2,j,k+1/2}^{n+1/2} - H_y|_{i+1/2,j,k-1/2}^{n+1/2}}{\Delta z},$$

$$\tag{2.19a}$$

$$\hat{\mathbf{y}} \cdot \nabla \times \mathbf{H}|_{i,j+1/2,k}^{n+1/2} = \frac{H_x|_{i,j+1/2,k+1/2}^{n+1/2} - H_x|_{i,j+1/2,k-1/2}^{n+1/2}}{\Delta z} - \frac{H_z|_{i+1/2,j+1/2,k}^{n+1/2} - H_z|_{i-1/2,j+1/2,k}^{n+1/2}}{\Delta x},$$

$$\tag{2.19b}$$

$$\hat{\mathbf{z}} \cdot \nabla \times \mathbf{H}|_{i,j,k+1/2}^{n+1/2} = \frac{H_y|_{i+1/2,j,k+1/2}^{n+1/2} - H_y|_{i-1/2,j,k+1/2}^{n+1/2}}{\Delta x} - \frac{H_x|_{i,j+1/2,k+1/2}^{n+1/2} - H_x|_{i,j-1/2,k+1/2}^{n+1/2}}{\Delta v},$$

$$\tag{2.19c}$$

$$\hat{\mathbf{x}} \cdot \nabla \times \mathbf{E}\big|_{i,j+1/2,k+1/2}^{n} = \frac{E_z\big|_{i,j+1,k+1/2}^{n} - E_z\big|_{i,j,k+1/2}^{n}}{\Delta y} - \frac{E_y\big|_{i,j+1/2,k+1}^{n} - E_y\big|_{i,j+1/2,k}^{n}}{\Delta z}, \quad (2.20a)$$

$$\hat{\mathbf{y}} \cdot \nabla \times \mathbf{E}\big|_{i+1/2,j,k+1/2}^{n} = \frac{E_x\big|_{i+1/2,j,k+1}^{n} - E_x\big|_{i+1/2,j,k}^{n}}{\Delta z} - \frac{E_z\big|_{i+1,j,k+1/2}^{n} - E_z\big|_{i,j,k+1/2}^{n}}{\Delta x}, \quad (2.20b)$$

$$\hat{\mathbf{z}} \cdot \nabla \times \mathbf{E}\big|_{i+1/2,j+1/2,k}^{n} = \frac{E_y\big|_{i+1,j+1/2,k}^{n} - E_y\big|_{i,j+1/2,k}^{n}}{\Delta x} - \frac{E_x\big|_{i+1/2,j+1,k}^{n} - E_x\big|_{i+1/2,j,k}^{n}}{\Delta y} \quad (2.20c)$$

Despite its overall robustness, this configuration is not very convenient because since it depends on the structure of the cell. In other words, if the definition of $\mathbf{E}$ and $\mathbf{H}$ on the element's edges and faces was the opposite, then (2.19) and (2.20) would be different. This is, indeed, the most significant side effect of a constant origin. On the other hand, the second choice is not constrained by the above weakness, thus exhibiting great flexibility and universality. Herein, each field component, under examination, is evaluated at a point (i, j, k), whereas the indices of the others are resolved in reference with this specific point and its neighboring stencils. This merely means that one moves the origin of the cell to the location of every grid quantity and extracts the approximating equations in a more straightforward and practical way, as

$$\hat{\mathbf{x}} \cdot \nabla \times \mathbf{H}\big|_{i,j,k}^{n+1/2} = \frac{H_z\big|_{i,j+1/2,k}^{n+1/2} - H_z\big|_{i,j-1/2,k}^{n+1/2}}{\Delta y} - \frac{H_y\big|_{i,j,k+1/2}^{n+1/2} - H_y\big|_{i,j,k-1/2}^{n+1/2}}{\Delta z}, \quad (2.21a)$$

$$\hat{\mathbf{y}} \cdot \nabla \times \mathbf{H}\big|_{i,j,k}^{n+1/2} = \frac{H_x\big|_{i,j,k+1/2}^{n+1/2} - H_x\big|_{i,j,k-1/2}^{n+1/2}}{\Delta z} - \frac{H_z\big|_{i+1/2,j,k}^{n+1/2} - H_z\big|_{i-1/2,j,k}^{n+1/2}}{\Delta x}, \quad (2.21b)$$

$$\hat{\mathbf{z}} \cdot \nabla \times \mathbf{H}\big|_{i,j,k}^{n+1/2} = \frac{H_y\big|_{i+1/2,j,k}^{n+1/2} - H_y\big|_{i-1/2,j,k}^{n+1/2}}{\Delta x} - \frac{H_x\big|_{i,j+1/2,k}^{n+1/2} - H_x\big|_{i,j-1/2,k}^{n+1/2}}{\Delta y}, \quad (2.21c)$$

$$\hat{\mathbf{x}} \cdot \nabla \times \mathbf{E}\big|_{i,j,k}^{n} = \frac{E_z\big|_{i,j+1/2,k}^{n} - E_z\big|_{i,j-1/2,k}^{n}}{\Delta y} - \frac{E_y\big|_{i,j,k+1/2}^{n} - E_y\big|_{i,j,k-1/2}^{n}}{\Delta z}, \quad (2.22a)$$

$$\hat{\mathbf{y}} \cdot \nabla \times \mathbf{E}\big|_{i,j,k}^{n} = \frac{E_x\big|_{i,j,k+1/2}^{n} - E_x\big|_{i,j,k-1/2}^{n}}{\Delta z} - \frac{E_z\big|_{i+1/2,j,k}^{n} - E_z\big|_{i-1/2,j,k}^{n}}{\Delta x}, \quad (2.22b)$$

$$\hat{\mathbf{z}} \cdot \nabla \times \mathbf{E}\big|_{i,j,k}^{n} = \frac{E_y\big|_{i+1/2,j,k}^{n} - E_y\big|_{i-1/2,j,k}^{n}}{\Delta x} - \frac{E_x\big|_{i,j+1/2,k}^{n} - E_x\big|_{i,j-1/2,k}^{n}}{\Delta y} \quad (2.22c)$$

Being entirely alike with the relations of the first approach, (2.21) and (2.22) also share identical indices to each other. These merits motivated us to adopt the preceding index notation, unless distinctly indicated.

A word of practicality should be mentioned here. It is clear that this kind of treatment exposes the half-cell positions of each **E** and **H** quantity. However, this arrangement cannot be promptly incorporated into a computer language, as popular programming compilers support arrays only with integer indices. So, whenever they include the FDTD expressions in their computer codes, most researchers prefer the rounding down technique that associates any lattice point $(i + 1/2, j + 1/2, k + 1/2)$ with an array location (i, j, k). Therefore, special attention must be paid toward the concise interconnection of electric or magnetic variables with the nodes of a particular cell and the original field location inside this cell.

2.2.3   Realization Assertions for EMC Problems

Having established the fundamental background regarding the discrete nature of temporal and spatial derivatives in time-dependent Maxwell's curl laws, the complete set of the FDTD update formulae for a general EMC application will be now composed. In the case of a computational domain occupied by materials with a constant variation of their properties as a function of space, the system of (2.17) and (2.18), together with (2.21) and (2.22), is developed in the subsequent form

Ampère's law:

$$E_x|_{i,j,k}^{n+1} = ep_{i,j,k}^{E_x} E_x|_{i,j,k}^{n} + eq_{i,j,k}^{E_x,\Delta y} \left[H_z|_{i,j+1/2,k}^{n+1/2} - H_z|_{i,j-1/2,k}^{n+1/2} \right]$$

$$- eq_{i,j,k}^{E_x,\Delta z} \left[H_y|_{i,j,k+1/2}^{n+1/2} - H_y|_{i,j,k-1/2}^{n+1/2} \right], \tag{2.23a}$$

$$E_y|_{i,j,k}^{n+1} = ep_{i,j,k}^{E_y} E_y|_{i,j,k}^{n} + eq_{i,j,k}^{E_y,\Delta z} \left[H_x|_{i,j,k+1/2}^{n+1/2} - H_x|_{i,j,k-1/2}^{n+1/2} \right]$$

$$- eq_{i,j,k}^{E_y,\Delta x} \left[H_z|_{i+1/2,j,k}^{n+1/2} - H_z|_{i-1/2,j,k}^{n+1/2} \right], \tag{2.23b}$$

$$E_z|_{i,j,k}^{n+1} = ep_{i,j,k}^{E_z} E_z|_{i,j,k}^{n} + eq_{i,j,k}^{E_z,\Delta x} \left[H_y|_{i+1/2,j,k}^{n+1/2} - H_y|_{i-1/2,j,k}^{n+1/2} \right]$$

$$- eq_{i,j,k}^{E_z,\Delta y} \left[H_x|_{i,j+1/2,k}^{n+1/2} - H_x|_{i,j-1/2,k}^{n+1/2} \right] \tag{2.23c}$$

Faraday's law:

$$H_x|_{i,j,k}^{n+1/2} = hp_{i,j,k}^{H_x} H_x|_{i,j,k}^{n-1/2} + hq_{i,j,k}^{H_x,\Delta z} \left[E_y|_{i,j,k+1/2}^{n} - E_y|_{i,j,k-1/2}^{n} \right]$$

$$- hq_{i,j,k}^{H_x,\Delta y} \left[E_z|_{i,j+1/2,k}^{n} - E_z|_{i,j-1/2,k}^{n} \right], \tag{2.24a}$$

$$
H_y\big|_{i,j,k}^{n+1/2} = hp_{i,j,k}^{H_y} H_y\big|_{i,j,k}^{n-1/2} + hq_{i,j,k}^{H_y,\Delta x}\left[E_z\big|_{i+1/2,j,k}^{n} - E_z\big|_{i-1/2,j,k}^{n}\right]
$$

$$
- hq_{i,j,k}^{H_y,\Delta z}\left[E\big|_{i,j,k+1/2}^{n} - E_x\big|_{i,j,k-1/2}^{n}\right], \tag{2.24b}
$$

$$
H_z\big|_{i,j,k}^{n+1/2} = hp_{i,j,k}^{H_z} H_z\big|_{i,j,k}^{n-1/2} + hq_{i,j,k}^{H_z,\Delta y}\left[E_x\big|_{i,j+1/2,k}^{n} - E_x\big|_{i,j-1/2,k}^{n}\right]
$$

$$
- hq_{i,j,k}^{H_z,\Delta x}\left[E_y\big|_{i+1/2,j,k}^{n} - E_y\big|_{i-1/2,j,k}^{n}\right], \tag{2.24c}
$$

$$
ep_{i,j,k}^{f} = \frac{2\varepsilon_{i,j,k} - \sigma_{i,j,k}\Delta t}{2\varepsilon_{i,j,k} + \sigma_{i,j,k}\Delta t}, \quad eq_{i,j,k}^{f,\Delta} = \frac{2\Delta t}{\Delta(2\varepsilon_{i,j,k} + \sigma_{i,j,k}\Delta t)}, \quad \text{for}\ \begin{cases} f = E_x, E_y, E_z \\ \Delta = \Delta x, \Delta y, \Delta z \end{cases},
$$

with $\tag{2.25}$

$$
hp_{i,j,k}^{f} = \frac{2\mu_{i,j,k} - \rho'_{i,j,k}\Delta t}{2\mu_{i,j,k} + \rho'_{i,j,k}\Delta t}, \quad hq_{i,j,k}^{f,\Delta} = \frac{2\Delta t}{\Delta(2\mu_{i,j,k} + \rho'_{i,j,k}\Delta t)}, \quad \text{for}\ \begin{cases} f = H_x, H_y, H_z \\ \Delta = \Delta x, \Delta y, \Delta z \end{cases}
$$

Coefficients (2.25) characterize each electric and magnetic quantity at point (i,j,k) in a unique manner, i.e., they can completely represent the profile of every field component anywhere on the FDTD lattice. Evidently, when a cubic mesh ($\Delta = \Delta x = \Delta y = \Delta z$) is considered, storage demands can be more efficiently handled. This parametric rendition permits the easier incorporation of material attributes and enhances the compactness of the analysis. Probing further, if the domain at hand comprises specific media of several electric and magnetic profiles, (2.23) and (2.24) may be written more elegantly. This is performed through the pertinent number (1 to 6) of integer arrays depending on the **E** and **H** components that interact with these media. For instance, the update of E_x and H_y becomes

$$
E_x\big|_{i,j,k}^{n+1} = \text{ep}[l_e] \cdot E_x\big|_{i,j,k}^{n} + \text{eqy}[l_e] \cdot \left[H_z\big|_{i,j+1/2,k}^{n+1/2} - H_z\big|_{i,j-1/2,k}^{n+1/2}\right]
$$

$$
- \text{eqz}[l_e] \cdot \left[H_y\big|_{i,j,k+1/2}^{n+1/2} - H_y\big|_{i,j,k-1/2}^{n+1/2}\right], \tag{2.26}
$$

$$
H_y\big|_{i,j,k}^{n+1/2} = \text{hp}[l_h] \cdot H_y\big|_{i,j,k}^{n-1/2} + \text{hqx}[l_h] \cdot \left[E_z\big|_{i+1/2,j,k}^{n} - E_z\big|_{i-1/2,j,k}^{n}\right]
$$

$$
- \text{hqz}[l_h] \cdot \left[E_x\big|_{i,j,k+1/2}^{n} - E_x\big|_{i,j,k-1/2}^{n}\right], \tag{2.27}
$$

with $l_e = \text{MAT\_Ex}[i][j][k]$ and $l_h = \text{MAT\_Hy}[i][j][k]$. In fact, MAT_Ex and MAT_Hy arrays determine the material type associated to the relevant field quantity, whereas indices i,j,k

describe the distinct grid cell occupied by this material. The parameters of (2.25) have now been turned into 1-D arrays. Hence, if the space region involves L media, then each of the arrays ep[.],eqy[.],eqz[.] and hp[.],hpx[.],hqz[.] will contain only L elements, which, via the l_e or l_h pointers, lead to the extraction of the analogous expressions. This simply implies that every set of constitutive parameters in the FDTD lattice is imposed once, with its coefficients computed and stored before the leapfrog integration is initiated. Moreover, the previous individual media assignment to every field component in a cell enables the successful representation of anisotropic materials with diagonal permittivity or permeability tensors in a rigorous manner. It is to be emphasized, however, that the spatial staggering of the FDTD scheme is also applicable to MAT_E and MAT_H arrays, thus enforcing the staircase perspective during the construction of a smoothly curved EMC object, accommodating both dielectric and magnetic properties. Even in the situation of full alignment with the Cartesian grid, problems can arise because of this inherent offset. A possible solution to alleviate, yet not eliminate, this shortcoming is to use average values of the constitutive parameters in the problematic areas. Conversely, from a programming outlook, a routine that sets dielectric and magnetic media can be devised, with the aim of building all sorts of objects without significant modifications. Anyhow, it must be always recalled that FDTD modeling does not fill cells with certain materials but rather appoints different materials at the field locations within a cell, so as not to leave void regions or partial simulations.

Up to this point, the method has been derived by means of the *differential* form of Maxwell's equations. Nonetheless, apart from this development, an alternative one can be obtained via the *integral* version of Ampère's and Faraday's laws [3, 20] implemented on electrically small orthogonal contours. Local spatial approximations to these expressions provide the suitable conditions for the treatment of thin wires, slots, or arbitrary curvatures, which otherwise would have entailed very fine grids. Figure 2.3 illustrates the mutually intersecting contours as a family of chainlike coupling paths that can associate all field variations, imposed by the previous objects, to the integrals of (2.1) and (2.2). This evidence, along with a careful contour deformation to capture any shape or surface peculiarity, extends the FDTD discretization capabilities and improves the comprehension of physical mechanisms that usual finite differencing is not able to deeply explain.

Starting from (2.1) for the time stepping of E_y and referring to Figure 2.3a, Ampère's law yields

$$\varepsilon \frac{\partial}{\partial t} \iint_{S_E} E_y \big|_{i,j,k}^{n+1/2} ds_E + \sigma \iint_{S_E} E_y \big|_{i,j,k}^{n+1/2} ds_E \simeq \sum_{C_H edges} H \big|_{edge}^{n+1/2} \text{length}_{edge}$$

$$= H_z \big|_{i-1/2,j,k}^{n+1/2} \Delta z + H_x \big|_{i,j,k+1/2}^{n+1/2} \Delta x - H_z \big|_{i+1/2,j,k}^{n+1/2} \Delta z - H_x \big|_{i,j,k-1/2}^{n+1/2} \Delta x , \quad (2.28)$$

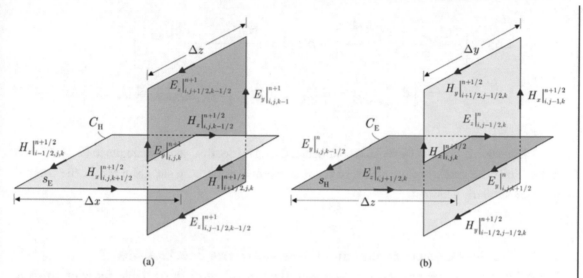

FIGURE 2.3: Integral formulation of the FDTD update equations. (a) Ampère's law for E_y and (b) Faraday's law for H_x.

where magnetic fields remain constant and equal to their average value along each edge. For the left-hand side of (2.28), temporal derivative may be approximated via central differences, as

$$\varepsilon \Delta x \Delta z \left[\frac{E_y \big|_{i,j,k}^{n+1} - E_y \big|_{i,j,k}^{n}}{\Delta t} \right] + \sigma \Delta x \Delta z \left[\frac{E_y \big|_{i,j,k}^{n+1} + E_y \big|_{i,j,k}^{n}}{2} \right]$$

$$= \left[H_x \big|_{i,j,k+1/2}^{n+1/2} - H_x \big|_{i,j,k-1/2}^{n+1/2} \right] \Delta x - \left[H_z \big|_{i+1/2,j,k}^{n+1/2} - H_z \big|_{i-1/2,j,k}^{n+1/2} \right] \Delta z , \quad (2.29)$$

under the constraint that E_y is invariable and equal to its average value over the surface of the cell. Dividing both sides of (2.29) by $\Delta x \Delta z$, the result is just the same as (2.23b). To complete the process, Faraday's law for H_x (Figure 2.3b) gives

$$\mu \frac{\partial}{\partial t} \iint_{S_H} H_x \big|_{i,j,k}^{n} d\mathbf{s}_H + \rho' \iint_{S_H} H_x \big|_{i,j,k}^{n} d\mathbf{s}_H \simeq - \sum_{C_E \, edges} E \big|_{edge}^{n} \mathrm{length}_{edge}$$

$$= E_y \big|_{i,j,k+1/2}^{n} \Delta y + E_z \big|_{i,j-1/2,k}^{n} \Delta z - E_y \big|_{i,j,k-1/2}^{n} \Delta y - E_z \big|_{i,j+1/2,k}^{n} \Delta z , \quad (2.30)$$

$$\mu \Delta y \Delta z \left[\frac{H_x|_{i,j,k}^{n+1/2} - H_x|_{i,j,k}^{n-1/2}}{\Delta t} \right] + \rho' \Delta y \Delta z \left[\frac{H_x|_{i,j,k}^{n+1/2} + H_y|_{i,j,k}^{n-1/2}}{2} \right]$$

$$= \left[E_y|_{i,j,k+1/2}^{n} - E_y|_{i,j,k+1/2}^{n} \right] \Delta y - \left[E_z|_{i,j+1/2,k}^{n} - E_z|_{i,j-1/2,k}^{n} \right] \Delta z \quad (2.31)$$

Likewise, if division by $\Delta y \Delta z$ is conducted, (2.31) is identical to its analogue in (2.24a), a fact that exhibits the absolute equivalence of the two approaches because since the rest of the components are similarly acquired.

2.2.4   Stability Considerations and Dispersion Error Mechanisms

The stability of a difference scheme deals with the unstable growth or stable decay of errors in the arithmetic operations required to solve the finite-difference equations. The FDTD algorithm exhibits, as already described, an explicit profile, which means that all unknowns at each time-step are evaluated from precalculated values at preceding time instants. In fact, electric and magnetic field components are updated and stored only for their latest values, an issue proven to be exceptionally versatile because it needs significantly less memory along with a decreased computational complexity. A limiting factor though (also holding for the majority of time-domain approaches) is their conditional stability, namely, it is essential to select a number of parameters—time step, size of spatial increments, etc.—in such a way that the difference scheme remains stable. Specifically, a certain bound for Δt relative to cell dimensions Δx, Δy, and Δz must be obtained to confirm the suppression of spurious waves as time-advancing continues. Hence, the derivation of closed-form analytical stability conditions is of great theoretical and practical interest.

The derivation of such a criterion for the 3-D FDTD technique, also known as the *Courant stability condition*, is performed by means of the von Neumann's method [2, 4]. In particular, the algorithm is decomposed into individual time and space eigenvalue problems to propagate plane-wave eigenvectors in the discretized domain. Both electric and magnetic field components are expressed via Fourier modes, whereas propagation is presumed to occur at arbitrary angles in the grid. The rationale is to specify the spectrum of eigenvalues for these modes after space approximation and subsequently compare the result to the stable spectrum of eigenvalues acquired by the time differentiation process. Thus, starting from

$$F_u|_{I,J,K}^{n} = F_{u0} e^{j(\omega n \Delta t - \tilde{k}_x I \Delta x - \tilde{k}_y J \Delta y - \tilde{k}_z K \Delta z)}, \quad \text{for } F = E, H, \ u = x, y, z, \ j^2 = -1, \quad (2.32)$$

and $\tilde{k}_x$, $\tilde{k}_y$, $\tilde{k}_z$, the components of the numerical wavevector $\tilde{\mathbf{k}} = (\tilde{k}_x, \tilde{k}_y, \tilde{k}_z)$ along x-, y-, and z-axis, respectively, update (2.23) and (2.24) are appropriately rewritten. For example, (2.23a), after the necessary substitutions from (2.32) and the expansion of the resulting complex exponentials, is modified to

$$\eta \, H_x|^{n+1/2} = \eta \, H_x|^{n-1/2} + 2jv\Delta t \left[\frac{E_z|^n}{\Delta y} \sin\left(\frac{\tilde{k}_y \Delta y}{2} \right) - \frac{E_y|^n}{\Delta z} \sin\left(\frac{\tilde{k}_z \Delta z}{2} \right) \right], \tag{2.33}$$

where $\eta = \sqrt{\mu/\varepsilon}$ is the characteristic impedance and $v = 1/\sqrt{\mu/\varepsilon}$ the speed of light (for ease, spatial stencils have been omitted). Let us now denote some variables that will facilitate the calculations, i.e.,

$$\rho_u = \frac{v\Delta t}{\Delta u} \quad \text{and} \quad P_u = \rho_u \sin\left(\frac{\tilde{k}_u \Delta u}{2} \right), \text{ for } u = x, y, z \tag{2.34}$$

Applying the treatment of (2.33) to the rest of (2.23) and (2.24), the system of Maxwell's equations becomes

$$E_x|^{n+1} = E_x|^n - 2j\eta \left(H_z|^{n+1/2} P_y - H_y|^{n+1/2} P_z \right), \tag{2.35a}$$

$$E_y|^{n+1} = E_y|^n - 2j\eta \left(H_x|^{n+1/2} P_z - H_z|^{n+1/2} P_x \right), \tag{2.35b}$$

$$E_z|^{n+1} = E_z|^n - 2j\eta \left(H_y|^{n+1/2} P_x - H_x|^{n+1/2} P_y \right), \tag{2.35c}$$

$$\eta \, H_x|^{n+1/2} = \eta \, H_x|^{n-1/2} - 2j\left(E_y|^n P_z - E_z|^n P_y \right), \tag{2.36a}$$

$$\eta \, H_y|^{n+1/2} = \eta \, H_y|^{n-1/2} - 2j\left(E_z|^n P_x - E_x|^n P_z \right), \tag{2.36b}$$

$$\eta \, H_z|^{n+1/2} = \eta \, H_z|^{n-1/2} - 2j\left(E_x|^n P_y - E_y|^n P_x \right) \tag{2.36c}$$

To synchronize $\mathbf{E}$ and $\mathbf{H}$ fields, it is also assumed that $E_u|^{n+1} = e^{j\omega\Delta t/2} E_u|^{n+1/2}$ and $H_u|^{n+1/2} = e^{j\omega\Delta t/2} H_u|^n$, which transform (2.35) to

$$E_x|^{n+1/2} - E_x|^{n-1/2} = 2j\eta \left(H_y|^n P_z - H_z|^n P_y \right), \tag{2.37a}$$

$$E_y\big|^{n+1/2} - E_y\big|^{n-1/2} = 2j\eta\left(H_z\big|^n P_x - H_x\big|^n P_z\right), \qquad (2.37b)$$

$$E_z\big|^{n+1/2} - E_z\big|^{n-1/2} = 2j\eta\left(H_x\big|^n P_y - H_y\big|^n P_x\right) \qquad (2.37c)$$

Next, the complex additions (2.36) + j(2.37) and definition $Q_u|^n = \eta H_u|^n + jE_u|^n$ are launched. So,

$$Q_x\big|^{n+1/2} = Q_x\big|^{n-1/2} + 2\left(Q_z\big|^n P_y - Q_y\big|^n P_z\right), \qquad (2.38a)$$

$$Q_y\big|^{n+1/2} = Q_y\big|^{n-1/2} + 2\left(Q_x\big|^n P_z - Q_z\big|^n P_x\right), \qquad (2.38b)$$

$$Q_z\big|^{n+1/2} = Q_z\big|^{n-1/2} + 2\left(Q_y\big|^n P_x - Q_x\big|^n P_y\right), \qquad (2.38c)$$

or in the more convenient matrix notation

$$\begin{bmatrix} Q_x\big|^{n+1/2} \\ Q_y\big|^{n+1/2} \\ Q_z\big|^{n+1/2} \end{bmatrix} = \begin{bmatrix} Q_x\big|^{n-1/2} \\ Q_y\big|^{n-1/2} \\ Q_z\big|^{n-1/2} \end{bmatrix} - 2 \begin{bmatrix} 0 & P_z & -P_y \\ -P_z & 0 & P_x \\ P_y & -P_x & 0 \end{bmatrix} \begin{bmatrix} Q_x\big|^n \\ Q_y\big|^n \\ Q_z\big|^n \end{bmatrix} \Rightarrow \mathbf{Q}^{n+1/2} = \mathbf{Q}^{n-1/2} - 2\mathbf{P}\mathbf{Q}^n$$

$$(2.39)$$

Equation (2.39) has not yet the proper form for a straightforward eigenvalue manipulation. In fact, the presence of the last term generates several difficulties regarding the well-posed nature of the whole scheme. This defect is evaded by introducing the auxiliary vector $\mathbf{V}^n = \mathbf{Q}^{n-1/2}$, which leads to the new system of

$$\begin{aligned} \mathbf{Q}^{n+1/2} &= \mathbf{V}^n - 2\mathbf{P}\mathbf{Q}^n \\ \mathbf{V}^{n+1/2} &= \mathbf{Q}^n \end{aligned} \quad \text{and via matrices to} \quad \begin{bmatrix} \mathbf{Q}^{n+1/2} \\ \mathbf{V}^{n+1/2} \end{bmatrix} = \begin{bmatrix} -2\mathbf{P} & \mathbf{I} \\ \mathbf{I} & \mathbf{0} \end{bmatrix} \begin{bmatrix} \mathbf{Q}^n \\ \mathbf{V}^n \end{bmatrix} \Rightarrow \Xi^{n+1/2} = \mathbf{G}\Xi^n$$

$$(2.40)$$

The mandatory condition for stability is the von Neumann constraint that enforces all eigenvalues λ_i of the amplification matrix $\mathbf{G}$ to satisfy inequality $|\lambda_i|\leq 1$. To this goal, one can set $\det(\lambda\mathbf{I} - \mathbf{G}) = 0$ and get

$$\det\begin{pmatrix} \lambda\mathbf{I} + 2\mathbf{P} & -\mathbf{I} \\ -\mathbf{I} & \lambda\mathbf{I} \end{pmatrix} = \det\begin{pmatrix} \lambda^2 - 1 & 2\lambda P_z & -2\lambda P_y \\ -2\lambda P_z & \lambda^2 - 1 & 2\lambda P_x \\ 2\lambda P_y & -2\lambda P_x & \lambda^2 - 1 \end{pmatrix}$$

$$= (\lambda^2 - 1)\left[(\lambda^2 - 1)^2 + 4\lambda^2(P_x^2 + P_y^2 + P_z^2)\right] = 0 \qquad (2.41)$$

The last relation of (2.41) is a fourth-order polynomial, and thus, the four distinct eigenvalues are

$$\lambda_{1,2} = \pm 1 \ \text{ and } \ \lambda_{3,4} = \pm j \sqrt{P_x^2 + P_y^2 + P_z^2} \pm \sqrt{1 - (P_x^2 + P_y^2 + P_z^2)} \qquad (2.42)$$

Overall stability is attained if and only if $P_x^2 + P_y^2 + P_z^2 \leq 1$ and $|\lambda_{3,4}| = 1$. Using (2.34), the limitation gives

$$P_x^2 + P_y^2 + P_z^2 = \rho_x^2 \sin^2\left(\frac{\tilde{k}_x \Delta x}{2}\right) + \rho_y^2 \sin^2\left(\frac{\tilde{k}_y \Delta y}{2}\right) + \rho_z^2 \sin^2\left(\frac{\tilde{k}_z \Delta z}{2}\right) \leq \rho_x^2 + \rho_y^2 + \rho_z^2$$

$$(2.43)$$

Finally, the Courant limit is expressed as

$$\rho_x^2 + \rho_y^2 + \rho_z^2 \leq 1 \Rightarrow v \Delta t \leq \left[\frac{1}{(\Delta x)^2} + \frac{1}{(\Delta y)^2} + \frac{1}{(\Delta z)^2} \right]^{-1/2}, \qquad (2.44)$$

which, for the frequently encountered cubic mesh, reduces to $v \Delta t \leq \Delta / \sqrt{3}$.

Of equivalently critical significance for the efficient implementation of the FDTD method, especially in EMC or immunity problems, is the intrinsic artificial effect of *numerical dispersion*. In general, this mesh weakness influences the precise propagation of electromagnetic waves in the computational space through an intense dependence of phase velocity on frequency (i.e., modal wavelength), direction, and lattice resolution [3, 21]. The costs of dispersion are diverse and frequently turn out to be inhibitive for the analysis. Among them, one may distinguish the phase errors and the unnatural refraction. Specifically, the former create oscillatory deviations from the typical speed of light, so leading to the amplification of waveforms and the generation of parasitical modes. In contrast, pseudorefraction appears because of the different cell shapes in curvilinear or nonorthogonal grids. Note that the dependence of phase velocity on grid direction is also known as *numerical anisotropy*, but its definition is frequently included in the meaning of dispersion. Therefore, the above defects alter the velocity of a traveling wave inside the discrete lattice and become the main source of accuracy limitations in the application of the FDTD technique, chiefly for electrically large EMC structures [16, 25].

The extraction of the FDTD dispersion relation follows the framework of stability investigation, where a set of Fourier numerical modes are assumed to propagate at arbitrary angles in the lattice for the electric and magnetic field components. By combining (2.32) with (2.35) and (2.36), using the auxiliary variables of (2.34) and expressing the complex exponential terms via some well-known trigonometric identities,

$$E_x \sin(\omega\Delta t/2) = \eta(H_y P_z - H_z P_y), \tag{2.45a}$$

$$E_y \sin(\omega\Delta t/2) = \eta(H_z P_x - H_x P_z), \tag{2.45b}$$

$$E_z \sin(\omega\Delta t/2) = \eta(H_x P_y - H_y P_x), \tag{2.45c}$$

$$\eta H_x \sin(\omega\Delta t/2) = -E_y P_z + E_z P_y, \tag{2.46a}$$

$$\eta H_y \sin(\omega\Delta t/2) = -E_z P_x + E_x P_z, \tag{2.46b}$$

$$\eta H_z \sin(\omega\Delta t/2) = -E_x P_y + E_y P_x \tag{2.46c}$$

If addition (2.46) + j(2.45) and definition $Q_u = \eta H_u + jE_u$ for $u = x, y, z$ are incorporated, one gets

$$Q_x \sin(\omega\Delta t/2) = j(Q_y P_z - Q_z P_y), \tag{2.47a}$$

$$Q_y \sin(\omega\Delta t/2) = j(Q_z P_x - Q_x P_z), \tag{2.47b}$$

$$Q_z \sin(\omega\Delta t/2) = j(Q_x P_y - Q_y P_x), \tag{2.47c}$$

whose matrix form results in the following system

$$\begin{bmatrix} j\sin(\omega\Delta t/2) & P_z & -P_y \\ -P_z & j\sin(\omega\Delta t/2) & P_x \\ P_y & -P_x & j\sin(\omega\Delta t/2) \end{bmatrix} \begin{bmatrix} Q_x \\ Q_y \\ Q_z \end{bmatrix} = 0 \Rightarrow \mathbf{PQ} = 0 \tag{2.48}$$

To have a nonzero solution of (2.48), the determinant of matrix $\mathbf{P}$ should be equal to zero, i.e.,

$$\det(\mathbf{P}) = 0 \Rightarrow -j\sin^3\left(\frac{\omega\Delta t}{2}\right) + j\sin\left(\frac{\omega\Delta t}{2}\right)(P_x^2 + P_y^2 + P_z^2) = 0 \Rightarrow \sin^2\left(\frac{\omega\Delta t}{2}\right)$$

$$= P_x^2 + P_y^2 + P_z^2, \tag{2.49}$$

or more elaborately and after the appropriate substitutions of P

$$\left[\frac{1}{\upsilon\Delta t}\sin\left(\frac{\omega\Delta t}{2}\right)\right]^2 = \left[\frac{1}{\Delta x}\sin\left(\frac{\tilde{k}_x \Delta x}{2}\right)\right]^2 + \left[\frac{1}{\Delta y}\sin\left(\frac{\tilde{k}_y \Delta y}{2}\right)\right]^2 + \left[\frac{1}{\Delta z}\sin\left(\frac{\tilde{k}_z \Delta z}{2}\right)\right]^2$$

$$\tag{2.50}$$

Equation (2.50) is the dispersion relation of the fields simulated by the FDTD method. In contrast to its numerical rendition, the analytical dispersion relation in a continuous medium without losses reads

$$\omega^2 = v^2 \left(k_x^2 + k_y^2 + k_z^2 \right) \qquad (2.51)$$

In spite of the superficial dissimilarity between (2.50) and (2.51), it is observed that as $\Delta x, \Delta y, \Delta z, \Delta t \rightarrow$ 0, the former approaches the latter. This remark justifies the use of finer grids for the suppression of dispersion errors and indicates the introduction of several phase discrepancies via the FDTD computations.

Finally, apart from phase, anisotropy, or dissipation errors, numerical dispersion can generate artificial refraction during the update of field vectors. This is principally frequent when variable cell complexes are implemented in subgridding techniques or in the vicinity of different material interfaces. As an outcome, mesh reflection occurs, even if these interfaces are located in free-space regions. The intense of such undesired phenomena depends on the magnitude and abruptness of the grid wave mode velocity distribution and can be deemed critical in several electrically large scattering or radiation problems involving phase cancellation of multiple surface waves to attain low levels of radar cross section. In fact, these flaws motivated the development of higher-order FDTD methods with various manifestations of lattice stencils.

2.2.5   Modeling Advantages and Structural Attributes

The formulation presented so far applies to uniform media in Cartesian coordinates with a constant phase velocity and free of any structural or mathematical constraints that commonly complement the FDTD technique. When the modeling of a geometry involves open-region areas, arbitrarily curved peculiarities or dissimilar materials, it becomes apparent that the ordinary FDTD procedure is not always sufficient to tackle with all these hurdles, thus obligating its interaction with other algorithms, capable of providing effective solutions. Elaborately discussed in the next chapters, these schemes affect the overall time-stepping as they bring in novel discretization strategies and update equations or even modify some of the inherent features concerning the second-order staggering procedure [17, 18]. Therefore, it goes without saying that as the complexity of an application augments, the stability of the entire FDTD method should take into account more factors or parameters apart from the usual ones. Amid the most frequently used algorithms, one can distinguish the absorbing boundary conditions, the nonorthogonal and unstructured lattice regimes as well as the mathematical realizations for the treatment of dispersive, nonlinear, or gain media.

Artificial boundary conditions furnish a widely used approach for the treatment of EMC problems initially formulated on unbounded domains and as a quantitative tool they need the processing of field data located nonlocally in space or time, a fact that may easily lead to severe instabilities. Actually, the total accuracy and performance of a numerical technique depends on the proper treatment of external boundaries. Because the choice of such conditions is never unique, the importance of their theoretical framework can be directly figured out. A widespread mistake is to conduct a selection based only on the solvability of the truncated problem. If analysis is restricted to this necessity only, then it is not guaranteed that simulations will be stable. In reality, a lot of issues should be carefully scrutinized and adapted to the specific problem. On the other hand, the use of body-conforming FDTD grids to capture the structure's curvatures or dissimilar material distribution, by varying the size of elementary cells, constitutes another source of concern. Because of the deformation of the elements in the vicinity of a geometric discontinuity or detail, their intrinsic attributes are likely to be distorted, hence affecting the consistency of the mesh.

To indicate the modeling capabilities of the FDTD method, consider the simple 3-D EMC problem of an H-bend waveguide with a uniform cross section along the y-axis. Such a structure is intended for the deliberate curving of magnetic dynamic lines, without affecting the electric ones. For an acceptable simulation, the domain is discretized into $120 \times 24 \times 120$ cells, and the excitation is launched at the $z = 0$ plane, whereas the open waveguide ends are terminated by the pertinent absorbing boundary conditions (see Chapter 4 for an elaborate explanation). Figure 2.4 presents the magnitude of the E_y component at the $y = 8$ plane for several time-step values, revealing the smoothness of propagation and the total absence of spurious modes. Moreover, Figure 2.5 provides a thorough inspection of the field at the critical sector of the bend. It can be observed that as soon as the wave approaches the specific area, it separates into two parts. The characteristic "tail" following the propagating wave is attributed to the reflections from the PEC walls of the waveguide.

A useful deduction from the previous remarks is that the option of spatial/temporal increments should always be in compromise with the specifications of the EMC problem. Accordingly, a suitable Δt is the one that offers the suitable grid resolution for the incorporation of every geometric detail. The essential limitation is that the cell size must be smaller than the smallest wavelength for precise solutions. A regular rule of thumb is 10 cells per wavelength λ, i.e., the side of each element should be $\lambda/10$ or less than the wavelength of the highest frequency. However, the most important side effect is the high increase of the overall memory and CPU requirements. As a matter of fact, the power and storage capacity of a computing system determines the upper limit of an FDTD simulation. Especially when large-scale microwave circuits and devices, operating at gigahertz frequencies, are studied, the ordinary FDTD method results in a prohibitive computational burden. Fortunately, these categories of problems are successfully solved via several efficient techniques that modify the original kernel of the algorithm and apply concepts of fast convergence. To realize why the cell size must be smaller than one wavelength, think that at any time instant, the lattice is a discrete spatial sample of field distribution. From the Nyquist theorem, there have to be at least two samples per

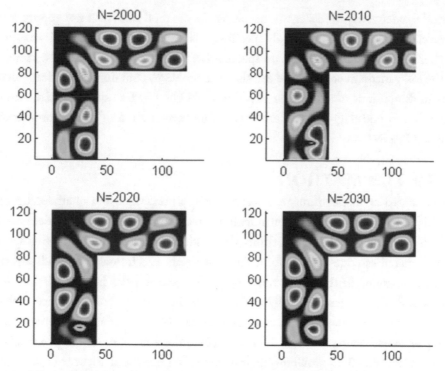

FIGURE 2.4: Magnitude of the E_y field component at different time steps. The numbers on the axes indicate the respective cell.

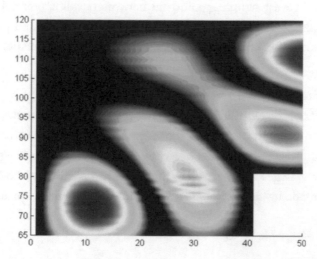

FIGURE 2.5: Magnitude of the E_y field component in the vicinity of the H-bend at $N = 2040$ time step.

spatial period to adequately model any information. As the total analysis is not approximate and the smallest wavelength cannot be punctually specified, a larger amount of samples is required. Finally, the researcher must focus its attention on spaces filled with penetrable materials. Here, the wavelength in the lossy medium controls the cell size. For problems containing electrically dense media, this results in dimensions much smaller than those obtained in free space or PEC arrangements. A possible solution (although not exclusive) is the implementation of nonuniform meshes in the diverse areas of interest.

2.3   THE TLM METHOD

Originally presented in the electromagnetics community around three and a half decades ago [6], the TLM technique has been proven a dynamic and versatile means for efficient field computations [6–9, 34–46]. In the explicit TLM algorithm, the continuous space is divided into a set of elementary cells, designated as nodes, which form the basic mesh. Electromagnetic fields are represented by wave pulses scattered in the nodes and propagating counterparts in transmission lines among adjacent nodes. Such a structure is fully analogous to the network concept; nevertheless, the introduction of wave amplitudes must be associated to transverse electric and magnetic components. As a consequence and contrary to the 1-D case where these amplitudes are practically a variable replacement, in 2-D and 3-D problems, their introduction necessitates a set of reference curves or planes, respectively, to specify tangential areas. As a matter of fact, field components are defined in terms of these planes, with the wave amplitudes being perpendicular to them. On each boundary plane separating two TLM cells, a sampling point for the tangential electric and magnetic quantities is defined, whereas in every sampling point, one network port is appointed to each polarization. This multiport configuration signifies the aforementioned TLM node and provides the pertinent flexibility for the discretization of the EMC structure or immunity application at hand.

2.3.1   Principal Properties of the TLM Network

The fundamental building block of a 2-D TLM network is a "shunt" node with four sections of transmission lines with a length of $\Delta h/2$, as shown in Figure 2.6a. This arrangement may be approximated by the lumped-element model of Figure 2.6b to give the basic equations. Conducting a comparison between voltages and currents in the equivalent circuit with the relations among the components of a *TE* mode inside a rectangular waveguide, the ensuing expressions are derived

$$E_y \equiv V_y, \; H_z \equiv I_{x_1} - I_{x_3}, \quad H_x \equiv I_{z_4} - I_{z_2}, \; \varepsilon \equiv 2C, \; \mu \equiv L, \qquad (2.52)$$

where V_y is the voltage, $I_{x_1}, I_{x_3}, I_{z_2}, I_{z_4}$ are the currents at the corresponding nodes, C is the capacitance, and L the inductance of the network. Moreover, for $\varepsilon_r = \mu_r = 1$, one obtains $1/\sqrt{LC} = 1/\sqrt{\mu_0 \varepsilon_0} = c$. It is

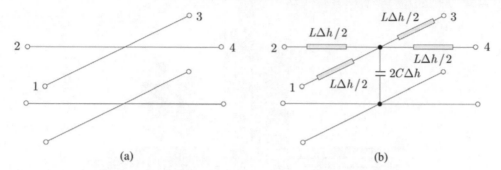

FIGURE 2.6: (a) Schematic description and (b) lumped-element equivalent of a shunt node in the TLM method.

mentioned that the 2-D TLM network can simulate an isotropic medium only as long as all frequencies are far below the network cutoff frequency, i.e., the overall propagation velocity is constant and equal to $c/\sqrt{2}$.

For the 3-D formulation, a $\Delta x \times \Delta y \times \Delta z$ elementary cell is considered to assign the characteristics of electromagnetic fields to the aspects of the TLM node. Toward this objective, the most efficient candidate is the symmetrical condensed node (SCN) [7–9, 34] of Figure 2.7, which comprises a set of interconnected lines such that on every cell face, two ports (orthogonal to each other) are applicable for the consistent modeling of any polarization pattern. Retaining the notation of the existing scientific literature, the voltages of the ports receive a consecutive labeling from 1 to 12. In the analysis herein, a cubical SCN, signifying a vacuum-filled cell, is studied at first place. Because of the higher dimensionality and the structure of the particular node, the extraction of the desired network counterparts is more complicated than the 2-D case. Scrutinizing Figure 2.7, two essential node "ingredients" may be discerned, namely, the so-called shunt and series nodes (Figure 2.8). As above, an equivalent voltage V_y and a current I_z are defined, whereas for the development of the suitable update equations, the principles of electric charge and magnetic flux conservation together with the continuity of electric and magnetic field quantities are strictly imposed. Enforcement of each of these conditions leads to three expressions among incident and reflected waves, hence, 12 equations in total.

Starting from electric charge conservation for the y-component of the electric field and consulting Figure 2.8a, the total incident (I) charge is equated to the total reflected (R) charge for the four half lines, as

$$\frac{C}{2}\left(V_3^I + V_4^I + V_{11}^I + V_8^I\right) = \frac{C}{2}\left(V_3^R + V_4^R + V_{11}^R + V_8^R\right), \tag{2.53}$$

with other two relations acquired in a similar fashion for the x and z terms.

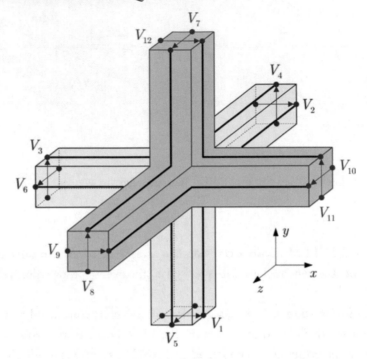

FIGURE 2.7: Description of the 3-D symmetrical condensed node for the TLM method.

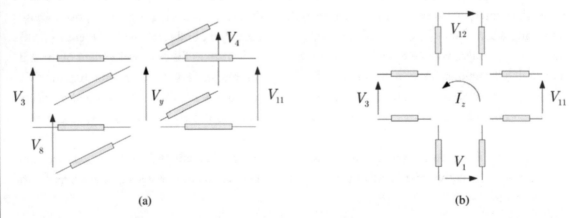

FIGURE 2.8: An example of a (a) shunt and (b) series node for the TLM method according to the numbering and orientation of the 3-D SCN model in Figure 2.7.

On the other hand, the conservation of magnetic flux—or evenly the fulfillment of (2.4)—implies that the total flux linked to all cell lines is zero,

$$\sum_m L_m I_m = \sum_m L_m (I_m^I - I_m^R) = 0\,, \tag{2.54}$$

which applied for the z-component of magnetic field (Figure 2.8b) gives

$$\frac{L}{2}\left(I_{11}^I - I_3^I + I_1^I - I_{12}^I\right) = \frac{L}{2}\left(I_{11}^R - I_3^R + I_1^R - I_{12}^R\right) \tag{2.55}$$

Substitution of $I^I = V^I/\eta$ and $I^R = V^R/\eta$ (for η the characteristic impedance) in (2.55) provides

$$V_{11}^I - V_3^I + V_1^I - V_{12}^I = -V_{11}^R + V_3^R - V_1^R + V_{12}^R\,, \tag{2.56}$$

and analogous equations for the x,y quantities.

Proceeding to the electric field continuity principle, one deduces the conclusion that y-components, regardless of being calculated on the x- or z-directed lines of the network, should be the same, namely,

$$V_{11} + V_3 = V_8 + V_4\,, \tag{2.57}$$

or via the decomposition with regard to incident and reflected voltages,

$$V_{11}^I + V_3^I + V_{11}^R + V_3^R = V_8^I + V_4^I + V_8^R + V_4^R \tag{2.58}$$

Finally, the continuity of magnetic fields entails that z-components must be the same, whether evaluated through the x- or y-directed lines of the network, i.e.,

$$I_1 - I_{12} = I_{11} - I_3\,, \tag{2.59}$$

or using incident and reflected voltages,

$$V_{11}^I - V_3^I + V_3^R - V_{11}^R = V_1^I - V_{12}^I + V_{12}^R - V_1^R \tag{2.60}$$

Similar expressions to (2.58) and (2.60) are obtained for the other components of electric and magnetic field, respectively [9]. In this manner, (2.53), (2.56), (2.58), (2.60), and their constituents for the remaining polarizations form a system of 12 equations, $\mathbf{V}^R = \mathbf{S}\mathbf{V}^I$, with

$$
\mathbf{S} = \frac{1}{2}
\begin{bmatrix}
0 & 1 & 1 & 0 & 0 & 0 & 0 & 0 & 1 & 0 & -1 & 0 \\
1 & 0 & 0 & 0 & 0 & 1 & 0 & 0 & 0 & -1 & 0 & 1 \\
1 & 0 & 0 & 1 & 0 & 0 & 0 & 1 & 0 & 0 & 0 & -1 \\
0 & 0 & 1 & 0 & 1 & 0 & -1 & 0 & 0 & 0 & 1 & 0 \\
0 & 0 & 0 & 1 & 0 & 1 & 0 & -1 & 0 & 1 & 0 & 0 \\
0 & 1 & 0 & 0 & 1 & 0 & 1 & 0 & -1 & 0 & 0 & 0 \\
0 & 0 & 0 & -1 & 0 & 1 & 0 & 1 & 0 & 1 & 0 & 0 \\
0 & 0 & 1 & 0 & -1 & 0 & 1 & 0 & 0 & 0 & 1 & 0 \\
1 & 0 & 0 & 0 & 0 & -1 & 0 & 0 & 0 & 1 & 0 & 1 \\
0 & -1 & 0 & 0 & 1 & 0 & 1 & 0 & 1 & 0 & 0 & 0 \\
-1 & 0 & 0 & 1 & 0 & 0 & 0 & 1 & 0 & 0 & 0 & 1 \\
0 & 1 & -1 & 0 & 0 & 0 & 0 & 0 & 1 & 0 & 1 & 0
\end{bmatrix}
\tag{2.61}
$$

that completely describes the SCN element. Specifically and after some algebra, the reflected voltages, as prescribed in Figure 2.7, are given by

$$
V_1^R = \frac{1}{2}\left(V_2^I + V_3^I + V_9^I - V_{11}^I\right), \qquad V_2^R = \frac{1}{2}\left(V_1^I + V_6^I + V_{12}^I - V_{10}^I\right), \tag{2.62a}
$$

$$
V_3^R = \frac{1}{2}\left(V_1^I + V_4^I + V_8^I - V_{12}^I\right), \qquad V_4^R = \frac{1}{2}\left(V_3^I + V_5^I + V_{11}^I - V_7^I\right), \tag{2.62b}
$$

$$
V_5^R = \frac{1}{2}\left(V_4^I + V_6^I + V_{10}^I - V_8^I\right), \qquad V_6^R = \frac{1}{2}\left(V_2^I + V_5^I + V_7^I - V_9^I\right), \tag{2.62c}
$$

$$
V_7^R = \frac{1}{2}\left(V_6^I + V_8^I + V_{10}^I - V_4^I\right), \qquad V_8^R = \frac{1}{2}\left(V_3^I + V_7^I + V_{11}^I - V_5^I\right), \tag{2.62d}
$$

$$
V_9^R = \frac{1}{2}\left(V_1^I + V_{10}^I + V_{12}^I - V_6^I\right), \qquad V_{10}^R = \frac{1}{2}\left(V_5^I + V_7^I + V_9^I - V_2^I\right), \tag{2.62e}
$$

$$
V_{11}^R = \frac{1}{2}\left(V_4^I + V_8^I + V_{12}^I - V_1^I\right), \qquad V_{12}^R = \frac{1}{2}\left(V_2^I + V_9^I + V_{11}^I - V_3^I\right) \tag{2.62f}
$$

Having determined the unknown voltages by means of (2.62), the subsequent step is the replacement of pulses with the suitable neighbors, as in the 2-D study (2.52). Note that the propagation velocity in the computational space is $\Delta t = \Delta / (2c)$, with $\Delta = \Delta x = \Delta y = \Delta z$ in the case of a cubical mesh. The aforementioned formulation indicates the versatility of the TLM rationale, which is proven to be more direct and robust than those of other time-domain techniques, especially when dealing with high-frequency EMC problems.

These assertions along with the fact that the validity of the discrete equations does not depend on the size and shape of the domain or on the materials its diverse regions are filled with, lead to a certain topological framework that is also transferred to the time-advancing procedure. Observe that as soon as the space-time domain quantities of the problem under investigation are properly discriminated and the correct node has been selected, the type of solution is uniquely determined. This is opposed to conventional practices that conduct two separate actions for the solution of a problem, i.e., they first discretize the domain in space and then create the system of equations. The topological interpretation of the TLM method has an additional advantage apart from the already enumerated ones; it reveals, in a "circuitous" way, the pitfalls of the classical electromagnetic outlook to construct—by combining arbitrary discretization schemes in space with diverse processes in time—rather unstable algorithms that do not adhere to the physics of the problem; a situation more frequently encountered in higher-order structural approximations.

2.3.2   Discretization of Maxwell's Equations

Apart from its circuit-oriented derivation, the TLM algorithm can be extracted directly from Maxwell's equations using certain concepts from Hilbert space theory and the method of moments [47] because the discrete field state may be described by an enumerable set of real/complex quantities and governed by linear mapping rules [7, 38]. Recollecting the geometry of Figure 2.7 and expanding the TLM node as shown in Figure 2.9, scattering at shunt nodes is shifted by half a temporal step with respect to the scattering at series nodes. Therefore, to obtain the fundamental TLM expressions, electric and magnetic components are represented by an expansion in subdomain base functions, which, in the present study, are triangular pulses in time and a product of 2-D triangular/rectangular pulses in space. Particularly, the expansions of $\mathbf{E}$ field quantities are shifted by half an increment in space and time relative to their $\mathbf{H}$ counterparts. In this manner, the resulting formulae preserve the consistency of the continuous state and guarantee accurate solutions.

Commencing from Maxwell's curl laws, (2.1) and (2.2), now written as

$$\nabla \times \mathbf{H} = \frac{1}{c\eta_0} \frac{\partial \mathbf{E}}{\partial t}, \quad \nabla \times \mathbf{E} = -\frac{\eta_0}{c} \frac{\partial \mathbf{H}}{\partial t}, \tag{2.63}$$

with $\eta_0 = \sqrt{\mu_0/\varepsilon_0}$ and $c = 1/\sqrt{\mu_0/\varepsilon_0}$, electric and magnetic field components are expanded in

$$E_x(x,y,z,t) = \sum_{i,j,k,n=-\infty}^{+\infty} E_x|_{i+1/2,j,k}^n P_{i+1/2}(x) q_j(y) q_k(z) q_n(t), \tag{2.64a}$$

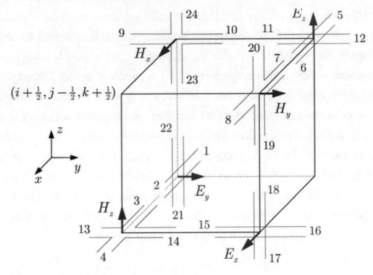

FIGURE 2.9: The 3-D elementary TLM node in an expanded form.

$$E_y(x,y,z,t) = \sum_{i,j,k,n=-\infty}^{+\infty} E_y\big|_{i,j+1/2,k}^{n} q_i(x) p_{j+1/2}(y) q_k(z) q_n(t), \qquad (2.64b)$$

$$E_z(x,y,z,t) = \sum_{i,j,k,n=-\infty}^{+\infty} E_z\big|_{i,j,k+1/2}^{n} q_i(x) q_j(y) p_{k+1/2}(z) q_n(t), \qquad (2.64c)$$

$$H_x(x,y,z,t) = \sum_{i,j,k,n=-\infty}^{+\infty} H_x\big|_{i,j+1/2,k+1/2}^{n+1/2} p_i(x) q_{j+1/2}(y) q_{k+1/2}(z) q_{n+1/2}(t), \qquad (2.65a)$$

$$H_y(x,y,z,t) = \sum_{i,j,k,n=-\infty}^{+\infty} H_y\big|_{i+1/2,j,k+1/2}^{n+1/2} q_{i+1/2}(x) p_j(y) q_{k+1/2}(z) q_{n+1/2}(t), \qquad (2.65b)$$

$$H_z(x,y,z,t) = \sum_{i,j,k,n=-\infty}^{+\infty} H_z\big|_{i+1/2,j+1/2,k}^{n+1/2} q_{i+1/2}(x) q_{j+1/2}(y) p_k(z) q_{n+1/2}(t), \qquad (2.65c)$$

where the rectangular pulse $p_m(u)$, with $m = i, j, k$ and $u = x, y, z$, is given by

$$p_m(u) = P\left(\frac{u}{\Delta u} - m\right), \text{ in which } P(u) = \begin{cases} 1 & \text{for } |u| \quad 1/2 \\ 1/2 & \text{for } |u| = 1/2 \\ 0 & \text{for } |u| \quad 1/2 \end{cases} \qquad (2.66)$$

and basis functions $q_m(u)$, with $u = x, y, z, t$, by

$$q_m(u) = Q\left(\frac{u}{\Delta u} - m\right), \quad \text{in which} \quad Q(u) = \begin{cases} 1 - |u| & \text{for} \quad |u| \quad 1 \\ 0 & \text{for} \quad |u| \geq 1 \end{cases} \qquad (2.67)$$

is a triangle scaling function. Observe that $p_m(u)$ and $q_m(u)$ attain the appropriate piecewise linear and step approximations of the exact solution of (2.63) along u-axis, respectively.

The next stage in the TLM formulation is the sampling of Maxwell's equations at the cell boundaries through delta functions in space and time. Thus, let us concentrate on the shunt nodes of Figure 2.9, for a cubical uniform grid, and opt for the $(i, j, k + 1/2)$ point. Such a choice requires the receipt of samples at $(i \pm 1/4, j, k + 1/2)$ and $(i, j \pm 1/4, k + 1/2)$ points at time steps $n + 1/4$ and $n + 3/4$. Because of the analogous treatment of all $\mathbf{E}$ and $\mathbf{H}$ quantities, the analysis deals with the z-component of Ampère's law (2.7c) only, for $\sigma = 0$. For example, the sampling of the $\partial E_z / \partial t$ partial derivative is conducted via

$$\text{Sampling}\left\{\frac{\partial E_z}{\partial t}\right\} \equiv \iiiint \frac{\partial E_z}{\partial t} \delta(x - i\Delta x)\delta(y - j\Delta y)\delta(z - k\Delta z)\delta(t - n\Delta t) \, \mathrm{d}x \, \mathrm{d}y \, \mathrm{d}z \, \mathrm{d}t$$

$$(2.68)$$

where for the pertinent mathematical manipulations, the following integrals are taken into account

$$\int_{-\infty}^{+\infty} \delta[u - (l + \tau)\Delta u] p_m(u) \mathrm{d}u = \delta_{l,m},$$

$$\int_{-\infty}^{+\infty} \delta(u - l\Delta u) q_m(u) \mathrm{d}u = \delta_{l,m},$$

$$\int_{-\infty}^{+\infty} \delta[u - (l + 1/4)\Delta u] q_m(u) \mathrm{d}u = \frac{1}{4}(3\delta_{l,m} + \delta_{l+1,m}),$$

$$\int_{-\infty}^{+\infty} \delta[u - (l + 3/4)\Delta u] q_m(u) \mathrm{d}u = \frac{1}{4}(\delta_{l,m} + 3\delta_{l+1,m}),$$

$$\int_{-\infty}^{+\infty} \delta[u - (l + 1/2 + \tau)\Delta u] \frac{\partial q_m(u)}{\partial u} \mathrm{d}u = \delta_{l+1,m} - \delta_{l,m},$$

with $\delta_{l,m}$ the Kronecker delta, $l, m = i, j, k, u = x, y, z, t$, and $\tau = 0, \pm 1/4$. This process, also applied to $\partial H_x / \partial y$ and $\partial H_y / \partial x$, leads to eight discrete expressions, which, if added, yield

$$E_z|_{i+1,j,k+1/2}^{n+1} + E_z|_{i-1,j,k+1/2}^{n+1} + E_z|_{i,j+1,k+1/2}^{n+1} + E_z|_{i,j-1,k+1/2}^{n+1} + 12\,E_z|_{i,j,k+1/2}^{n+1}$$

$$- E_z|_{i+1,j,k+1/2}^{n} - E_z|_{i-1,j,k+1/2}^{n} - E_z|_{i,j+1,k+1/2}^{n} - E_z|_{i,j-1,k+1/2}^{n} - 12\,E_z|_{i,j,k+1/2}^{n}$$

$$= \kappa\eta\left(H_y|_{i+1/2,j,k+1/2}^{n+3/2} + 6\,H_y|_{i+1/2,j,k+1/2}^{n+1/2} + H_y|_{i+1/2,j,k+1/2}^{n-1/2}\right)$$

$$- \kappa\eta\left(H_y|_{i-1/2,j,k+1/2}^{n+3/2} + 6\,H_y|_{i-1/2,j,k+1/2}^{n+1/2} + H_y|_{i-1/2,j,k+1/2}^{n-1/2}\right)$$

$$- \kappa\eta\left(H_x|_{i,j+1/2,k+1/2}^{n+3/2} + 6\,H_x|_{i,j+1/2,k+1/2}^{n+1/2} + H_x|_{i,j+1/2,k+1/2}^{n-1/2}\right)$$

$$+ \kappa\eta\left(H_x|_{i,j-1/2,k+1/2}^{n+3/2} + 6\,H_x|_{i,j-1/2,k+1/2}^{n+1/2} + H_x|_{i,j-1/2,k+1/2}^{n-1/2}\right) \tag{2.69}$$

for $\kappa = 2\eta_0 c\Delta t\,/\,(\eta\Delta)$, with Δ as the spatial increment of the cubic mesh and η as an arbitrary impedance. To correctly relate wave amplitudes to tangential electric and magnetic field quantities at the boundary surfaces that isolate the six 2-D TLM nodes of Figure 2.9, an instructive cell boundary mapping is introduced. According to this efficient approach, each component is described as the result of its series expansion at the boundary of the cell [37]. Hence, for the E_z terms of (2.69), it holds that

$$E_z|_{i\pm1/4,j,k+1/2}^{n} = \frac{1}{4}\left(E_z|_{i\pm1,j,k+1/2}^{n} + 3\,E_z|_{i,j,k+1/2}^{n}\right), \tag{2.70a}$$

$$E_z|_{i,j\pm1/4,k+1/2}^{n} = \frac{1}{4}\left(E_z|_{i,j\pm1,k+1/2}^{n} + 3\,E_z|_{i,j,k+1/2}^{n}\right), \tag{2.70b}$$

whereas for the H_x and H_y ones

$$H_x|_{i,j+1/4,k+1/2}^{n+1/2} = \frac{1}{4}\left(3\,H_x|_{i,j+1/2,k+1/2}^{n+1/2} + H_x|_{i,j-1/2,k+1/2}^{n+1/2}\right), \tag{2.71a}$$

$$H_x|_{i,j-1/4,k+1/2}^{n+1/2} = \frac{1}{4}\left(H_x|_{i,j+1/2,k+1/2}^{n+1/2} + 3\,H_x|_{i,j-1/2,k+1/2}^{n+1/2}\right), \tag{2.71b}$$

$$H_y|_{i+1/4,j,k+1/2}^{n+1/2} = \frac{1}{4}\left(3\,H_x|_{i+1/2,j,k+1/2}^{n+1/2} + H_x|_{i-1/2,j,k+1/2}^{n+1/2}\right), \tag{2.71c}$$

$$H_y|_{i-1/4,j,k+1/2}^{n+1/2} = \frac{1}{4}\left(H_x|_{i+1/2,j,k+1/2}^{n+1/2} + 3\,H_x|_{i-1/2,j,k+1/2}^{n+1/2}\right) \tag{2.71d}$$

To this end, similar values are defined for the temporal variation in (2.69), as

$$E_z|_{i,j,k}^{n+1/4} = \frac{1}{4}\left(E_z|_{i,j,k}^{n+1} + 3\,E_z|_{i,j,k}^{n}\right), \quad E_z|_{i,j,k}^{n+3/4} = \frac{1}{4}\left(3\,E_z|_{i,j,k}^{n+1} + E_z|_{i,j,k}^{n}\right), \tag{2.72a}$$

$$H_{x,y}\big|_{i,j,k}^{n+1/4} = \frac{1}{4}\left(3\,H_{x,y}\big|_{i,j,k}^{n+1/2} + H_{x,y}\big|_{i,j,k}^{n-1/2}\right),\ H_{x,y}\big|_{i,j,k}^{n+3/4} = \frac{1}{4}\left(H_{x,y}\big|_{i,j,k}^{n+3/2} + 3\,H_{x,y}\big|_{i,j,k}^{n+1/2}\right) \quad (2.72b)$$

Substitution of (2.70) and (2.71) in (2.69) gives

$$E_z\big|_{i+1/4,j,k+1/2}^{n+3/4} + E_z\big|_{i-1/4,j,k+1/2}^{n+3/4} + E_z\big|_{i,j+1/4,k+1/2}^{n+3/4} + E_z\big|_{i,j-1/4,k+1/2}^{n+3/4}$$

$$- E_z\big|_{i+1/4,j,k+1/2}^{n+1/4} - E_z\big|_{i-1/4,j,k+1/2}^{n+1/4} - E_z\big|_{i,j+1/4,k+1/2}^{n+1/4} - E_z\big|_{i,j-1/4,k+1/2}^{n+1/4}$$

$$= \kappa\eta\left(H_y\big|_{i+1/4,j,k+1/2}^{n+3/4} + H_y\big|_{i+1/2,j,k+1/2}^{n+1/4} - H_y\big|_{i-1/4,j,k+1/2}^{n+3/4} - H_y\big|_{i-1/2,j,k+1/2}^{n+1/4}\right)$$

$$- \kappa\eta\left(H_x\big|_{i,j+1/4,k+1/2}^{n+3/4} + H_x\big|_{i,j+1/4,k+1/2}^{n+1/4} - H_x\big|_{i,j-1/4,k+1/2}^{n+3/4} - H_x\big|_{i,j-1/4,k+1/2}^{n+1/4}\right), \quad (2.73)$$

with equivalent expressions derived for the rest of (2.9).

The boundary mapping of wave amplitudes is completed by defining an electric $|\mathcal{E}_{sh}\rangle$ and a magnetic $|\mathcal{H}_{sh}\rangle$ field vector in the Hilbert space, which, respectively, combine all **E** and **H** components at the three shunt nodes of Figure 2.9 in suitable 12×1 column vectors. Thus, one acquires

$$|\mathcal{E}_{sh}\rangle = \sum_{i,j,k,n=-\infty}^{+\infty}\begin{bmatrix} E_y\big|_{pa-}^{n+3/4} & E_y\big|_{pa+}^{n+3/4} & E_z\big|_{pb-}^{n+3/4} & E_z\big|_{pb+}^{n+3/4} & E_z\big|_{pc-}^{n+3/4} & E_z\big|_{pc+}^{n+3/4} \\ E_x\big|_{pd-}^{n+3/4} & E_x\big|_{pd+}^{n+3/4} & E_x\big|_{pe-}^{n+3/4} & E_x\big|_{pe+}^{n+3/4} & E_y\big|_{pf-}^{n+3/4} & E_y\big|_{pf+}^{n+3/4} \end{bmatrix}^T |n;i,j,k\rangle, \quad (2.74)$$

$$|\mathcal{H}_{sh}\rangle = \eta\sum_{i,j,k,n=-\infty}^{+\infty}\begin{bmatrix} H_z\big|_{pa-}^{n+3/4} & H_z\big|_{pa+}^{n+3/4} & H_y\big|_{pb-}^{n+3/4} & H_y\big|_{pb+}^{n+3/4} & H_x\big|_{pc-}^{n+3/4} & H_x\big|_{pc+}^{n+3/4} \\ H_z\big|_{pd-}^{n+3/4} & H_z\big|_{pd+}^{n+3/4} & H_y\big|_{pe-}^{n+3/4} & H_y\big|_{pe+}^{n+3/4} & H_x\big|_{pf-}^{n+3/4} & H_y\big|_{pf+}^{n+3/4} \end{bmatrix}^T |n;i,j,k\rangle, \quad (2.75)$$

where

$$pa\mp = (i\mp1/4,j-1/2,k), \qquad pb\mp = (i\mp1/4,j,k+1/2),$$

$$pc\mp = (i,j\mp1/4,k+1/2), \qquad pd\mp = (i+1/2,j\mp1/4,k),$$

$$pe\mp = (i+1/2,j,k\mp1/4), \qquad pf\mp = (i,j-1/2,k\mp1/4),$$

and $|n;i,j,k\rangle$ as the ket-vector notation that assigns a base vector $|i,j,k\rangle$ to every (i,j,k) discrete node of the TLM lattice at time-step n. In the same context, $|\mathcal{E}_{sh}\rangle$ and $|\mathcal{H}_{se}\rangle$ vectors merge all electric and magnetic field quantities at the three series nodes of Figure 2.9, as

$$|\mathcal{E}_{se}\rangle = \sum_{i,j,k,n=-\infty}^{+\infty} \begin{bmatrix} E_y\big|_{qa(1)}^{n+1/4} & E_y\big|_{qa(3)}^{n+1/4} & E_z\big|_{qb(1)}^{n+1/4} & E_z\big|_{qb(3)}^{n+1/4} & E_z\big|_{qc(3)}^{n+1/4} & E_z\big|_{qc(1)}^{n+1/4} \\ E_x\big|_{qd(3)}^{n+1/4} & E_x\big|_{qd(1)}^{n+1/4} & E_x\big|_{qe(1)}^{n+1/4} & E_x\big|_{qe(3)}^{n+1/4} & E_y\big|_{qf(1)}^{n+1/4} & E_y\big|_{qf(3)}^{n+1/4} \end{bmatrix}^T |n;i,j,k\rangle, \qquad (2.76)$$

$$|\mathcal{H}_{se}\rangle = \eta \sum_{i,j,k,n=-\infty}^{+\infty} \begin{bmatrix} H_z\big|_{qa(1)}^{n+1/4} & H_z\big|_{qa(3)}^{n+1/4} & H_y\big|_{qb(1)}^{n+1/4} & H_y\big|_{qb(3)}^{n+1/4} & H_x\big|_{qc(3)}^{n+1/4} & H_x\big|_{qc(1)}^{n+1/4} \\ H_z\big|_{qd(3)}^{n+1/4} & H_z\big|_{qd(1)}^{n+1/4} & H_y\big|_{qe(1)}^{n+1/4} & H_y\big|_{qe(3)}^{n+1/4} & H_x\big|_{qf(1)}^{n+1/4} & H_y\big|_{qf(3)}^{n+1/4} \end{bmatrix}^T |n;i,j,k\rangle$$

$$(2.77)$$

where

$$qa(m) = (i + m/4, j - 1/2, k), \qquad qb(m) = (i + m\ 4, j, k + 1/2),$$

$$qc(m) = (i, j - m/4, k + 1/2), \qquad qd(m) = (i + 1/2, j - m\ 4, k),$$

$$qe(m) = (i + 1/2, j, k + m/4), \qquad qf(m) = (i, j - 1/2, k + m/4),$$

Then, the incident and reflected wave amplitudes in the Hilbert space are represented by means of

$$|\mathcal{A}_d\rangle = \sum_{i,j,k,n=-\infty}^{+\infty} A_d\big|_{i,j,k}^{n} |n,i,j,k\rangle \;\; \text{and} \;\; |\mathcal{B}_d\rangle = \sum_{i,j,k,n=-\infty}^{+\infty} B_d\big|_{i,j,k}^{n} |n,i,j,k\rangle \qquad (2.78)$$

If $d = sh$, vectors $|\mathcal{A}_{sh}\rangle$ and $|\mathcal{B}_{sh}\rangle$ refer to the wave amplitudes at the three shunt nodes, and if $d = se$, vectors $|\mathcal{A}_{se}\rangle$ and $|\mathcal{B}_{se}\rangle$ incorporate the corresponding terms at the three series nodes of the mesh. Specifically considering the numbering of Figure 2.9, one deduces that nodes 1, 2, 5, 6, 11, 12, 15, 16, 17, 18, 21, 22 belong to $\mathbf{A}_{sh}$ and $\mathbf{B}_{sh}$, whereas nodes 3, 4, 7, 8, 9, 10, 13, 14, 19, 20, 23, 24 to $\mathbf{A}_{se}$ and $\mathbf{B}_{se}$.

In this framework and after some calculus, the desired mapping for the TLM method is expressed as

$$|\mathcal{A}_d\rangle = \frac{1}{2}\left(|\mathcal{E}_d\rangle + \Xi|\mathcal{H}_d\rangle\right), \qquad |\mathcal{B}_d\rangle = \frac{1}{2}\left(|\mathcal{E}_d\rangle - \Xi|\mathcal{H}_d\rangle\right), \qquad (2.79a)$$

or

$$|\mathcal{E}_d\rangle = |\mathcal{A}_d\rangle + |\mathcal{B}_d\rangle, \qquad |H_d\rangle = \Lambda\left(|A_d\rangle - |B_d\rangle\right), \qquad (2.79b)$$

with

$$\Xi = \begin{bmatrix} \Lambda & \mathbf{0} & \mathbf{0} \\ \mathbf{0} & \Lambda & \mathbf{0} \\ \mathbf{0} & \mathbf{0} & \Lambda \end{bmatrix} \text{ and } \Lambda = \begin{bmatrix} 1 & 0 & 0 & 0 \\ 0 & -1 & 0 & 0 \\ 0 & 0 & -1 & 0 \\ 0 & 0 & 0 & 1 \end{bmatrix}$$

It is significant to stress that on every boundary surface, incident wave amplitudes into a 2-D TLM node are identical to their counterparts reflected from the adjacent nodes, namely,

$$|\mathcal{A}_{sh}\rangle = \mathbf{M}_{sh}|\mathcal{B}_{se}\rangle \text{ and } |\mathcal{A}_{se}\rangle = \mathbf{M}_{se}|\mathcal{B}_{sh}\rangle \tag{2.80}$$

in which the connection matrices $\mathbf{M}_{sh}$ and $\mathbf{M}_{se}$ are obtained by

$$\mathbf{M}_{sh} = \mathbf{X}(\Phi_{1,2} + \Phi_{3,4}) + \mathbf{Y}^{\dagger}(\Phi_{6,5} + \Phi_{8,7}) + \mathbf{Z}(\Phi_{9,10} + \Phi_{11,12})$$
$$+ \Phi_{2,1} + \Phi_{4,3} + \Phi_{5,6} + \Phi_{7,8} + \Phi_{10,9} + \Phi_{12,11} \tag{2.81}$$

and $\mathbf{M}_{se} = \mathbf{M}_{sh}^{\dagger}$, for $(\Phi_{i,j})_{l,m} = \delta_{i,l}\delta_{j,m}$ a set of 12×12 (l, m)-matrices and shift operators

$$\mathbf{X}|n; i, j, k\rangle \equiv |n; i+1, j, k\rangle \qquad \mathbf{X}^{\dagger}|n; i, j, k\rangle \equiv |n; i-1, j, k\rangle, \tag{2.82a}$$

$$\mathbf{Y}|n; i, j, k\rangle \equiv |n; i, j+1, k\rangle \qquad \mathbf{Y}^{\dagger}|n; i, j, k\rangle \equiv |n; i, j-1, k\rangle, \tag{2.82b}$$

$$\mathbf{Z}|n; i, j, k\rangle \equiv |n; i, j, k+1\rangle \qquad \mathbf{Z}^{\dagger}|n; i, j, k\rangle \equiv |n; i, j, k-1\rangle, \tag{2.82c}$$

with $\dagger$ denoting the Hermitian conjugate of the respective matrix.

Finally, (2.73), for $d = sh$, can be reformulated in terms of Hilbert space representation as

$$[0\ 0\ 1\ 1\ 1\ 1\ 0\ 0\ 0\ 0\ 0\ 0](1 - \mathbf{T})|\mathcal{E}_{sh}\rangle = [0\ 0\ -1\ 1\ 1\ -1\ 0\ 0\ 0\ 0\ 0\ 0](1 + \mathbf{T})\kappa|\mathcal{H}_{sh}\rangle, \tag{2.83}$$

where $\mathbf{T}|n; i, j, k\rangle \equiv |n+1; i, j, k\rangle$ is the corresponding time shift operator as those of (2.82).

Repeating the sampling and development procedure of (2.68)–(2.82) for (2.8c)—the dual time-dependent Maxwell's equation of (2.8c)—one concludes to

$$[1\ 1\ 0\ 0\ 0\ 0\ 1\ 1\ 0\ 0\ 0\ 0](1 - \mathbf{T})|\mathcal{H}_{sh}\rangle = [-1\ 1\ 0\ 0\ 0\ 0\ -1\ 1\ 0\ 0\ 0\ 0](1 + \mathbf{T})\nu|\mathcal{E}_{sh}\rangle, \tag{2.84}$$

with $v = 2\eta c\Delta t / (\eta_0\Delta)$. If κ and v are set to unity, (2.83) and (2.84), through (2.79), become

$$[0\ 0\ 1\ 1\ 1\ 1\ 0\ 0\ 0\ 0\ 0\ 0]\,|\mathcal{B}_{sh}\rangle = [0\ 0\ 1\ 1\ 1\ 1\ 0\ 0\ 0\ 0\ 0\ 0]\,\mathbf{T}\,|\mathcal{A}_{sh}\rangle \qquad (2.85)$$

$$[-1\ 1\ 0\ 0\ 0\ 0\ -1\ 1\ 0\ 0\ 0\ 0]\,|\mathcal{B}_{sh}\rangle = [-1\ 1\ 0\ 0\ 0\ 0\ -1\ 1\ 0\ 0\ 0\ 0]\,\mathbf{T}\,|\mathcal{A}_{sh}\rangle \qquad (2.86)$$

An equivalent couple to (2.85), (2.86) is derived for $d = se$ and the series nodes, by employing the derivatives of spatial delta functions as test functions. In an akin manner, the other four Cartesian components (2.7a), (2.7b), and (2.8a), (2.8b) may be adequately processed to provide the other four couples of expressions. The 12 relations constitute the fundamental TLM equations, which in matrix notation are given by

$$|\mathcal{B}\rangle = \mathbf{T}\Pi\,|\mathcal{A}\rangle, \qquad (2.87)$$

for $|\mathcal{A}\rangle = \left[\,|\mathcal{A}_{sh}\rangle\,|\mathcal{A}_{se}\rangle\,\right]$, $|\mathcal{B}\rangle = \left[\,|\mathcal{B}_{sh}\rangle\,|\mathcal{B}_{se}\rangle\,\right]$, and Π the scattering matrix acquired via

$$\Pi = \begin{bmatrix} 0 & \Pi_0 & \Pi_0^T \\ \Pi_0^T & 0 & \Pi_0 \\ \Pi_0 & \Pi_0^T & 0 \end{bmatrix}, \quad \text{where} \quad \Pi_0 = \begin{bmatrix} 0 & 0 & 1/2 & -1/2 \\ 0 & 0 & -1/2 & 1/2 \\ 1/2 & 1/2 & 0 & 0 \\ 1/2 & 1/2 & 0 & 0 \end{bmatrix} \qquad (2.88)$$

is a real, symmetric, and Hermitian operator, defined in the initial Hilbert space. It is to be stressed that the dispersion relation of (2.87) is proven to be, after some algebra, analogous to that of the FDTD method, thus indicating a similar behavior of the TLM technique in the case of lattice artificialities [35, 36, 40].

2.3.3   Profits and EMC Realization Issues

The TLM method can be adapted to a wide area of EMC problems such as high frequency circuit boards, coupled structures or biomedical applications to name a few. Not only is it an expedient numerical tool, but owing to its affinity with the physical mechanism of wave propagation, it has the potential to offer reliable insights into the nature and behavior of electromagnetic fields. The major merit of the technique is its efficiency to handle involved devices. Distinguished by an indisputable flexibility, the TLM network encompasses the properties of $\mathbf{E}$ and $\mathbf{H}$ vectors accompanied by their interactions with all boundaries and material interfaces. Therefore, it is not mandatory to

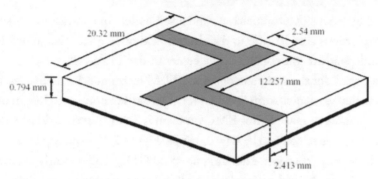

FIGURE 2.10: A 3-D microstrip filter realized on a RT/Duroid substrate.

reformulate the problem under study for every minor modification of its parameters because no loss of convergence, instabilities, or spurious modes is generated.

Another profit of the explicit algorithm rests on the notable amount of evidence obtained during a single evaluation. Amid them, one may perceive the extraction of rigorous impulse responses, leading in sequence to successful responses to any excitation scheme, as well as the determination of the dominant and higher-order modes, calculated in the frequency domain through the Fourier transform [37, 43, 45]. To this aim, enhanced performance is attained by the elaborate classification of the problem's topology and the correct association of its diverse regions with the proper amount of cells. Actually, the use of SCN for the formation of the fundamental TLM equations can be

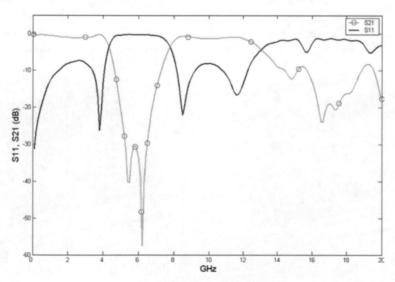

FIGURE 2.11: *S*-parameter calculation for the 3-D microstrip filter.

evenly conducted by means of asymmetrical condensed nodes. In contrast to the symmetrical case, magnetic field expansions are identical to the electric field ones, a fact that allows the definition of the six linearly independent components at the center of the TLM cell.

As an illustration for the competences of the TLM technique, let us analyze the filter of Figure 2.10, implemented via a microstrip line on a RT/Duroid substrate with height of 0.794 mm and $\varepsilon_r = 2.2$. The computational space of this EMC problem is divided into a grid of $80 \times 108 \times 22$ cells, and the time increment is selected to be $\Delta t = 0.441$ ps. Figure 2.11 displays the S-parameters of the filter for a wide frequency spectrum, extending up to 20 GHz, which clearly exhibit its function in the range of 4 to 8 GHz. Also, the magnitude of E_z component at a plane in the substrate located one cell beneath the microstrip line is depicted in Figure 2.12. Again, the modeling efficiency of the method is easily discernible.

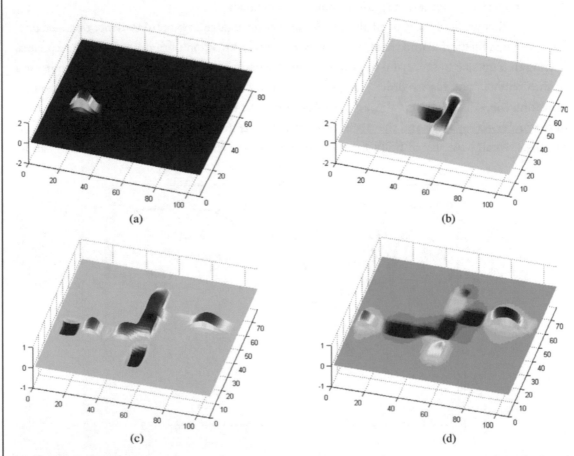

(a)

(b)

(c)

(d)

FIGURE 2.12: Snapshots of the magnitude of E_z component at (a) 12.853 ps, (b) 35.677 ps, (c) 72.351 ps, and (d) 150.398 ps.

The main limitation of the method is its computational overhead that depends on the complexity of the structures and the nonuniformity of fields. Generally, the finest feature—compared with the smallest wavelength λ of the application—must at least include three nodes (virtually $\lambda / 10$) for an acceptable resolution. As a consequence, the total storage may be specified by bearing in mind that every 2-D node needs five computer memory places along with an extra amount for the description of the output impulse response. To this extent, a 3-D TLM node requires 12 storage positions, whereas for losses or diverse constitutive parameters, it opts for up to 26. So, the construction of optimal grids is, indeed, a critical issue for the algorithm, even in the presence of modern computing systems, which, apparently, augment its universality.

2.4   THE FINITE INTEGRATION TECHNIQUE

The FIT has become a promising numerical method for the simulation of macroscopic electromagnetic phenomena and coupled problems [10, 11]. Being of similar chronological vintage to the TLM method, the algorithm establishes its principal concepts by a discrete reformulation of Maxwell's equations in their integral rendition, rather than the differential one, and incorporates six components of electric/magnetic voltages and fluxes on a dual lattice tessellation. More specifically, the former quantities are assigned to contours, whereas the latter are assigned to surfaces, thus permitting the construction of a dynamic matrix system, pertinent also for irregular and nonorthogonal grids [48–52]. This consistent finite-volume-based discretization strategy rests on the use of integral balances, which lead to long-term stable and energy-conserving analyses devoid of spurious parasites, even before commencing time-domain numerical computations. Taking into account the constantly augmenting requirements of modern EMC industry for highly accurate fabrication standards and cost-effective prototypes, it becomes apparent that the aforesaid characteristics appoint to the FIT the competency to provide promising solutions for realistic applications with convoluted geometries [53–60]. Among them, one can discern waveguide devices, antennas, shielding or immunity structures, test facilities, biomedical models or more recent configurations such as PCBs, integrated circuit components, and specialized dispersive/anisotropic media arrangements.

2.4.1   Basic Attributes and Grid Equations

The first stage of the FIT comprises the confinement of the electromagnetic field problem, which typically relates an open boundary region to a simply connected as well as bounded computational domain $\Omega \in \mathbb{R}^3$. To this objective, the method creates the algebraic analogues of Maxwell's equations that guarantee the preservation of all physical field properties in the discretized space and the most significant; they offer a unique and well-posed solution [10]. Then, Ω is decomposed into a finite number of N_V cells $V_m (m = 1, 2, \ldots, N_V)$ under the convention that all candidates are exactly adapted to each other, i.e., the intersection of two different cells is either empty or a 2-D polygon face, an 1-D edge shared by both cells or a node. Each cell edge and polygonal face is properly oriented toward a prefixed direction. This kind of decomposition forms G complex; the numerical grid of the algorithm, which may be tetrahedral, hexahedral, or any other (non)orthogonal coordinate mesh. Nonetheless and for simplicity throughout this section, Ω consists of bricks, and the discretization is attained via the tensor product Cartesian cell complex

$$G \equiv \left\{ V_m \in \mathbb{R}^3 \Rightarrow V_m \equiv [x_i, x_{i+1}] \times \left[y_j, y_{j+1}\right] \times [z_k, z_{k+1}] \right.$$
$$\left. \text{for } i = 1, 2, \quad , N_x, \ j = 1, 2, \quad , N_y, \ k = 1, 2, \quad , N_z \right\},$$

(2.89)

where (x_i, y_j, z_k) points are denoted by the i, j, k coordinate indices along the x-, y-, and z-axis, respectively. Moreover, in G, the nonempty faces $A_m (m = 1, 2, \ldots, N_A)$ are defined as the intersections of two cells, edges $L_m (m = 1, 2, \ldots, N_L)$ represent the intersections of faces, and nodes $P_m (m = 1, 2, \ldots, N_p)$ are defined by the intersection of edges.

Next, Maxwell's laws and the related constitutive equations are transformed from the continuous to discrete status by allocating electric voltages, **e**, on the edges and magnetic fluxes, **b**, on the faces of a lattice—called the primary grid—and magnetic voltages, **h**, on the edges and electric fluxes, **d**, on the faces of a dual lattice—called the secondary grid (Figure 2.13). It is stressed that the use of voltages and fluxes as state variables instead of field components directly (as in other

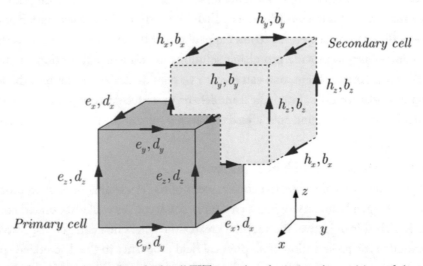

FIGURE 2.13: A primary/secondary dual-cell FIT complex, depicting the position of electric and magnetic field components.

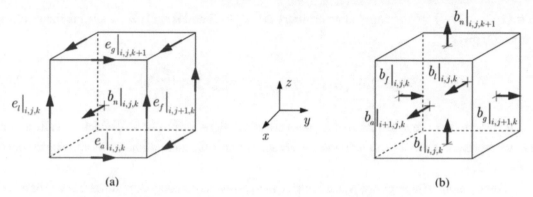

FIGURE 2.14: Graphical allocation of (a) four electric voltages on the edges of an elementary FIT cell for the evaluation of Faraday's law and (b) six magnetic fluxes on the faces of an elementary FIT cell for the computation of Ampère's law.

techniques) gives the fairly convenient option to account for measurable quantities of integral type, without having to conduct nonphysical assumptions.

In this context, expressing the electric voltage e_m along edge L_m and the corresponding magnetic flux b_m through face A_m, via

$$e_m = \oint_{L_m} \mathbf{E} \cdot d\mathbf{l} \quad \text{and} \quad b_m = \iint_{A_m} \mathbf{B} \cdot d\mathbf{s}, \tag{2.90}$$

Faraday's law in G receives the form of

$$\oint_{L_m} \mathbf{E} \cdot d\mathbf{l} = -\iint_{A_m} \frac{\partial \mathbf{B}}{\partial t} \cdot d\mathbf{s}, \tag{2.91}$$

or in accordance with the previous theoretical aspects, the ordinary differential notation of

$$\sum_p c_{mp} e_p = -\frac{\partial b_m}{\partial t}, \tag{2.92}$$

with p indicating the number of edges required for the circulation. Because the choice for the elements of Ω is the hexahedral brick, faces A_m are rectangular and p involves four electric voltage quantities. Moreover, coefficients c_{mp} contain the appropriate topological information on the incidence relation of cell edges within G as well as on their orientation and Hence, they share solely the values of -1, 0, or $+1$. For illustration, from the graphical depiction of Figure 2.14a, where $a, f,$

$g, l \in \{1, 2, \ldots, N_L\}$ are the edges of an arbitrary cell $V_m \in G$ and $n \in \{1, 2, \ldots, N_A\}$ is the resulting face, (2.92) gives

$$e_a|_{i,j,k} - e_g|_{i,j,k+1} + e_f|_{i,j+1,k} - e_l|_{i,j,k} = -\frac{\partial}{\partial t} b_n|_{i,j,k} \qquad (2.93)$$

Observe that the orientation of L_m does affect the signs within (2.56), which actually attains an exact representation of (2.91), due to the absence of any approximation or discretization procedure [11, 48, 49].

Furthermore, Ampère's law is manipulated in a completely analogous manner, but on the aforementioned fully dual (secondary) cell complex $\tilde{G}$, where $\tilde{G}$ is built by taking the foci of cells V_m as lattice points for the mesh cells $\tilde{V}_m$ of $\tilde{G}$. In the case of more generalized or unstructured G complexes, it is also viable to use the cell barycenters as boundary vertices for the construction of $\tilde{G}$. So, by means of this policy, a one-to-one relation is established between the edges of G and the faces of $\tilde{G}$ and vice versa. Retaining the same geometrical association concept, magnetic voltages h_m are defined on the $\tilde{L}_m$ edges of $\tilde{G}$, whereas electric fluxes d_m and current densities j_m are assigned to dual faces $\tilde{A}_m$, in terms of

$$h_m = \oint_{\tilde{L}_m} \mathbf{H} \cdot d\mathbf{l}, \, d_m = \iint_{\tilde{A}_m} \mathbf{D} \cdot d\mathbf{s}, \, j_m = \iint_{\tilde{A}_m} \mathbf{J} \cdot d\mathbf{s}, \qquad (2.94)$$

Therefore, Ampère's law becomes

$$\oint_{\tilde{L}_m} \mathbf{H} \cdot d\mathbf{l} = \iint_{\tilde{A}_m} \left(\frac{\partial \mathbf{D}}{\partial t} + \mathbf{J} \right) \cdot d\mathbf{s}, \qquad (2.95)$$

or in differential form

$$\sum_p \tilde{c}_{mp} h_p = \frac{\partial d_m}{\partial t} + j_m, \qquad (2.96)$$

with $\tilde{c}_{mp}$ describing the dual counterparts of c_{mp}.

Similarly, the third and fourth equations (for $\rho_m = 0$) of Maxwell's system are transformed to

$$\sum_p s_{mp} b_p = 0 \quad \text{and} \quad \sum_p \tilde{s}_{mp} d_p = q_m, \quad \text{with } q_m = \iiint_{\tilde{V}_m} \rho dV, \qquad (2.97)$$

$s_{mp}, \tilde{s}_{mp} \in \{-1, 0, +1\}$ the suitable topological coefficients for the primary/secondary cell volumes and p involving six magnetic fluxes for the computation of the closed surface integral within the cell volume V_m owing to its hexahedral structure. As an example, let us focus on Figure 2.14b, where $a, f, g, l, n, t \in \{1, 2, \ldots, N_A\}$, and prove the validity of the first relation in (2.97) through

$$b_a|_{i+1,j,k} - b_l|_{i,j,k} + b_g|_{i,j+1,k} - b_f|_{i,j,k} + b_n|_{i,j,k+1} - b_t|_{i,j,k} = 0, \qquad (2.98)$$

which again constitutes an exact formulation, as the only deviation from the physical law is the substitution of magnetic fluxes by their integrals over faces A_m.

The discretization process, presented above, results in the subsequent set of explicit matrix expressions,

$$\mathbf{Ce} = -\frac{\mathrm{d}}{\mathrm{d}t}\mathbf{b}, \qquad \tilde{\mathbf{C}}\mathbf{h} = \frac{\mathrm{d}}{\mathrm{d}t}\mathbf{d} + \mathbf{j}, \qquad (2.99)$$

$$\mathbf{Sb} = \mathbf{0}, \qquad \tilde{\mathbf{S}}\mathbf{d} = \mathbf{q}, \qquad (2.100)$$

which enclose the pertinent lattice evidence from the dual-cell complex pair $\{G, \tilde{G}\}$ and are known as *Maxwell grid equations (MGE)* [10]. The elements of "curl" matrices $\mathbf{C}, \tilde{\mathbf{C}}$ are coefficients c_{mp}, $\tilde{c}_{mp}$ of (2.92) and (2.96), whereas those of "divergence" matrices $\mathbf{S}, \tilde{\mathbf{S}}$ are coefficients s_{mp}, $\tilde{s}_{mp}$ of (2.97). Their contribution to the robust development of FIT is indeed crucial, because they act as structural degrees of freedom. Hence, the former represent the edges-to-faces incidence on the primary and secondary mesh G and $\tilde{G}$, respectively, and the latter the faces-to-volume boundary in the interior of the entire cell.

Having acquired the matrix mapping of Maxwell's equations, the next step is to derive the correct material relations between electric and magnetic voltages and fluxes or differently speaking to discretize the well-known constitutive laws. This is exactly the actual source of numerical errors in FIT, as MGE contain only topological information. Therefore, considering the one-to-one correspondence of the faces and their penetrating dual edges anywhere in Ω, such relations lead to

$$\mathbf{d} = \mathbf{M}_\varepsilon \mathbf{e} + \mathbf{p}, \qquad \mathbf{j} = \mathbf{M}_\sigma \mathbf{e}, \qquad \mathbf{h} = \mathbf{M}_\mu \mathbf{b} + \mathbf{m}, \qquad (2.101)$$

which are completely independent of the local field approximation method from an algebraic consistency point of view. In (2.101), $\mathbf{M}_\varepsilon$ and $\mathbf{M}_\mu$ are the positive definite permittivity and permeability matrices, whereas $\mathbf{M}_\sigma$ is the semipositive matrix of conductivities. Particularly, on Cartesian lattices where material tensors are isotropic, these sparse matrices have a diagonal form. Furthermore, vectors $\mathbf{p}$ and $\mathbf{m}$ arise from the existence of permanent electric and magnetic polarizations. This kind of discretization is deemed advantageous, as the approximation of field components comes through the media relations rather than the usual finite-difference expressions for spatial/temporal derivatives, such as the FDTD technique. For instance, with the definition of a maximal length l_{max} of the mesh cell edges in the Cartesian complex pair $\{G, \tilde{G}\}$, the outcome for the coupling of densities j_m and voltages e_m yields an entry of $\mathbf{M}_\sigma$

$$\frac{j_m}{e_m} = \frac{\iint_{\tilde{A}_m} \mathbf{J} \cdot \mathrm{d}s}{\int_{L_m} E \cdot \mathrm{d}\mathbf{l}} = \frac{\iint_{\tilde{A}_m} \sigma \, \mathrm{d}s}{\int_{L_m} \mathrm{d}l} + O\left[l_{max}^\lambda\right], \tag{2.102}$$

for an $\lambda = 2$ value of the discretization error exponent λ in the case of nonuniform grid stencils.

2.4.2 Algebraic Assets and Operator Functionality

Before proceeding to the extraction of the time-domain FIT formulae for the solution of general EMC applications, it would be fairly instructive to deal with the description of harmonic models basically because of the indisputable flexibility achieved by combined frequency- and time-domain implementations.

A closer inspection of matrices $\mathbf{C}$, $\tilde{\mathbf{C}}$ and $\mathbf{S}$, $\tilde{\mathbf{S}}$ reveals a set of important properties from the duality of $\{G, \tilde{G}\}$. In fact, these algebraic features are in complete accordance with the corresponding analytical (continuous state) relations. Consequently, for the primary grid matrices, one obtains

$$\underbrace{\mathbf{C}\tilde{\mathbf{S}}^T = \mathbf{0}}_{\text{FIT state}} \Leftrightarrow \underbrace{\mathrm{curl}\{\mathrm{grad}\} = \mathbf{0}}_{\text{Continuous state}} \quad \text{and} \quad \underbrace{\mathbf{SC} = 0}_{\text{FIT state}} \Leftrightarrow \underbrace{\mathrm{div}\{\mathrm{curl}\} = 0}_{\text{Continuous state}}, \tag{2.103}$$

with superscript T denoting the transpose matrix. Likewise, for the secondary mesh constituents, it holds

$$\underbrace{\tilde{\mathbf{C}}\mathbf{S}^T = \mathbf{0}}_{\text{FIT state}} \Leftrightarrow \underbrace{\mathrm{curl}\{\mathrm{grad}\} = \mathbf{0}}_{\text{Continuous state}} \quad \text{and} \quad \underbrace{\tilde{\mathbf{S}}\tilde{\mathbf{C}} = 0}_{\text{FIT state}} \Leftrightarrow \underbrace{\mathrm{div}\{\mathrm{curl}\} = 0}_{\text{Continuous state}} \tag{2.104}$$

To validate the previous aspects, let us reexpress MGE, (2.99) and (2.100), in the frequency domain as

$$\left.\begin{array}{r} \mathbf{Ce} = -j\omega\mathbf{b} \\ \tilde{\mathbf{C}}\mathbf{M}_\mu^{-1}\mathbf{b} = j\omega\mathbf{M}_\varepsilon\mathbf{e} \end{array}\right\} \Rightarrow \tilde{\mathbf{C}}\mathbf{M}_\mu^{-1}\mathbf{Ce} = \omega^2\mathbf{M}_\varepsilon\mathbf{e}, \tag{2.105}$$

where temporal derivatives have been substituted by $j\omega$, under the assumption of a lossless and source-free medium. Normalizing the curl–curl (2.105) by the $\mathbf{e}_N = \mathbf{M}_\varepsilon^{1/2}\mathbf{e}$ energy quantity,

$$\mathbf{K}\mathbf{K}^T\mathbf{e}_N = \omega^2\mathbf{e}_N, \quad \text{for } \mathbf{K} = \mathbf{M}_\varepsilon^{-1/2}\tilde{\mathbf{C}}\mathbf{M}_\mu^{-1/2}, \tag{2.106}$$

which signifies a linear problem with real and positive eigenvalues ω^2, equipped with the potential of decomposition into a linear combination of undamped lossless eigensolutions that do not oscillate in time.

Apart from its consistent profile in harmonic problems, FIT is proven to be fairly reliable in the key issue of energy conservation. Indeed, if this condition is violated, the development of any criterion for a late-time stable time integration of electromagnetic fields is hardly possible. For this purpose, recall (2.99) and multiply its first equation with the complex conjugate matrix $[\mathbf{h}^T]^*$ and its second equation with $[\mathbf{e}^T]^*$ to obtain

$$[\mathbf{h}^T]^* \mathbf{Ce} = -j\omega[\mathbf{h}^T]^* \mathbf{b}, \qquad (2.107)$$

$$[\mathbf{e}^T]^* \tilde{\mathbf{C}}\mathbf{h} = j\omega[\mathbf{e}^T]^* \mathbf{d} \qquad (2.108)$$

If (2.108) is transposed and the complex conjugate of the outcome is received,

$$[\mathbf{h}^T]^* \mathbf{Ce} = -j\omega[\mathbf{d}^T]^* \mathbf{e}, \qquad (2.109)$$

which along with (2.107) shows that $[\mathbf{h}^T]^* \mathbf{b} = [\mathbf{d}^T]^* \mathbf{e}$ and certifies the equality (i.e., preservation) of electric and magnetic energy in G and $\tilde{G}$. These energies, practically time-independent, can be easily expressed, through scalar ϕ and vector $\mathbf{a}$ potential, as

$$W_{\text{electric}} = \underbrace{\frac{1}{2}\mathbf{q}\phi}_{\text{FIT state}} \Leftrightarrow \underbrace{\frac{1}{2}\iiint_\Omega \rho\Phi dV}_{\text{Continuous state}}, \qquad (2.110)$$

$$W_{\text{magnetic}} = \underbrace{\frac{1}{2}\mathbf{j}\mathbf{a}}_{\text{FIT state}} \Leftrightarrow \underbrace{\frac{1}{2}\iiint_\Omega \mathbf{A} \cdot \mathbf{J} dV}_{\text{Continuous state}}, \qquad (2.111)$$

with $\underbrace{\mathbf{e} = \tilde{\mathbf{S}}^T \phi}_{\text{FIT state}} \Leftrightarrow \underbrace{\mathbf{E} = -\text{grad}\{\Phi\}}_{\text{Continuous state}}$ and $\underbrace{\mathbf{b} = \mathbf{Ca}}_{\text{FIT state}} \Leftrightarrow \underbrace{\mathbf{B} = -\text{curl}\{\mathbf{A}\}}_{\text{Continuous state}}$, and $\Phi, \mathbf{A}$ are the continuous analogues of $\phi, \mathbf{a}$.

Another critical attribute of the FIT is extracted by multiplying both sides of (2.105) with matrix $\tilde{\mathbf{S}}$ and

$$\tilde{\mathbf{S}}\tilde{\mathbf{C}}\mathbf{M}_\mu^{-1}\mathbf{Ce} = \omega^2 \tilde{\mathbf{S}}\mathbf{M}_\varepsilon \mathbf{e} \longrightarrow 0 = \omega^2 \tilde{\mathbf{S}}\mathbf{M}_\varepsilon \mathbf{e} \qquad (2.112)$$

Moreover, if the computational domain Ω is terminated by simply connected metallic boundaries, (2.112) enables two different solutions for $\mathbf{d}$ and, through (2.101), for $\mathbf{e}$, namely,

$$\mathbf{e}, \mathbf{d} = \begin{cases} \omega^2 \neq 0, & \text{when } \tilde{\mathbf{S}}\mathbf{d} = \mathbf{0} \\ \omega^2 = 0, & \text{when } \tilde{\mathbf{S}}\mathbf{d} \neq \mathbf{0} \end{cases} \tag{2.113}$$

Because the initial problem, launched via (2.105), is real-valued and symmetric, an orthonormal set of eigenvectors $\mathbf{e}_m$, such that $\mathbf{e}_m^{\mathrm{T}}\mathbf{e}_n = \delta_{m,n}$ ($\delta_{m,n}$ the Kronecker delta) can be constructed [51, 52]. This set permits the definition of two orthogonal vector subspaces Ω_ω, Ω_0, which span the solution space of (2.105). So,

$$\Omega = \Omega_\omega \cup \Omega_0, \tag{2.114}$$

demonstrating that all rotational dynamic solutions in Ω_ω, for $\omega^2 \neq 0$, are orthogonal to the irrotational static discrete modes in Ω_0, for $\omega^2 = 0$. In an effort to manipulate the latter modes, (2.105) is generalized to a discrete Helmholtz-type equation by means of the curl{curl} - grad{div} $= -\nabla^2$ analytical identity. An immediate use of the resulting matrix is not workable because real-world problems encompass irregular media distributions. In this way, Helmholtz-grid equation is written as

$$\left(\tilde{\mathbf{C}}\mathbf{M}_\mu^{-1}\mathbf{C} + \mathbf{D}_A\tilde{\mathbf{S}}^T\mathbf{D}_B\tilde{\mathbf{S}}\mathbf{D}_B\right)\mathbf{e} = \omega^2\mathbf{M}_\varepsilon\mathbf{e}, \tag{2.115}$$

with $\mathbf{D}_A$ and $\mathbf{D}_B$ diagonal matrices, suitably picked to attain a well-posed discretization of (2.105) in homogeneous domains. A practical preference could be the following: $\mathbf{D}_A = \mathbf{M}_\varepsilon$ and $\mathbf{D}_B = v\mathbf{D}_{\tilde{G}}^{-1}$, with $\mathbf{D}_{\tilde{G}}$ as the diagonal matrix of dual-cell volumes in $\tilde{G}$ and v a scaling coefficient. For a more detailed clarification of $\mathbf{D}_{\tilde{G}}$ role in (2.115), an alternative set

$$\mathbf{D}_A = \mathbf{M}_\varepsilon + (j\omega)^{-1}\mathbf{M}_\sigma, \tag{2.116}$$

$$\mathbf{D}_B = \mathbf{D}\mathbf{D}_{\tilde{G}}^{-1}\mathbf{D}_{c\mu}^{-1}\mathbf{D}_{c\varepsilon}^{-1}\left[\mathbf{D}_{c\varepsilon}^{\mathrm{T}}\right]^{-1}, \tag{2.117}$$

is considered, where $\mathbf{D}$ is diagonal, with element values 0 or 1, specifying the cell vertices (particularly those in perfectly electric conducting regions) of G on which the ∇^2 operator performs. Furthermore, matrices $\mathbf{D}_{c\varepsilon}$ and $\mathbf{D}_{c\mu}$ provide structural data of complex permittivities and permeabilities in average form over the $\tilde{G}$ cells. The above aspects prove that (2.115) in homogeneous areas becomes a rigorous discrete approximation for ∇^2. However, in the case of nonisotropic and inhomogeneous distributions of dielectric and permeable media, the selection turns into a more cumbersome task because of the presence of vector parasites.

Therefore, the eigenvalue problem associated with operator ∇^2 reads

$$\mathbf{D}_A\tilde{\mathbf{S}}^T\mathbf{D}_B\tilde{\mathbf{S}}\mathbf{D}_B\mathbf{e} = \omega^2\mathbf{M}_\varepsilon\mathbf{e}, \tag{2.118}$$

and may be treated like that of (2.105), however, its eigenvectors are not a solution to Maxwell's equations, despite the fact that they are resolved concurrently with the original ones. These nonphysical products are usually designated as spurious modes. Fortunately, the concept of MGE circumvents this serious hindrance and accomplishes accurate simulations, without any perturbations or misleading results.

2.4.3   Generalized Time-Domain FIT Models for EMC Applications

To construct the appropriate time-domain expressions for the update of electromagnetic fields via FIT, the topic of temporal integration should be carefully analyzed. Thus, for a given function $f(t)$, the time axis is discretized in terms of equidistant increments Δt, with $t = n\Delta t$ and $f|^n = f(n\Delta t)$. Using the well-known forward and backward finite-difference approximators

$$\frac{\partial f}{\partial t}\bigg|^n_{\text{forward}} \simeq \frac{f|^{n+1} - f|^n}{\Delta t} \quad \text{and} \quad \frac{\partial f}{\partial t}\bigg|^n_{\text{backward}} \simeq \frac{f|^n - f|^{n-1}}{\Delta t}, \tag{2.119}$$

for the temporal derivatives, the MGE system is expressed in the more compact form of

$$\Xi \mathbf{F} = \Lambda \frac{\partial \mathbf{F}}{\partial t} + \Pi, \tag{2.120}$$

where

$$\Xi = \begin{bmatrix} \mathbf{0} & \tilde{\mathbf{C}} \\ -\mathbf{C} & \mathbf{0} \end{bmatrix}, \qquad \Lambda = \begin{bmatrix} \eta_0 \mathbf{M}_\varepsilon & \mathbf{0} \\ \mathbf{0} & \eta_0^{-1}\mathbf{M}_\mu \end{bmatrix}$$

and

$$\mathbf{F} = \begin{bmatrix} \eta_0^{-1/2}\mathbf{e} \\ \eta_0^{1/2}\mathbf{h} \end{bmatrix}, \qquad \Pi = \begin{bmatrix} \eta_0^{1/2}\mathbf{j} \\ \mathbf{0} \end{bmatrix}, \qquad \eta_0 = \sqrt{\frac{\mu_0}{\varepsilon_0}}$$

The decision on the optimal time-update mechanism is a point of principal importance, as an incorrect option may lead to totally unstable solutions. Hence, application of the forward finite-difference formulas (2.119) to (2.120) gives

$$\mathbf{F}^{n+1} = \left(\mathbf{I} + \Delta t\Lambda^{-1}\Xi\right)\mathbf{F}^n - \Delta t\Lambda^{-1}\Pi^n, \tag{2.121}$$

which is proven to be unstable for any Δt. On the other hand, the backward operator leads to

$$\left(\mathbf{I} - \Delta t\Lambda^{-1}\Xi\right)\mathbf{F}^{n+1} = \mathbf{F}^n - \Delta t\Lambda^{-1}\Pi^{n+1},$$

which, through the required matrix inversion, becomes

$$\mathbf{F}^{n+1} = \left(\mathbf{I} - \Delta t\Lambda^{-1}\Xi\right)^{-1}\left(\mathbf{F}^n - \Delta t\Lambda^{-1}\Pi^{n+1}\right) \tag{2.122}$$

Although (2.122) is unconditionally stable, it is literally ineffective for modern EMC problems, whose large dimensions and detailed structure stipulate the generation of extremely big matrices—i.e., huge systems of equations for each time step—with unaffordable inversion procedures.

Nonetheless, the case of the leapfrog scheme, firstly encountered in the FDTD technique, seems to offer the best compromise between stability and algorithmic implementation because it needs only simple matrix multiplications to advance from one time step to another [50, 60]. So, sampling $\mathbf{b}$ and $\mathbf{e}$ vectors in an interleaving sense with a temporal shift of half a time step, MGE may be written as

$$\mathbf{b}^{n+1} = \mathbf{b}^n - \Delta t \mathbf{C} \mathbf{e}^{n+1/2}$$
$$\mathbf{e}^{n+1/2} = \mathbf{e}^{n-1/2} + \Delta t \mathbf{M}_\varepsilon^{-1} \left(\tilde{\mathbf{C}} \mathbf{M}_\mu^{-1} \mathbf{b}^n - \mathbf{j}^n \right) \tag{2.123}$$

To validate the stability of (2.123), a general rendition of von Neumann method is used to provide

$$\mathbf{F}^{n+1} = \Theta \mathbf{F}^n + \Pi^n, \tag{2.124}$$

with

$$\Theta = \begin{bmatrix} \mathbf{I} & -\Delta t \mathbf{C} \\ \Delta t \mathbf{M}_\varepsilon^{-1} \tilde{\mathbf{C}} \mathbf{M}_\mu^{-1} & \mathbf{I} - (\Delta t)^2 \mathbf{M}_\varepsilon^{-1} \tilde{\mathbf{C}} \mathbf{M}_\mu^{-1} \mathbf{C} \end{bmatrix}$$

and

$$\mathbf{F}^n = \begin{bmatrix} \mathbf{b}^n \\ \mathbf{e}^{n+1/2} \end{bmatrix}, \qquad \Pi^n = \begin{bmatrix} \mathbf{0} \\ -\Delta t \mathbf{M}_\varepsilon^{-1} \mathbf{j}^{n+1/2} \end{bmatrix}.$$

Extracting the eigenvalues λ_m of Θ, it can be easily shown that all of them are located inside the unit circle in the complex plane, and therefore, (2.124) is found to be conditionally stable as long as the time-step remains smaller than a prefixed value (i.e., the magnitudes of λ_m do not exceed the unit circle), determined by a certain criterion. The most versatile, yet somewhat stringent, one is the Courant condition of (2.44),

$$\upsilon \Delta t \leq \left(\sqrt{\frac{1}{(\Delta x)^2} + \frac{1}{(\Delta y)^2} + \frac{1}{(\Delta z)^2}} \right)^{-1}, \tag{2.125}$$

which is basically applicable to homogeneous regions, with Δx, Δy, Δz as the spatial increments along the respective Cartesian axes. Consequently, for the solution of (2.123), one must first derive the largest eigenvalues of Θ, examine their position relative to the unit circle, and then proceed to the simulation. Such a procedure is not so laborious because eigenvalue computation can nowadays be conducted via efficient algorithms.

Additionally, for media with conductivity profiles such that their skin depth is larger than the minimum spatial increment, the leapfrog scheme is replaced by an exponentially based one, as

$$\mathbf{b}^{n+1} = \mathbf{b}^n - \Delta t \mathbf{C} \mathbf{e}^{n+1/2}$$
$$\mathbf{e}^{n+1/2} = e^{-\Delta t \mathbf{M}_\varepsilon^{-1} \mathbf{M}_\sigma} \mathbf{e}^{n-1/2} + \mathbf{M}_\sigma^{-1} \left(\mathbf{I} - e^{-\Delta t \mathbf{M}_\varepsilon^{-1} \mathbf{M}_\sigma} \right) \left(\tilde{\mathbf{C}} \mathbf{M}_\mu^{-1} \mathbf{b}^n - \mathbf{j}^n \right), \tag{2.126}$$

which, however, is more prone to round off arithmetic errors, owing to the larger number of computer operations per time step. In contrast, when the skin depth is smaller than the minimum spatial increment, a surface impedance approach with the appropriate local plane wave considerations may be possibly adopted.

Attempting a comparison between (2.123) and (2.126), it is concluded that the former is pertinent for the calculation of EMC quantities in a time scale close to the order of spatial dimensions, whereas the latter is best suited for smoothly varying functions and free of abrupt temporal oscillations and elongated investigations. As far as charge conservation issues are concerned, (2.86) are multiplied with divergence matrices $\mathbf{S}$ and $\tilde{\mathbf{S}}\mathbf{M}_\varepsilon$

$$\mathbf{Sb}^{n+1} = \mathbf{Sb}^n - \Delta t \mathbf{SCe}^{n+1/2} \longrightarrow \mathbf{Sb}^{n+1} = \mathbf{Sb}^n$$

$$\tilde{\mathbf{S}}\mathbf{M}_\varepsilon \mathbf{e}^{n+1/2} = \tilde{\mathbf{S}}\mathbf{M}_\varepsilon \mathbf{e}^{n-1/2} + \Delta t \tilde{\mathbf{S}}\mathbf{M}_\varepsilon \mathbf{M}_\varepsilon^{-1} \left(\tilde{\mathbf{C}}\mathbf{M}_\mu^{-1}\mathbf{b}^n - \mathbf{j}^n \right) \longrightarrow \tilde{\mathbf{S}}\mathbf{M}_\varepsilon \mathbf{e}^{n+1/2} = \tilde{\mathbf{S}} \left(\mathbf{M}_\varepsilon \mathbf{e}^{n-1/2} - \Delta t \mathbf{j}^n \right)$$

From the previous manipulations, it can be easily discerned that this significant property is inherently built in the leapfrog scheme. Equivalent deductions are drawn for the exponential update method (2.126) as well.

At this point, it is noteworthy to explore the competence of the time-domain FIT to model the continuity equation. To this aim, the second MGE in (2.99) is multiplied with $\tilde{\mathbf{S}}$, as

$$\tilde{\mathbf{S}}\tilde{\mathbf{C}}\mathbf{h} = \frac{\mathrm{d}}{\mathrm{d}t}\tilde{\mathbf{S}}\mathbf{d} + \tilde{\mathbf{S}}\mathbf{j}, \tag{2.127}$$

and by means of (2.105), becomes

$$\underbrace{\tilde{\mathbf{S}}\mathbf{j} = -\frac{\partial}{\partial t}\mathbf{q}}_{\text{FIT state}} \Leftrightarrow \underbrace{\mathrm{div}\{\mathbf{J}\} = -\rho}_{\text{Continuous state}}, \tag{2.128}$$

which is the FIT continuity equation, under the mandatory assumption of charge-conserving sources $\mathbf{j}$.

2.4.4   Merits and Implementation Aspects

The FIT is probably among the most rapidly developing numerical methods for electromagnetic field simulations. Based on its competences to offer accurate solutions for EMC problems, with a variety of material properties, in both the time and frequency domain, FIT has been turned into a strong contender for the analysis of a wide range of devices. Actually, its significant performance can be attributed to three reasons. First, it is the profound theoretical background that achieves a sufficient late-time stability profile, a robust orthogonality in modal evaluation, the dynamic suppression of spurious parasites, as well as the conservation of energy and charges anywhere in the domain. Second, FIT may be easily applied to electrically large or complicated structures. Lastly, it works efficiently not only for any lattice type but may also employ diverse discretization strategies in the modeling of dissimilar media interfaces.

Nevertheless, the most important superiority of FIT over traditional numerical techniques lies in the system of MGE. Being a consistent discrete representation of the original field equations, they preserve all the physical properties of the continuous problem. These characteristics are deemed instructive in a twofold manner. One issue is the drastic elimination of numerical lattice artificialities owing to the absence of approximation schemes, whereas the other is the algebraic nature of numerical solutions that enhances the precision of the outcomes. The resulting degrees of freedom collected in the vectors of the MGE system, typically consist of measurable integral state variables, i.e., voltages, fluxes, currents, and charges. Moreover, the operation matrices are sparse and contain critical data on the incidence relations of the dual grids, thus allowing the choice of the correct time integration process, even for nongauged cases that yield singular equations.

To give an indicative impression of FIT capabilities, the frequency selective surface of Figure 2.15, comprising diverse crosslike dipoles, is investigated. The dimensions of this characteristic EMC problem are $T_x = T_y = 1.5$ cm, whereas the metallic parts of each dipole have a length of 1.125 cm and a width of 0.187 cm. Furthermore, the structure is excited by a transverse electric incident wave. Figure 2.16 shows the variation of the reflection coefficient versus the incidence angle and various frequency values. As can be easily observed, the FIT manages to cope successfully with the

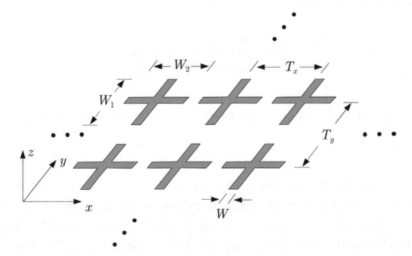

FIGURE 2.15: A frequency selective surface consisting of crosslike dipoles.

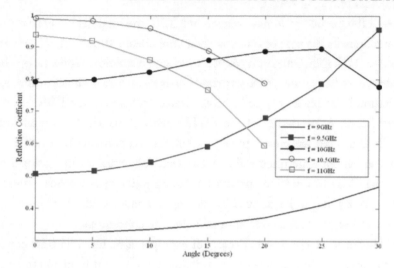

FIGURE 2.16: Reflection coefficient versus angle of incidence for different frequencies.

periodic nature of the above configuration, without inducing any instabilities or grid discrepancies. Similar deductions are obtained by the electric field snapshots of Figure 2.17.

Regarding the time-domain treatment of curved surfaces and termination conditions, FIT has established the so-called *perfect boundary approximation (PBA)* [11, 51]. According to this framework, plane waves can be realized as inhomogeneous boundary conditions and therefore play the role of stable low-reflection regions. Additionally, PBA is equally applied to waveguide wall conditions. After obtaining a rigorous eigenmode solution across the wave port (namely, known propagation

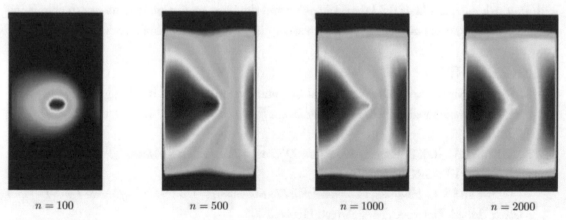

$n = 100$ $n = 500$ $n = 1000$ $n = 2000$

FIGURE 2.17: Reflection coefficient versus angle of incidence for different frequencies.

constants) through the frequency domain version of MGE, the total field may be expressed as a sum of all waveguide modes. In the case of transversely homogeneous structures, it is adequate to acquire a single-frequency set of eigenmodes and adopt a wideband excitation. Conversely, when inhomogeneous waveguides are to be handled, such as printed boards, microstrip lines, or filters, the preceding broadband excitation is simply an approximation. Hence, in terms of the PBA concept, FIT incorporates the multifrequency properties of the FDTD method, which, for a remarkable number of contemporary EMC devices, are proven to be very helpful and particularly versatile.

Moreover, extensive experiments illustrated that for actual computations, a Δt given by the equality in the Courant criterion suffices for the majority of situations, whereas more accurate outcomes are not going to be derived by picking a smaller value. Indeed, when the equality holds, the discretized wave most closely approximates the continuous propagation and grid errors are minimized. However, exceptions do occur. For example, a material with its conductivity much greater than zero asks for a time increment fairly below the Courant limit. As the phase velocity in the conducting region is less than that of free space, the selected time step will fulfill the stability condition everywhere in the domain. Once cell dimensions are determined, one may estimate the total FIT space together with any supplementary modeling modifications.

Finally, from a numerical implementation perspective, FIT offers considerable memory and CPU time savings. Toward this direction contributes the sparse form of "curl" and "divergence" matrices, which contain integer-valued elements and require a confined amount of floating point operations affordable by any computer system. Also, for large-scale EMC problems, such as biomedical applications with detailed models of the human body or characterization studies of test chambers, the mesh-independent construction of MGE leads to optimal discretizations and as a natural consequence to rapid simulations [57, 61]. These aspects have broadened the universality of FIT to new design horizons and especially to coupled problems (electromagnetic systems coupled with thermal, mechanical or fluid dynamic ones) and double-negative metamaterial arrangements, whose demanding traits cannot be taken into account by other time-domain realizations.

REFERENCES

1. K. S. Yee, "Numerical solution of initial boundary value problems involving Maxwell's equations in isotropic media," *IEEE Trans. Antennas Propagat.*, vol. AP-14, no. 3, pp. 302–307, May 1966.
2. K. S. Kunz and R. J. Luebbers, *The Finite Difference Time Domain Method for Electromagnetics*. Boca Raton, FL: CRC Press, 1993.
3. A. Taflove and S. C. Hagness, *Computational Electrodynamics: The Finite-Difference Time-Domain Method*, 3rd ed. Norwood, MA: Artech House, 2005.

4. A. Taflove, Ed., *Advances in Computational Electrodynamics: The Finite-Difference Time-Domain Method*. Norwood, MA: Artech House, 1998.

5. D. M. Sullivan, *Electromagnetic Simulation Using the FDTD Method*. Piscataway, NJ: IEEE Press, 2000.

6. P. B. Johns and R. L. Beurle, "Numerical solutions of 2-dimensional scattering problems using a transmission-line matrix," *Proc. IEEE*, vol. 118, pp. 1203–1208, 1971.

7. W. J. R. Hoefer, "The transmission line matrix (TLM) method," in *Numerical Techniques for Microwave and Millimeter Wave Passive Structures*, T. Itoh, Ed. New York: Wiley, 1989, pp. 496–451.

8. M. Krumpholz and P. Russer, "A field theoretical derivation of TLM," *IEEE Trans. Microwave Theory Tech.*, vol. 42, no. 9, pp. 1660–1668, Sept. 1994.

9. C. Christopoulos, *The Transmission-Line Modeling Method: TLM*. New York: IEEE Press, 1995.

10. T. Weiland, "A discretization method for the solution of Maxwell's equations for six-component fields," *Electron. Commun. (AEÜ)*, vol. 31, no. 3, pp. 116–120, 1977.

11. T. Weiland, "Time domain electromagnetic field computation with finite difference methods," *Int. J. Numer. Model.: Electron. Networks Devices Fields*, vol. 3, pp. 295–319, 1996.

12. R. Holland, "Finite-difference solution of Maxwell's equations in generalized nonorthogonal coordinates," *IEEE Trans. Nuclear Sci.*, vol. NS-30, no. 6, pp. 4589–4591, Dec. 1983.

13. A. C. Cangellaris, "Time-domain finite methods for electromagnetic wave propagation and scattering," *IEEE Trans. Magn.*, vol. 27, no. 5, pp. 3780–3785, Sept. 1991.

14. T. Itoh and B. Houshmand, Eds., *Time-Domain Methods for Microwave Structures: Analysis and Design*. Piscataway, NJ: IEEE Press, 1998.

15. W. Sui, *Time-Domain Computer Analysis of Nonlinear Hybrid Systems*. Boca Raton, FL: CRC Press, 2002.

16. N. V. Kantartzis and T. D. Tsiboukis, *Higher-Order FDTD Schemes for Waveguide and Antenna Structures*, San Rafael, CA: Morgan & Claypool Publishers, 2006, doi:10.2200/S00018ED1V 01Y200604CEM003.

17. G. Mur, "Absorbing boundary conditions for the finite-difference approximation of the time-domain electromagnetic field equations," *IEEE Trans. Electromagn. Compat.*, vol. 23, no. 4, pp. 377–382, Nov. 1981.

18. J.-P. Bérenger, "A perfectly matched layer for the absorption of electromagnetic waves," *J. Comp. Phys.*, vol. 114, no. 2, pp. 185–200, Oct. 1994, doi:10.1006/jcph.1994.1159.

19. J. A. Roden, C. R. Paul, W. T. Smith, and S. D. Gedney, "Finite-difference time-domain analysis of lossy transmission lines," *IEEE Trans. Electromagn. Compat.*, vol. 38, no. 1, pp. 15–24, Feb. 1996, doi:10.1109/15.485691.

20. S. Dey and R. Mittra, "A locally conformal finite-difference time-domain algorithm for modeling three-dimensional perfectly conducting objects," *IEEE Microw. Guided Wave Lett.*, vol. 7, pp. 273–275, 1997.

21. C. L. Wagner and J. B. Schneider, "Divergent fields, charge, and capacitance in FDTD simulations," *IEEE Trans. Microwave Theory Tech.*, vol. 46, no. 12, pp. 2131–2136, Dec. 1998, doi:10.1109/22.739294.

22. M. Celuch-Marcysiak and W. K. Gwarek, "On the nature of solutions produced by the finite difference schemes in the time domain," *Int. J. Numer. Model.: Electron. Networks Devices Fields*, vol. 12, pp. 23–40, Dec. 1999.

23. S. Grivet-Talocia and F. Canavero, "Wavelet-based high-order, adaptive modeling of lossy interconnects," *IEEE Trans. Electromagn. Compat.*, vol. 43, no. 4, pp. 471–484, Nov. 2001, doi:10.1109/15.974626.

24. W. Wallyn, D. De Zutter, and H. Rogier, "Prediction of the shielding and resonant behaviour of multisection enclosures based on magnetic current modelling," *IEEE Trans. Electromagn. Compat.*, vol. 44, no. 1, pp. 130–138, Feb. 2002, doi:10.1109/15.990719.

25. S. Wang and F. L. Teixeira, "Dispersion-relation-preserving FDTD algorithms for large-scale three-dimensional problems," *IEEE Trans. Antennas Propagat.*, vol. 51, no. 8, pp. 1818–1828, Aug. 2003.

26. S. V. Georgakopoulos, C. R. Birtcher, C. A. Balanis, and R. A. Renaut, "HIRF penetration and PED coupling analysis for scaled fuslage models using a hybrid subgrid FDTD(2,2)/ FDTD(2,4) method," *IEEE Trans. Electromagn. Compat.*, vol. 45, no. 2, pp. 293–305, May 2003, doi:10.1109/TEMC.2003.811308.

27. M. Sarto, F. Sarto, M. C. Larciprete, M. Scalora, M. D'Amore, C. Sibilia, and M. Bertolotti, "Nanotechnology of transparent metals for radiofrequency electromagnetic shielding," *IEEE Trans. Electromagn. Compat.*, vol. 45, no. 4, pp. 586–594, Nov. 2003, doi:10.1109/TEMC.2003.819057.

28. C. L. Holloway, D. A. Hill, J. M. Ladbury, and G. Koepke, "Requirements for an effective reverberation chamber: Unloaded or loaded," *IEEE Trans. Electromagn. Compat.*, vol. 48, no. 1, pp. 187–194, Feb. 2006, doi:10.1109/TEMC.2006.870709.

29. C. E. Baum, "Reminiscences of high-power electromagnetics," *IEEE Trans. Electromagn. Compat.*, vol. 49, no. 2, pp. 211–218, May 2007.

30. Y. Baba, N. Nagaoka, and A. Ametani, "Modeling of thin wires in a lossy medium for FDTD simulations," *IEEE Trans. Electromagn. Compat.*, vol. 47, no. 1, pp. 54–60, Feb. 2005, doi:10.1109/TEMC.2004.842115.

31. T. Kasuga and H. Inoue, "Novel FDTD simulation method using multiple-analysis-space for electromagnetic far field," *IEEE Trans. Electromagn. Compat.*, vol. 47, no. 2, pp. 274–284, May 2005, doi:10.1109/TEMC.2005.847413.

32. Y. Yi, B. Chen, D.-G. Fang, and B.-H. Zhou, "A new 2-D FDTD method applied to scattering by infinite objects with oblique incidence," *IEEE Trans. Electromagn. Compat.*, vol. 47, no. 4, pp. 749–756, Nov. 2005, doi:10.1109/TEMC.2005.860559.

33. A. Christ, S. Benkler, J. Fröhlich, and N. Kuster, "Analysis of the accuracy of the numerical reflection coefficient of the finite-difference time-domain method at planar material interfaces," *IEEE Trans. Electromagn. Compat.*, vol. 48, no. 2, pp. 264–272, May 2006, doi:10.1109/TEMC.2006.873850.

34. P. B. Johns, "A symmetrical condensed node for the TLM-method," *IEEE Trans. Microwave Theory Tech.*, vol. MTT-35, no. 4, pp. 370–377, Apr. 1987, doi:10.1109/TMTT.1987.1133658.

35. W. J. Hoefer, "The transmission-line matrix method—Theory and applications," *IEEE Trans. Microwave Theory Tech.*, vol. MTT-33, no. 10, pp. 882–893, Oct. 1995.

36. P. Russer, *Electromagnetics, Microwave Circuit and Antenna Design for Communications Engineering*, 2nd ed. Boston, MA: Artech House, 2006.

37. C. Christopoulos, *The Transmission-Line Modeling (TLM) Method in Electromagnetics*. San Rafael, CA: Morgan & Claypool Publishers, 2006, doi:10.2200/S00027ED1V01Y200605CEM007.

38. H. Jin and R. Vahldiek, "The frequency-domain TLM-method—A new concept," *IEEE Trans. Microwave Theory Tech.*, vol. 40, no. 12, pp. 2207–2218, Dec. 1992.

39. V. Trenkic, C. Christopoulos, and T. Benson, "On the time step in hybrid symmetrical condensed TLM nodes," *IEEE Trans. Microwave Theory Tech.*, vol. 43, no. 9, pp. 2172–2174, Sept. 1995, doi:10.1109/22.414558.

40. L. de Menezes and W. J. R. Hoefer, "Accuracy of TLM solutions of Maxwell's equations," *IEEE Trans. Microwave Theory Tech.*, vol. 44, no. 12, pp. 2512–2518, Dec. 1996.

41. S. Le Maguer, A. Peden, D. Bourreau, and M. M. Ney, "Split-step TLM (SS-TLM)—A new scheme for accelerating electromagnetic-field simulation," *IEEE Trans. Microwave Theory Tech.*, vol. 52, no. 4, pp. 1182–1190, Apr. 2004.

42. J. D. Paul, C. Christopoulos, D. W. P. Thomas, and X. Liu, "Time-domain modelling of electromagnetic wave interaction with thin-wires using TLM," *IEEE Trans. Electromagn. Compat.*, vol. 47, no. 3, pp. 447–455, Aug. 2005, doi:10.1109/TEMC.2005.852217.

43. P. Sewell, T. M. Benson, C. Christopoulos, D. W. P. Thomas, A. Vukovic, and J. Wykes, "Transmission-line modelling (TLM) based upon tetrahedral meshes," *IEEE Trans. Electromagn. Compat.*, vol. 53, no. 4, pp. 1919–1928, Nov. 2005.

44. P. P. M. So, D. Huilian, and W. J. R. Hoefer, "Modeling of metamaterials with negative refractive index using 2-D shunt and 3-D SCN TLM networks," *IEEE Trans. Microwave Theory Tech.*, vol. 53, no. 4, pp. 1496–1505, Apr. 2005, doi:10.1109/MWSYM.2004.1338945.

45. P. Lorenz, J. V. Vital, B. Biscontini, and P. Russer, "TLM-G—A grid-enabled time-domain transmission-line-matrix system for the analysis of complex electromagnetic structures," *IEEE Trans. Microwave Theory Tech.*, vol. 53, no. 11, pp. 3631–3637, Nov. 2005.

46. D. Lukashevich, A. C. Cangellaris, and P. Russer, "Oblique–oblique projection in TLM-MOR for high-Q structures," *IEEE Trans. Microwave Theory Tech.*, vol. 54, no. 7, pp. 1472–1483, July 2006.

47. R. F. Harrington, *Field Computation by Moment Methods*. New York: IEEE Press, 1968.

48. E. Tonti, "On the geometrical structure of electromagnetism," in *Gravitation, Electromagnetism and Geometrical Structures*, G. Ferraese, Ed. Bologna: Pitagora, 1996, pp. 281–308.

49. A. Bossavit and L. Kettunnen, "Yee-like schemes on tetrahedral mesh with diagonal lumping," *Int. J. Numer. Model.: Electron. Networks Devices Fields*, vol. 12, no. 1–2, pp. 129–142, 1999, doi:10.1002/(SICI)1099-1204(199901/04)12:1/2<129::AID-JNM327>3.0.CO;2-G.

50. U. van Rienen, M. Günter, and D. Hecht, Eds., *Scientific Computing in Electrical Engineering*. Berlin: Springer Verlag, 2001.

51. R. Schuhmann and T. Weiland, "The nonorthogonal finite integration technique applied to 2D- and 3D-eigenvalue problems," *IEEE Trans. Magn.*, vol. 36, no. 4, pp. 807–901, July 2000.

52. O. Podebrad, M. Clemens, and T. Weiland, "New flexible subgridding scheme for the finite integration technique," *IEEE Trans. Magn.*, vol. 39, no. 3, pp. 1662–1665, May 2003.

53. J. E. Lebaric and D. Kajfez, "Analysis of dielectric resonator cavities using the finite integration technique," *IEEE Trans. Microwave Theory Tech.*, vol. 37, no. 11, pp. 1740–1748, Nov. 1989.

54. H. Kruger, F. Mayer, and T. Weiland, "Time domain simulation of ferrite materials using the finite integration technique," *Proc. IEEE Int. Symp. Electromagn. Compat.*, vol. 2, pp. 858–863, 2001, doi:10.1109/ISEMC.2001.950490.

55. S. Caniggia, E. Leroux, A. Orlandi, and L Sassi, "Finite integration technique numerical modeling for EMC and signal integrity problems," *Proc. IEEE Int. Symp. Indust. Electron.*, vol. 4, pp. 1404–1409, 2002, doi:10.1109/ISIE.2002.1026000.

56. G. Spadacini, S. P. Pignari, and F. Marliani, "Closed-form transmission line model for radiated susceptibility in metallic enclosures," *IEEE Trans. Electromagn. Compat.*, vol. 47, no. 4, pp. 701–708, Nov. 2005, doi:10.1109/TEMC.2005.857875.

57. M. Walter and I. Munteanu, "FIT for EMC," *Proc. 17th Int. Zurich Symp. Electromagn. Compat.*, pp. 15–17, 2006.

58. A. Barchanski, T. Steiner, H. de Gersem, M. Clemens, and T. Weiland, "Local grid refinement for low-frequency current computations in 3-D human anatomy models," *IEEE Trans. Magn.*, vol. 42, no. 4, pp. 1371–1374, Apr. 2006.

59. D. Ioan, G. Ciuprina, M. Radulescu, and E. Seebacher, "Compact modeling and fast simulation of on-chip interconnect lines," *IEEE Trans. Magn.*, vol. 42, no. 4, pp. 547–550, Apr. 2006.

60. S. Caniggia and F. Maradei, "Numerical prediction and measurement of ESD radiated fields by free-space field sensors," *IEEE Trans. Electromagn. Compat.*, vol. 49, no. 3, pp. 494–503, Aug. 2007.

61. C. R. Paul, "A brief history of work in transmission lines for EMC problems," *IEEE Trans. Electromagn. Compat.*, vol. 49, no. 2, pp. 237–252, May 2007.

CHAPTER 3

Alternative Time-Domain Techniques in EMC Modeling

3.1 INTRODUCTION

With the theoretical background of the three prime time-domain methods firmly established, it becomes obvious that for a substantial variety of problems, they are expected to act as very skillful and trustworthy workhorses. Yet, the underlying approximations of physical quantities restrain their use in some specialized situations. Furthermore, the incessant proliferation of advanced devices in the area of EMC engineering with high complexity and cutting-edge qualities advocates the construction of discrete models not readily supported by existing practices. Hence, the need for customized techniques became a thought-provoking motive for academics and practitioners with a considerable exposure to computational electromagnetics. The outgrowth of these endeavors has, indeed, elevated our knowledge on numerical analysis because several of the developed algorithms appointed new initiatives and multiple directions of research. Although not so prevalent as the basic time-domain approaches, they soon earned notable appreciation because of their overwhelming performance in diverse tough cases, where alternative design and optimization perspectives are necessitated.

The aim of this chapter is to present the fundamentals of the most regularly encountered methods together with some useful hints for their implementation. Starting from the *finite-element time-domain (FETD)* technique both in its nodal- and edge-element variant, the description proceeds to the *finite-volume time-domain (FVTD)* algorithm with an emphasis on its topological and conservation law formulation. As a matter of fact, the two schemes are the first ones that incorporated the concept of temporal integration in a leapfrog-like regime. Next, the contribution of *wavelet representation* functions in generalized time-domain configurations, through elements of multiresolution analysis, comes into scrutiny, and explicit field approximations are derived. Recognizing the detrimental effect of dispersion errors in spatial-derivative evaluations, the chapter then investigates the properties of *pseudospectral time-domain (PSTD)* methods and elucidates on the incorporation of Fourier transforms for the reduction of the appropriate lattice resolutions. The issue of constructing efficient algorithms free of any preventive stability condition is subsequently explored via the *alternating-direction implicit*

(ADI) FDTD technique. In addition, the potential of minimizing the intrinsic reflection discrepancies of finite-difference discretizations is discussed during the development of the *nonstandard FDTD* approach, whereas the extraction of *higher-order time-domain* forms and *weighted essentially nonoscillatory (WENO)* counterparts completes the motivating theme of grid optimality. Finally, the chapter closes with a detailed reference on the state-of-the-art topic of *hybridizations*, which attempt to merge the best aspects of time-dependent algorithms into one powerful framework, under the suitable boundary continuity conditions at their interface.

3.2   THE FAMILY OF FETD ALGORITHMS

The principal notion of combining the unique capabilities of the finite element method in the development of flexible meshes with powerful temporal integration concepts was proposed for EMC modeling almost 20 years ago [1–4]. Nevertheless, the prominence of the so-called FETD technique is still rather restricted as compared with its main competitor—the FDTD algorithm— although various efficient implementations have been until now (and continue to be) presented. This may be attributed to the minimum complexity and straightforward programming effort of the latter, especially for high-frequency problems. Unfortunately, in its most popular version, the FDTD method is based on totally structured lattices conforming to Cartesian coordinates, which make the overall application difficult when more convoluted geometries are to be tackled.

In fact, there exist two principal approaches to devise FETD schemes for Maxwell's curl laws. The first one stems from the *second-order vector wave equation*, where either the electric or the magnetic field is eliminated by means of certain basis functions. Specifically, edge elements are practically selected, as these basis functions because of their conformity to the discrete version of the de Rham cohomology and the suppression of spurious parasites at nonzero frequencies. Contrarily, the second discipline lies on the *direct discretization* of the coupled first-order Maxwell's equations and involves both fields as simultaneous unknowns. Such a technique lends itself to the combination of edge elements for the expansion of electric quantities with face elements for the representation of magnetic components. Amid its most prominent merits, one may discern the preservation of de Rham cohomology properties in the discrete state, the absence of artificial gradient-like oscillatory modes, the relatively easy development of energy-conserving forms, the larger effectiveness in hybridizations with the FDTD algorithm, and the notably enhanced capabilities in the treatment of dispersive or anisotropic media owing to their individual implementation, without the need for spatial differentiations. Concerning the computational burden both methods are deemed comparable, despite that the latter calculates and stores two field unknowns. This is because the discretization of second-order temporal derivatives mandates the storage of two past time steps, unlike the only one value required by the direct first-order formulation, the sparse linear system solution of which is applied solely in the update of electric fields.

From the preceding remarks, it becomes apparent that the second classification of FETD schemes provides some serious advantages over other traditional time-domain constituents as well. Literally, the most perceptible one is its superior versatility in the generation of adaptive simplicial grids that attain a very natural way of manipulating field/flux continuity conditions at material interfaces. Therefore, this section will be chiefly devoted to it and, in the first place, will deal with the fundamental Faedo–Galerkin strategy for the development of semidiscrete approximations to Maxwell's time-dependent laws and vector wave [5–9]. Then, analysis will carry on to the more elaborate description of explicit/implicit nodal and edge (Whitney) element realizations [10–15]. To this aim and setting as a prerequisite a firm knowledge on the theoretical background and abstractions of conventional finite elements in the frequency domain, the important issue of test/trial functions along with the topology of the resulting degrees of freedom is also explored.

3.2.1   The General Concept of Faedo–Galerkin Formulation

First-order curl equations case. Let us consider a region of space Ω occupied by a medium with position-dependent relative constitutive parameters ε_r and μ_r, amenable to Maxwell's curl laws (2.1) and (2.2), for $\mathbf{M}_c = \mathbf{0}$. All electromagnetic interactions inside Ω must fulfill the well-known boundary conditions $\hat{\mathbf{n}} \times \mathbf{E} = \mathbf{0}$ and $\hat{\mathbf{n}} \times \mathbf{H} = \mathbf{0}$ on perfect electric and magnetic surfaces Γ_E and Γ_H, respectively, with $\hat{\mathbf{n}}$ as their normal unit vector. Moreover, in the case of an unbounded domain, outgoing waves should satisfy an appropriate radiation or boundary condition, whereas for the accomplishment of a unique solution, all fields in Ω at $t = 0$ have to be known. In this context, two weak forms[1] can be obtained through the Galerkin or weighted residual method [1, 4]. The first one employs a set of vector functions $\mathbf{u}$, which are actually square-integrable quantities with finite energy. Hence, for every $\mathbf{u}$, one gets

$$\int_{\Omega} \mathbf{u} \cdot \left(\nabla \times \mathbf{H} - \mathbf{J}_c - \varepsilon_0 \varepsilon_r \frac{\partial \mathbf{E}}{\partial t} \right) d\Omega = 0, \tag{3.1}$$

$$\int_{\Omega} \mathbf{u} \cdot \left(\nabla \times \mathbf{E} + \mu_0 \mu_r \frac{\partial \mathbf{H}}{\partial t} \right) d\Omega = 0, \tag{3.2}$$

in an effort to find electric $\mathbf{E}(\mathbf{r}, t) \in \mathcal{H}_E(\nabla \times; \Omega)$ and magnetic $\mathbf{H}(\mathbf{r}, t) \in \mathcal{H}_H(\nabla \times; \Omega)$ fields, which belong to the corresponding curl Hilbert spaces and are defined in the temporal interval $[0, T]$.

---

[1]In the finite element technique, it is required that the interpolation/expansion functions, and *not* their derivatives, are continuous. Solutions extracted in this manner are called *weak* because they satisfy the governing differential equations only in a weak sense.

On the other hand, the second weak form requires that (2.1) is multiplied by a set of square-integrable functions χ and integrated over Ω. Similarly, multiplication of (2.2) by analogous functions ψ and integration of the curl term by parts, leads to

$$\int_{\Omega} \chi \cdot \left(\nabla \times \mathbf{H} - \mathbf{J}_c - \varepsilon_0 \varepsilon_r \frac{\partial \mathbf{E}}{\partial t} \right) d\Omega = 0, \tag{3.3}$$

$$\int_{\Omega} \left(\nabla \times \psi \cdot \mathbf{E} + \psi \cdot \mu_0 \mu_r \frac{\partial \mathbf{H}}{\partial t} \right) d\Omega = 0, \tag{3.4}$$

for every χ and ψ, which again looks for the pertinent electric and magnetic vectors, with the only exception that a natural boundary condition stems from the variational principle, i.e., $\hat{\mathbf{n}} \times (\nabla \times \mathbf{E}) = 0$ on Γ_{H}.

Second-order vector wave equation case. Although the preceding formulation seems adequate for the majority of EMC setups, it would be very instructive to extract the mathematical framework for the Faedo–Galerkin formulation in terms of the second-order vector wave equation with regard to $\mathbf{E}$ (likewise for $\mathbf{H}$)

$$\nabla \times \frac{1}{\mu_r} \nabla \times \mathbf{E} + \frac{\varepsilon_r}{c^2} \frac{\partial^2 \mathbf{E}}{\partial t^2} = -\mu_0 \frac{\partial \mathbf{J}_c}{\partial t}, \tag{3.5}$$

that strictly fulfills $\hat{\mathbf{n}} \times \mathbf{E} = 0$ on Γ_{E} and $\hat{\mathbf{n}} \times (\nabla \times \mathbf{E}) = 0$ on Γ_{H}, together with a properly tailored absorbing boundary condition for the termination of the infinite domain into a finite region.

The Galerkin version of the problem in (3.5) reads: determine vector function $\mathbf{u}(t) \in U$ such that

$$\int_{\Omega} \mathbf{v} \cdot \left(\nabla \times \frac{1}{\mu_r} \nabla \times \mathbf{u} + \frac{\varepsilon_r}{c^2} \frac{\partial^2 \mathbf{u}}{\partial t^2} + \mu_0 \frac{\partial \mathbf{J}_c}{\partial t} \right) d\Omega = 0, \tag{3.6}$$

for every $\mathbf{v} \in V$ and $t > 0$, with U and V as the trial and test spaces, respectively, which in the weighted residual method should not be necessarily the same. The most typical implementation of (3.6) is to balance the option of $\mathbf{u}$ and $\mathbf{v}$ by means of Green's theorem. More specifically, a weak formulation of (3.5) is given by

$$\int_{\Omega} \left(\nabla \times \mathbf{v} \cdot \frac{1}{\mu_r} \nabla \times \mathbf{u} \right) d\Omega + \frac{1}{c^2} \int_{\Omega} \mathbf{v} \cdot \varepsilon_r \frac{\partial^2 \mathbf{u}}{\partial t^2} d\Omega$$

$$+ \mu_0 \int_{\Omega} \mathbf{v} \cdot \frac{\partial \mathbf{J}_c}{\partial t} d\Omega + \frac{1}{c} \oint_{\Gamma_{\Omega}} \mathbf{v} \cdot \left(\frac{\partial \mathbf{u}}{\partial t} \times \hat{\mathbf{n}} \right) d\Gamma_{\Omega} = 0, \tag{3.7}$$

where Γ_Ω is the outer termination boundary of Ω. According to the finite-element basic abstraction, the intention is to find the optimal solution, $\mathbf{u}_p$, of (3.7) with the appropriate number of degrees of freedom. After some calculus, the Faedo–Galerkin approach generates a system of ordinary differential equations, namely,

$$\Lambda\frac{d^2\mathbf{u}_c}{dt^2} + \mathbf{M}\frac{d\mathbf{u}_c}{dt} + \mathbf{N}\mathbf{u}_c + g = 0, \qquad (3.8)$$

for $\mathbf{u}_c$, the coefficient vector of $\mathbf{u}_p$, and concerning the matrices,

$$\Lambda_{ij} = \frac{1}{c^2}\int_\Omega \mathbf{w}_i \cdot \varepsilon_r \mathbf{w}_j \, d\Omega, \qquad \mathbf{M}_{ij} = \frac{1}{c}\oint_{\Gamma_\Omega} \mathbf{w}_i \cdot (\mathbf{w}_j \times \hat{\mathbf{n}})d\Gamma_\Omega,$$

$$\mathbf{N}_{ij} = \int_\Omega \nabla \times \mathbf{w}_i \cdot \frac{1}{\mu_r}\nabla \times \mathbf{w}_j \, d\Omega, \qquad g_i = \mu_0\int_\Omega \mathbf{w}_i \cdot \frac{\partial\mathbf{J}_c}{\partial t}d\Omega,$$

with $\mathbf{w}_i$ as a set of vector basis functions that span the entire 2-D or 3-D space of interest.

3.2.2   Nodal Element Realizations

Among different approaches, the technique of point-matched finite elements has been inarguably the first one to employ nodal finite-element interpolation notions for the derivation of explicit time-advancing schemes for Maxwell's equations. In particular, the algorithm retains the spatial staggering of electric and magnetic fields; yet, as opposed to the FDTD method, all three components of each vector are accommodated at the same node [1, 5–8]. Therefore, two complementary lattices are used for numerical calculations. Both meshes are built in such a way that every electric field node is enclosed within the volume of an element belonging to the magnetic field grid and every magnetic field node is limited within the volume of an element positioned in the electric field grid. The combination of the aforesaid lattices with nodal finite elements creates the pertinent subspaces intended for the finite-element approximation. In this context, one has

$$\mathbf{E}(\mathbf{r}, t) = \sum_{i=1}^{M_E} \phi_i(\mathbf{r})\mathbf{E}_i(t) \quad \text{and} \quad \mathbf{H}(\mathbf{r}, t) = \sum_{j=1}^{M_H} \theta_j(\mathbf{r})\mathbf{H}_j(t), \qquad (3.9)$$

with M_E, M_H as the number of electric and magnetic field nodes and $\phi_i(\mathbf{r})$, $\theta_j(\mathbf{r})$ as the scalar basis functions for the accomplishment of rigorous computations in the two meshes, respectively. As a result, the temporal evolutions of $\mathbf{E}_i(\mathbf{r})$, $\mathbf{H}_j(\mathbf{r})$ play the role of the technique's degrees of freedom at the nodes of the cell complexes. For the completion of the Faedo–Galerkin formulation, a variety of test functions can be picked, thus yielding either implicit or explicit integration procedures.

Actually, the latter scheme is easily implemented via the point-matching concept that sets as collocation points the nodes of the two complementary meshes. This is equivalent to the use of delta functions appointed to electric field nodes $\delta(\mathbf{r}-\mathbf{r}_i)$ and associated with magnetic field ones $\delta(\mathbf{r}-\mathbf{r}_j)$ as test functions. Hence, the substitution of (3.7) in (3.1) and (3.2) gives the subsequent system of state equations

$$\varepsilon_0\varepsilon_r(\mathbf{r}_i)\frac{d\mathbf{E}_i(t)}{dt} = \sum_{j=1}^{M_H} \nabla\theta_j(\mathbf{r}_i) \times \mathbf{H}_j(t) - \mathbf{J}_c(\mathbf{r}_i), \text{ for } i = 1, 2, \quad , M_E \qquad (3.10)$$

$$\mu_0\mu_r(\mathbf{r}_j)\frac{d\mathbf{H}_j(t)}{dt} = -\sum_{i=1}^{M_E} \nabla\phi_i(\mathbf{r}_j) \times \mathbf{E}_i(t), \text{ for } j = 1, 2, \quad , M_H \qquad (3.11)$$

To derive the update expressions for the FETD method, (3.10) and (3.11) must be formulated in a more convenient matrix form. To this objective,

$$\bar{\mathbf{E}} = \left[\bar{\mathbf{E}}_1(t), \bar{\mathbf{E}}_2(t), \quad , \bar{\mathbf{E}}_{M_E}(t)\right]^{\mathrm{T}}, \ \bar{\mathbf{J}}_c = \left[\bar{\mathbf{J}}_{c,1}(t), \bar{\mathbf{J}}_{c,2}(t), \quad , \bar{\mathbf{J}}_{c,M_E}(t)\right]^{\mathrm{T}},$$

$$\bar{\mathbf{H}} = \left[\bar{\mathbf{H}}_1(t), \bar{\mathbf{H}}_2(t), \quad , \bar{\mathbf{H}}_{M_H}(t)\right]^{\mathrm{T}}$$

vectors are defined, where

$$\bar{\mathbf{E}}_i(t) = [E_{ix}(t), E_{iy}(t), E_{iz}(t)]^{\mathrm{T}}, \bar{\mathbf{J}}_{c,i}(t) = [J_{c,ix}(t), J_{c,iy}(t), J_{c,iz}(t)]^{\mathrm{T}}, \bar{\mathbf{H}}_j = [H_{jx}(t), H_{jy}(t), H_{jz}(t)]^{\mathrm{T}}$$

In this sense, the expansion of cross-products in (3.10) and (3.11) provides

$$\begin{bmatrix} 0 & -\partial_z\theta_j & \partial_y\theta_j \\ \partial_z\theta_j & 0 & -\partial_x\theta_j \\ -\partial_y\theta_j & \partial_x\theta_j & 0 \end{bmatrix} \begin{bmatrix} H_{jx} \\ H_{jy} \\ H_{jz} \end{bmatrix} = \mathbf{S}_{\mathbf{H}_j}^i \bar{\mathbf{H}}_j \text{ and } \begin{bmatrix} 0 & -\partial_z\phi_j & \partial_y\phi_j \\ \partial_z\phi_j & 0 & -\partial_x\phi_j \\ -\partial_y\phi_j & \partial_x\phi_j & 0 \end{bmatrix} \begin{bmatrix} E_{ix} \\ E_{iy} \\ E_{iz} \end{bmatrix} = \mathbf{S}_{\mathbf{E}_i}^j \bar{\mathbf{E}}_i,$$

$$(3.12)$$

with the superscripts i, j in the 3×3 operators $\mathbf{S}$ denoting that their elements are calculated at points $\mathbf{r}_i$ for $i = 1, 2, \ldots, M_E$ and $\mathbf{r}_j$ for $j = 1, 2, \ldots, M_H$, respectively. This compact notation leads to the construction of

$$\frac{d\bar{\mathbf{E}}}{dt} = \mathbf{A}^{-1}\left(\mathbf{R}\bar{\mathbf{H}} - \bar{\mathbf{J}}_c\right) \text{ and } \frac{d\bar{\mathbf{H}}}{dt} = -\mathbf{C}^{-1}\mathbf{T}\bar{\mathbf{E}}, \qquad (3.13)$$

for $R_{ij} = \mathbf{S}_{H_j}^i$, $\mathbf{T}_{ji} = \mathbf{S}_{E_i}^j$. Moreover in (3.13), the electromagnetic properties of all materials involved in the computational domain are incorporated in the diagonal permittivity $\mathbf{A}$ and permeability $\mathbf{C}$ matrices with elements $\mathbf{A}_{ii} = \varepsilon_0 \varepsilon_r(\mathbf{r}_i)\mathbf{I}$ and $\mathbf{C}_{jj} = \mu_0 \mu_r(\mathbf{r}_j)\mathbf{I}$, where $\mathbf{I}$ is the corresponding 3×3 identity matrix. The final step for the completion of the FETD method focuses on the discretization of temporal derivatives in (3.13) through a conditionally stable leapfrog-like integration scheme, in such a way that the temporal increments of $\bar{\mathbf{E}}$ are interleaved one-half time-step with regard to those of $\bar{\mathbf{H}}$. Consequently, according to the von Neumann analysis, the stability condition for the system of (3.13) is given by

$$\Delta t \leq \frac{2}{|\lambda_{\max}(\Xi)|}, \quad \text{with} \quad \Xi = \mathbf{A}^{-1}\mathbf{RTC}^{-1}, \quad (3.14)$$

and $\lambda_{\max}$ is the maximum eigenvalue of matrix Ξ. A more careful inspection of (3.12) unveils that if the regular finite-element interpolation functions are chosen for the evaluation of field spatial variations, the resultant matrices are very sparse because of the basis functions' compact support and the temporal update is indeed notably competent. As far as the topology of the two complementary grids is concerned, the option of quadrilateral elements for 2-D and hexahedral elements for 3-D applications seems to be a fairly effective practice, despite its superficial complexity. Nevertheless, the nodal FETD algorithm suffers from some defects, the most significant of which lies on the collocated profile of electric and magnetic components, so imposing several difficulties regarding the derivation of the suitable boundary conditions at dissimilar media interfaces. A drastic solution to this shortcoming comes from Whitney elements, described in the following paragraph.

3.2.3   Whitney Element Time-Domain Techniques

Acknowledged as a breakthrough in the arsenal of numerical techniques, vector finite elements offer an excellent background for the solution of high-frequency electromagnetic problems [2]. Although their basic contribution concentrates on the significant amelioration of spurious modes, there are other practical reasons that make these elements extremely universal [10–15]. First, for the lowest order class, the degrees of freedom are proportional to the tangential electric field along an edge, hence rendering tangential continuity very easy to impose. Second, flux continuity is launched as a natural boundary condition devoid of any cumbersome assumptions, whereas finally geometrical oddities with singular fields, such as corners or abrupt discontinuities, can be efficiently modeled because no nodal value exists at the singularity. Consonant with the fundamental principles of the method, Maxwell's equations are optimally discretized if $\mathbf{E},\mathbf{H}$ fields are approximated by edge elements (Whitney 1-forms) and **B,D** fluxes by facet elements (Whitney 2-forms).[2]

Regarding the former elements, the basis functions are given by

$$\mathbf{w}_{ij}^{(1)} = \zeta_i \nabla \zeta_j - \zeta_j \nabla \zeta_i, \qquad (3.15)$$

along edge $\{ij\}$ (where the circulation of (3.15) is equal to unity and zero along all other edges), for ζ_i the standard linear nodal basis functions (Lagrange interpolation polynomials) connected to vertex i. In this manner, any vector field **F**, which is twice integrable, can be expressed in terms of Whitney 1-forms as

$$\mathbf{F} = \sum_{\text{edge}\{ij\}} \mathcal{F}_{\text{circ}}^{ij} \mathbf{w}_{ij}^{(1)}, \qquad (3.16)$$

with $\mathcal{F}_{\text{circ}}^{ij}$ denoting the circulation of **F** along edge $\{ij\}$. Likewise **F**, related to a certain facet $\{ijk\}$, reads

$$\mathbf{F} = \sum_{\text{facet}\{ijk\}} \mathcal{F}_{\text{flux}}^{ijk} \mathbf{w}_{ijk}^{(2)}, \qquad (3.17)$$

by means of Whitney 2-forms, where $\mathbf{w}_{ijk}^{(2)}$ are the respective basis functions

$$\mathbf{w}_{ijk}^{(2)} = 2 \left(\zeta_i \nabla \zeta_j \times \nabla \zeta_k + \zeta_j \nabla \zeta_k \times \nabla \zeta_i + \zeta_k \nabla \zeta_i \times \nabla \zeta_j \right), \qquad (3.18)$$

and $\mathcal{F}_{\text{flux}}^{ijk}$ the flux of **F**, equal to unity through facet $\{ijk\}$ and zero through all other facets. It is stressed that (3.17) associates the previous fluxes with the degrees of freedom of a tetrahedral element.

A dual-field explicit algorithm. For the distinct scheme, one has to conduct the succeeding interpolations

$$\mathbf{F}_A = \sum_{\text{edge}\{ij\}} \mathcal{F}_{A,\text{circ}}^{ij} \mathbf{w}_{ij}^{(1)} \quad \text{and} \quad \mathbf{F}_B = \sum_{\text{facet}\{ijk\}} \mathcal{F}_{B,\text{flux}}^{ijk} \mathbf{w}_{ijk}^{(2)}, \qquad (3.19)$$

---

[2] An important issue in the theory of vector finite elements is the involvement of Whitney (differential) forms, which actually constitute a family of geometrical objects—sharing the structure of a totally antisymmetric tensor—that establish the correct functional framework for Maxwell's equations. More specifically, the following kinds of forms are applicable to electromagnetic quantities:

- *Whitney 0-forms:* scalar functions with functional but not derivative continuity, such as the electric static potential ϕ.
- *Whitney 1-forms:* vector functions with tangential but not normal continuity, like field intensities **E** and **H**.
- *Whitney 2-forms:* vector functions with normal but not tangential continuity, such as flux densities **D** and **B**.
- *Whitney 3-forms:* discontinuous scalar functions, like $\nabla \times \mathbf{D}$.

with $\mathbf{F}_A = [\mathbf{E}, \mathbf{H}]$, $\mathbf{F}_B = [\mathbf{D}, \mathbf{B}, \mathbf{J}_c]$, and $\mathcal{F}^{ij}_{A,\text{circ}} = [\mathcal{E}^{ij}_{\text{circ}}, \mathcal{H}^{ij}_{\text{circ}}]$, $\mathcal{F}^{ijk}_{B,\text{flux}} = [\mathcal{D}^{ijk}_{\text{flux}}, \mathcal{B}^{ijk}_{\text{flux}}, \mathcal{J}^{ijk}_{c,\text{flux}}]$, as evidently derived from (3.16) and (3.17) [10–12]. Plugging (3.19) into Ampère's and Faraday's laws, (2.1) and (2.2),

$$\sum_{\text{edge}\{ij\}} \mathcal{H}^{ij}_{\text{circ}} \nabla \times \mathbf{w}^{(1)}_{ij} = \sum_{\text{facet}\{ijk\}} \left(\frac{\partial \mathcal{D}^{ijk}_{\text{flux}}}{\partial t} + \mathcal{J}^{ijk}_{c,\text{flux}} \right) \mathbf{w}^{(2)}_{ijk}, \qquad (3.20)$$

$$\sum_{\text{edge}\{ij\}} \mathcal{E}^{ij}_{\text{circ}} \nabla \times \mathbf{w}^{(1)}_{ij} = - \sum_{\text{facet}\{ijk\}} \frac{\partial \mathcal{B}^{ijk}_{\text{flux}}}{\partial t} \mathbf{w}^{(2)}_{ijk} \qquad (3.21)$$

To derive more convenient relations from an algebraic point of view, the collocation algorithm is applied. For instance, on facet $\{ijk\}$ of area Ω_{ijk}, (3.20) and (3.21) lead to

$$\iint_{\Omega_{ijk}} \left[\sum_{\text{edge}\{ij\}} \mathcal{H}^{ij}_{\text{circ}} \nabla \times \mathbf{w}^{(1)}_{ij} \right] \cdot d\mathbf{s} = \iint_{\Omega_{ijk}} \left[\sum_{\text{facet}\{ijk\}} \left(\frac{\partial \mathcal{D}^{ijk}_{\text{flux}}}{\partial t} + \mathcal{J}^{ijk}_{c,\text{flux}} \right) \mathbf{w}^{(2)}_{ijk} \right] \cdot d\mathbf{s}, \quad (3.22)$$

$$\iint_{\Omega_{ijk}} \left[\sum_{\text{edge}\{ij\}} \mathcal{E}^{ij}_{\text{circ}} \nabla \times \mathbf{w}^{(1)}_{ij} \right] \cdot d\mathbf{s} = - \iint_{\Omega_{ijk}} \left[\sum_{\text{facet}\{ijk\}} \frac{\partial \mathcal{B}^{ijk}_{\text{flux}}}{\partial t} \mathbf{w}^{(2)}_{ijk} \right] \cdot d\mathbf{s} \qquad (3.23)$$

Simplification of the above expressions is achieved via the properties of Whitney 1 and 2 forms, as

$$\frac{\partial \mathcal{D}^{ijk}_{\text{flux}}}{\partial t} = \mathcal{H}^{ij}_{\text{circ}} + \mathcal{H}^{jk}_{\text{circ}} + \mathcal{H}^{ki}_{\text{circ}} - \mathcal{J}^{ijk}_{c,\text{flux}} \quad \text{and} \quad \frac{\partial \mathcal{B}^{ijk}_{\text{flux}}}{\partial t} = - \left(\mathcal{E}^{ij}_{\text{circ}} + \mathcal{E}^{jk}_{\text{circ}} + \mathcal{E}^{ki}_{\text{circ}} \right), \quad (3.24)$$

or in matrix notation

$$\frac{\partial \boldsymbol{\mathcal{D}}}{\partial t} = \mathbf{Z}\boldsymbol{\mathcal{H}} - \boldsymbol{\mathcal{J}}_c \qquad \text{and} \qquad \frac{\partial \boldsymbol{\mathcal{B}}}{\partial t} = -\mathbf{Z}\boldsymbol{\mathcal{E}}, \qquad (3.25)$$

where $\mathbf{Z}$ is the circulation array that is filled either by unity or zero entries. To complete the system of Maxwell's equations and assign fields to fluxes or vice versa, the ensuing constitutive relations are derived

$$\sum_{\text{facet}\{ijk\}} \mathcal{D}^{ijk}_{\text{flux}} \mathbf{w}^{(2)}_{ijk} = \varepsilon \sum_{\text{edge}\{ij\}} \mathcal{E}^{ij}_{\text{circ}} \mathbf{w}^{(1)}_{ij}, \qquad (3.26)$$

$$\sum_{\text{facet}\{ijk\}} \mathcal{B}^{ijk}_{\text{flux}} \mathbf{w}^{(2)}_{ijk} = \mu \sum_{\text{edge}\{ij\}} \mathcal{H}^{ij}_{\text{circ}} \mathbf{w}^{(1)}_{ij}, \qquad (3.27)$$

and with the aid of the collocation approach

$$\mathcal{E} = \mathbf{K}^{\mathrm{E}}\mathcal{D} \quad \text{and} \quad \mathcal{H} = \mathbf{K}^{\mathrm{H}}\mathcal{B}, \tag{3.28}$$

where

$$K^{\mathrm{E}}_{\{ij\}\{ijk\}} = \int_{\mathrm{edge}\{ij\}} \varepsilon^{-1}\mathbf{w}^{(2)}_{ijk} \cdot \mathrm{d}\mathbf{l} \quad \text{and} \quad K^{\mathrm{H}}_{\{ij\}\{lms\}} = \int_{\mathrm{edge}\{ij\}} \mu^{-1}\mathbf{w}^{(2)}_{ijk} \cdot \mathrm{d}\mathbf{l},$$

with $\{ij\}$ and $\{ijk\}$ as the numbers that correspond to the unknown quantities of edge $\{ij\}$ and facet $\{ijk\}$.

Finally, if temporal derivatives in (3.25) are evaluated through a central finite-difference scheme, staggered in time according to a leapfrog-oriented discipline, the 3-D Whitney elements time domain (WETD) update formulae are obtained. So,

$$\mathcal{D}^{n+1} = \mathcal{D}^n + \Delta t \mathbf{Z}\mathcal{H}^{n+1/2} - \Delta t \mathcal{J}^{n+1/2}_c, \quad \mathcal{E}^{n+1} = \mathbf{K}^{\mathrm{E}}\mathcal{D}^{n+1}, \tag{3.29}$$

$$\mathcal{B}^{n+1/2} = \mathcal{B}^{n-1/2} - \Delta t \mathbf{Z}\mathcal{E}^n, \quad \mathcal{H}^{n+1/2} = \mathbf{K}^{\mathrm{H}}\mathcal{B}^{n+1/2} \tag{3.30}$$

with Δt the temporal increment. The conditional stability of (3.29) and (3.30) is investigated via the amplification matrix technique, which, after eliminating $\mathcal{E}$ and $\mathcal{H}$ vectors, yields the compact notation of

$$\mathcal{X}^{n+1} = \mathbf{G}\mathcal{X}^n, \quad \text{where} \quad \mathcal{X}^{n+1} = \left[\mathcal{B}^{n+1/2}, \mathcal{D}^{n+1}\right], \tag{3.31}$$

$$\mathbf{G} = \begin{bmatrix} \mathbf{I} & -\Delta t \mathbf{Z}\mathbf{K}^{\mathrm{E}} \\ \Delta t \mathbf{Z}\mathbf{K}^{\mathrm{H}} & \mathbf{I} - (\Delta t)^2 \mathbf{Z}\mathbf{K}^{\mathrm{H}}\mathbf{Z}\mathbf{K}^{\mathrm{E}} \end{bmatrix}, \tag{3.32}$$

$\mathbf{I}$ is the identity matrix and $\mathbf{J}_c = \mathbf{0}$. Considering the necessary condition of $|\lambda_s(\mathbf{G})| \leq 1$, for every s, the stability criterion for the explicit formulation is

$$\Delta t \leq \frac{2}{\left|\lambda_{\max}\left(\mathbf{Z}\mathbf{K}^{\mathrm{H}}\mathbf{Z}\mathbf{K}^{\mathrm{E}}\right)\right|}, \tag{3.33}$$

with $\lambda_{\max}(\mathbf{N})$ as the largest eigenvalue of matrix $\mathbf{N}$. This limit guarantees the selection of the suitable Δt, even for the case of curvilinear structures, without the need of the inhibiting FDTD staircase approximation.

As an example for the aforementioned explicit FETD methodology, assume the rectangular microstrip antenna of Figure 3.1, which is placed over a RT/Duroid ($\varepsilon_r = 2.2$) substrate with a height of 0.698 mm. The computational domain is divided into 102,456 elements, whereas for the insertion of the excitation signal, a 50-Ω microstrip line is used. Figure 3.2 illustrates four snapshots

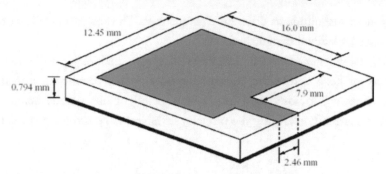

FIGURE 3.1: A rectangular microstrip antenna on a RT/Duroid substrate.

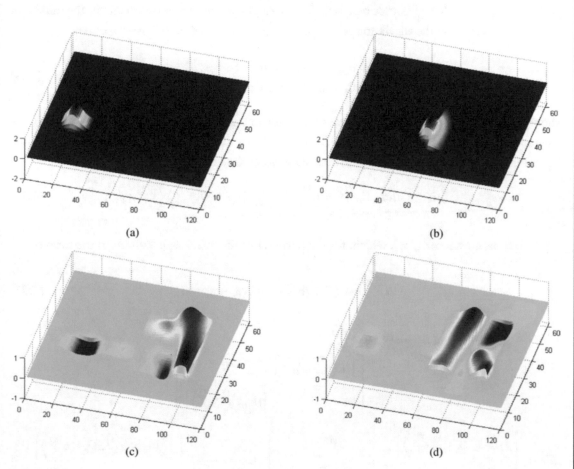

FIGURE 3.2: Snapshots of the magnitude of E_z component at (a) 32.392 ps, (b) 82.461 ps, (c) 174.085 ps, and (d) 258.972 ps.

of the E_z component magnitude at different time instants. The smooth profile of the propagating waves can be, indeed, readily distinguished.

An implicit two-stage approach. As already stated, basis functions $\mathbf{w}_{ij}^{(1)}$ in (3.15) ensure the tangential continuity of fields across the element interfaces while permitting normal discontinuity [13, 14]. Nonetheless, $\mathbf{w}_{ij}^{(1)}$ do not satisfy the pertinent orthogonality requirements. To render them orthogonal and concurrently retaining the property of tangential continuity, the set of vector basis functions,

$$\mathbf{z}_{ij}^{(1)} = \begin{cases} \left(\hat{\mathbf{t}}_{ij} \cdot \mathbf{w}_{ij}^{(1)}\right)\hat{\mathbf{t}}_{ij}, & \text{in the middle of edge}\{ij\} \\ \mathbf{0}, & \text{in the middle of other edges} \end{cases} \tag{3.34}$$

is constructed, where $\hat{\mathbf{t}}_{ij}$ is the unit vector tangential to edge $\{ij\}$. Inspecting (3.34), one may come up with the deduction that $\mathbf{z}_{ij}^{(1)}$ cannot represent fields that possess normal components in the middle of the element edges. For the alleviation of this problem, an additional set of basis functions

$$\mathbf{y}_{ij}^{(1)} = \begin{cases} \hat{\mathbf{n}}_{ij}, & \text{in the middle of edge}\{ij\} \\ \mathbf{0}, & \text{in the middle of other edges} \end{cases} \tag{3.35}$$

should be introduced, with $\hat{\mathbf{n}}_{ij}$ the normal unit vector to edge $\{ij\}$. In this way, $\mathbf{y}_{ij}^{(1)}$ are not only orthogonal to each other, but also orthogonal with regard to all $\mathbf{z}_{ij}^{(1)}$. Because the previous conventions do not uniquely determine $\mathbf{z}_{ij}^{(1)}$ and $\mathbf{y}_{ij}^{(1)}$, these are reestablished as

$$\mathbf{y}_{ij}^{(1)} = \zeta_i \zeta_j \hat{\mathbf{n}}_{ij} \quad \text{and} \quad \mathbf{z}_{ij}^{(1)} = \mathbf{w}_{ij}^{(1)} - \sum_{p=\text{edge}\{ij\}} \left\langle \mathbf{w}_{ij}^{(1)}, \mathbf{y}_p^{(1)} \right\rangle \left\langle \mathbf{y}_p^{(1)}, \mathbf{y}_p^{(1)} \right\rangle^{-1} \mathbf{y}_p^{(1)} \tag{3.36}$$

Let us, now, recall (3.8), which, for the purposes of the analysis, is written in the form of

$$\mathbf{\Lambda}\frac{d^2\mathbf{u}_c}{dt^2} + (\mathbf{\Gamma} + \mathbf{\Phi})\frac{d\mathbf{u}_c}{dt} + \mathbf{N}\mathbf{u}_c + g = 0, \tag{3.37}$$

with the element of the matrices given by

$$\mathbf{\Lambda}_{ij} = \begin{bmatrix} \left[\mu\left\langle \mathbf{z}_{jk}^{(1)}, \mathbf{z}_{ki}^{(1)} \right\rangle\right] & \mathbf{0} \\ \mathbf{0} & \left[\mu\left\langle \mathbf{y}_{jk}^{(1)}, \mathbf{y}_{ki}^{(1)} \right\rangle\right] \end{bmatrix},$$

$$\mathbf{\Gamma}_{ij} = \begin{bmatrix} \left[\mu\sigma\varepsilon^{-1}\left\langle \mathbf{z}_{jk}^{(1)}, \mathbf{z}_{ki}^{(1)} \right\rangle\right] & \mathbf{0} \\ \mathbf{0} & \left[\mu\sigma\varepsilon^{-1}\left\langle \mathbf{y}_{jk}^{(1)}, \mathbf{y}_{ki}^{(1)} \right\rangle\right] \end{bmatrix}, \mathbf{\Phi}_{ij} = \frac{1}{\sqrt{\mu\varepsilon}}\begin{bmatrix} \left[\left\langle \hat{\mathbf{n}}_\Omega \times \mathbf{z}_{jk}^{(1)}, \hat{\mathbf{n}}_\Omega \times \mathbf{z}_{ki}^{(1)} \right\rangle\right] & \mathbf{0} \\ \mathbf{0} & \mathbf{0} \end{bmatrix},$$

$$\mathbf{N}_{ij} = \begin{bmatrix} \left[\varepsilon^{-1} \left\langle \nabla \times \mathbf{z}_{jk}^{(1)}, \nabla \times \mathbf{z}_{ki}^{(1)} \right\rangle \right] & \left[\varepsilon^{-1} \left\langle \nabla \times \mathbf{z}_{jk}^{(1)}, \nabla \times \mathbf{y}_{ki}^{(1)} \right\rangle \right] \\ \left[\varepsilon^{-1} \left\langle \nabla \times \mathbf{y}_{jk}^{(1)}, \nabla \times \mathbf{z}_{ki}^{(1)} \right\rangle \right] & \left[\varepsilon^{-1} \left\langle \nabla \times \mathbf{y}_{jk}^{(1)}, \nabla \times \mathbf{y}_{ki}^{(1)} \right\rangle \right] \end{bmatrix},$$

where $\hat{\mathbf{n}}_{\Omega}$ is the outgoing unit vector normal to the Γ_{Ω} outer boundary that surrounds computational space Ω. If a typical central finite-difference scheme is applied to (3.37), one derives the final update expression

$$\mathbf{A}\mathbf{u}_c^{n+1} = \left[\frac{2}{(\Delta t)^2} \mathbf{\Lambda} - \mathbf{N} \right] \mathbf{u}_c^n + \left[\frac{1}{2\Delta t}(\mathbf{\Gamma} + \mathbf{\Phi}) - \frac{1}{(\Delta t)^2} \mathbf{\Lambda} \right] \mathbf{u}_c^{n-1} - g^n, \qquad (3.38)$$

with $\mathbf{A} = \dfrac{1}{(\Delta t)^2}\mathbf{\Lambda} + \dfrac{1}{2\Delta t}(\mathbf{\Gamma} + \mathbf{\Phi})$

Although the preceding scheme is, in general, implicit—because of the nonorthogonal nature of (3.34) and (3.35)—it may be easily transformed to an explicit one if the vector basis functions of (3.36) are selected, and hence, $\mathbf{\Lambda}$, $\mathbf{\Gamma}$, $\mathbf{\Phi}$, $\mathbf{A}$ become diagonal. Toward this direction contributes the sparse profile of matrix N. To point out the equivalent efficiency of the implicit two-stage FETD algorithm, as compared with its explicit counterpart, previously described, the S_{11} parameter of the

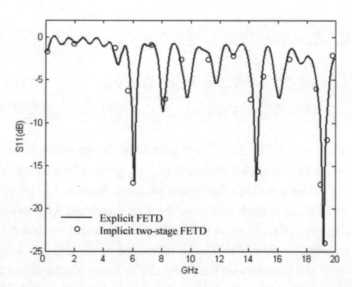

FIGURE 3.3: Comparison of the S_{11}-parameter evaluation through the explicit and the two-stage implicit FETD methods.

rectangular microstrip antenna (Figure 3.1) is computed in Figure 3.3 via both techniques. Evidently, the agreement of the two plots is very satisfactory.

Higher-order WETD implementations. For the realization of the specific time-domain approaches, a more precise approximation of the constitutive relations should be pursued. Such a concept is based on the derivation of higher-order Whitney elements, particularly tangentially and normally continuous second-order ones [15]. Additionally, the definition of degrees of freedom must be conducted in a compatible policy, as a 1-form field intensity is associated with a 2-form flow via a curl equation in the Maxwell's system.

In this context, the second-order tetrahedral tangentially continuous Whitney elements are generated by means of a systematic procedure, founded on the decoupling property of degrees of freedom and the correct modeling of the curl operator in the final update expressions. The resultant basis functions are given by

$$\mathbf{w}_{ij}^{(1)} = \left(8\zeta_i^2 - 4\zeta_i\right)\nabla\zeta_j + \left(-8\zeta_i\zeta_j + 2\zeta_j\right)\nabla\zeta_i, \tag{3.39}$$

$$\mathbf{w}_{ijk}^{(2)} = 16\zeta_i\zeta_j\nabla\zeta_k - 8\zeta_j\zeta_k\nabla\zeta_i - 8\zeta_k\zeta_i\nabla\zeta_j, \tag{3.40}$$

for the edge and facet case, respectively. Working along a similar fashion, the normal counterpart may be successfully constructed. For the sake of completeness, three degrees of freedom on each face and three extra ones, $\mathbf{w}_{ijkl}^{(3)}$, in the volume of the element are resolved. After some algebra, the relevant basis functions read

$$\mathbf{w}_{ijk}^{(2)} = (30\zeta_i\zeta_k - 6\zeta_k)\nabla\zeta_i \times \nabla\zeta_j + \left(30\zeta_i^2 - 12\zeta_i\right)\nabla\zeta_j \times \nabla\zeta_k + (30\zeta_i\zeta_j - 6\zeta_j)\nabla\zeta_k \times \nabla\zeta_i, \tag{3.41}$$

$$\begin{aligned}\mathbf{w}_{ijkl}^{(3)} = &- 30\zeta_k\zeta_l\nabla\zeta_i \times \nabla\zeta_j + 30\zeta_j\zeta_l\nabla\zeta_i \times \nabla\zeta_k \\ &+ 30\zeta_j\zeta_k\nabla\zeta_i \times \nabla\zeta_l + 60\zeta_i\zeta_l\nabla\zeta_j \times \nabla\zeta_k - 60\zeta_i\zeta_k\nabla\zeta_j \times \nabla\zeta_l\end{aligned} \tag{3.42}$$

Observe that the definition of (3.39)–(3.42) is consistent in the sense that if the 2-form field is the curl of the 1-form field, explicit expressions regarding the degrees of freedom for all electric and magnetic components are obtained through the Stokes theorem. Actually, this property constitutes the most serious issue for an efficient discretization of the time-dependent Ampère's and Faraday's law.

Tracking the notions of (3.15)–(3.23), higher-order elements are launched during the discretization of the constitutive equations. The development of the leapfrog scheme initiates from the approximation of **E** and **B** vectors via the first-order Whitney elements and the basis functions of (3.15) and (3.18), as by now described. Hence, time-advancing equations like the second one of (3.24) are acquired, without any other complicated manipulations. The next step focuses on the computation of flux density **D** and intensity **H**. Nevertheless, if the same order of elements is used, a sizeable loss of accuracy is experienced. Mitigation of the latter weakness comes from the establishment of a dual lattice, which accomplishes notable precision enhancements in the sense that it produces finer resolution values in the space domain. To this preference, another possible justification comes from the fact that electromagnetic laws are, in fact, second-order partial differential equations in space and time, and thus, their modeling stipulates advanced discretizations. Unfortunately, when general unstructured primary meshes are dealt with, the dual ones do not comprise tetrahedral elements only, making it extremely strenuous to define WETD configurations on them, unless specialized grid structures are taken into account. For this reason, **H** and **D** are calculated in terms of *second-order* Whitney elements tangential and normal, respectively, leading to

$$\mathbf{H}^{n+1/2} = \mu^{-1}\mathbf{PB}^{n+1/2}, \qquad \mathbf{D}^n = \varepsilon\mathbf{QE}^n, \qquad \mathbf{E}^{n+1} = \varepsilon^{-1}\mathbf{RD}^{n+1}, \qquad (3.43)$$

constitutive equations, with **P**, **Q**, **R** matrices involving integrations of basis functions of a normal element according to the definition of degrees of freedom for the tangential elements. The coefficients of **P**, **Q**, **R** depend on the geometric features of the problem under study; yet they remain the same as time evolves. Furthermore, it can be verified that the approximation order for the tangentially continuous fields determined from (3.43), is reduced by one compared with the order of the normally continuous counterparts. This is *exactly* why the algorithm will fail, if first-order elements are the only chosen.

Because Faraday's law is going to be derived from (3.24), the last stage for the completion of the higher-order WETD method is the discretization of Ampère's law, this time involving second-order degrees of freedom. So, if one applies Stokes theorem for suitably picked fields on any facet $\{ijk\}$, a set of explicit expressions for the electric flux density and the magnetic field intensity is devised. Note that these update equations are not *identical*, due to the lack of symmetry in the definition of facet degrees of freedom for the tangential elements. Regarding the conditional stability of the algorithm, a prefixed temporal increment equal to half or one third of the smallest dimension in the domain, divided by the propagation velocity, is deemed sufficient.

In conclusion, the above features of WETD techniques reveal their promising potential in the solution of difficult EMC applications. Even so, there are several issues that need to be further investigated for the specific simulators. Amid them, the use of collocation approaches and different kinds of basis functions in diverse regions of the computational domain is the most prominent. It becomes apparent that such functions should be carefully established to be fully consistent with the expected field smoothness in the various areas. Finally, the implementation of more rigorous spatial interpolation schemes for the reduction of dispersion errors seems to be a favorable sector of improvement, especially in terms of wavelet expansions for the fulfillment of demanding field continuity conditions, as will be discussed in the succeeding sections.

3.3   THE FVTD METHOD

Soon after the initial application of the traditional FDTD algorithm in real-world EMC problems, a significant impediment arose; the incapacity of the "staircase" approximation to treat curved surfaces, abrupt geometric details, or irregular discontinuities, without the unaffordable increase of mesh resolution. Obviously, the idea for the direct discretization of the integral form of Maxwell's equations to derive flexible numerical models had started to mature and several successful realizations appeared. A typical representative of this class is, without any doubt, the FVTD method that evolved as an instructive, yet more laborious, tool for the solution of challenging structures [16–22] and the most important: it became the primary vehicle for the development of other conspicuous time-domain approaches for handling generalized nonorthogonal or mixed polyhedral unstructured grids, as elaborately exemplified in Chapter 4. Essentially, the FVTD approach uses a leapfrog integration scheme—conditionally stable and nondissipative—that conserves the total charge in a global and local sense. Thus, the canonical staggered-meshed FDTD formalism is efficiently circumvented with the computational overhead limited in reasonable levels. The goal of this section is the formulation of 3- and reduced 2-D FVTD variants by means of a topological (use of dually interleaved lattices) and a conservation law discipline, tailored for structure-conforming discretizations.

3.3.1   Topological Construction of FVTD Schemes

Development of 3-D update equations. According to the topological point of view, the extraction of the FVTD technique requires the division of the computational domain into a set of dual cells of arbitrary size, shape, and curvature; however, throughout this paragraph, analysis will assume the presence of hexahedral elements. The corners of every primary cell, whose origin is at the (i, j, k) node, are defined as $N^{P}_{i,j,k}$ and those of the secondary cells as $N^{S}_{i,j,k}$. Inspecting Figure 3.4, one may promptly notice that the two grids share the ordinary offset of 1/2 along each direction in relevance to the primary grid. Also, let the lengths of the matching edges—e.g., toward the y-coordinate—be $L^{P,2}_{i,j,k} = |N^{P,y}_{i,j+1,k} - N^{P,y}_{i,j,k}|$ and $L^{S,2}_{i,j,k} = |N^{S,y}_{i,j+1,k} - N^{S,y}_{i,j,k}|$, respectively, with $L^{P,1}_{i,j,k}$, $L^{P,3}_{i,j,k}$ and $L^{S,1}_{i,j,k}$, $L^{S,3}_{i,j,k}$ denoted in an analogous manner along x- and z-coordinate. Because, in the general case, the four corners of each cell face do not necessarily lie on the same plane, it is mandatory to describe the geometrical aspects of all edges and faces in both meshes [16]. So, regarding the former, direction cosines $\mathbf{C}^{P,m}_{i,j,k}$ and $\mathbf{C}^{S,m}_{i,j,k}$, for $m = 1, 2, 3$, of unit vector form are required. On the other hand, primary and secondary faces are characterized via their vector areas

$$\mathbf{A}^{P,m}_{i,j,k} = A^{P,m}_{i,j,k}\,\hat{\mathbf{n}}^{P,m}_{i,j,k} \qquad \text{and} \qquad \mathbf{A}^{S,m}_{i,j,k} = A^{S,m}_{i,j,k}\,\hat{\mathbf{n}}^{S,m}_{i,j,k}, \qquad (3.44)$$

referring to the +1 direction, where $A^{P,m}_{i,j,k}$, $A^{S,m}_{i,j,k}$ are the corresponding scalar areas and $\hat{\mathbf{n}}^{P,m}_{i,j,k}$, $\hat{\mathbf{n}}^{S,m}_{i,j,k}$ the average inward-pointing unit normals of the relative faces. The purpose of introducing (3.44) is that

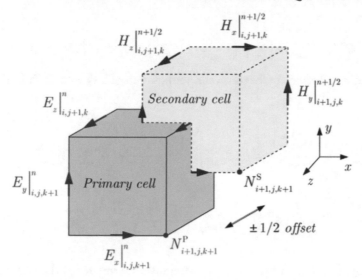

FIGURE 3.4: The spatial staggering offset in the FVTD elementary primary/secondary cell complex.

albeit the scalar area may not have a unique value, expressly when the face is not planar, its vector counterpart is always unique, as long as the surface it prescribes is limited by straight lines connecting the four corners.

Because of the involvement of curvilinear (or in extent of nonorthogonal) coordinates, $\mathbf{C}_{i,j,k}^{P,m}$ and $\mathbf{C}_{i,j,k}^{S,m}$ will not mandatorily be parallel or perpendicular to $\hat{\mathbf{n}}_{i,j,k}^{P,m}$ and $\hat{\mathbf{n}}_{i,j,k}^{S,m}$. Such a deduction implies that the 3×3 matrices formed by the dot products of $\hat{\mathbf{n}}_{i,j,k}^{S,m}$ and $\mathbf{C}_{i,j,k}^{P,m}$ or vice versa will not be unitary and sometimes not even diagonal. As a consequence, to complete the description of the domain, the diagonal elements of these matrices should be obtained in a preprocessing stage. For illustration,

$$\hat{\mathbf{n}}_{i,j,k}^{S,1} \cdot \mathbf{C}_{i,j,k}^{P,1} = n_{i,j-1,k-1}^{S,1x} C_{i,j,k}^{P,1x} + n_{i,j-1,k-1}^{S,1y} C_{i,j,k}^{P,1y} + n_{i,j-1,k-1}^{S,1z} C_{i,j,k}^{P,1z}, \qquad (3.45)$$

with equivalent expressions holding for $\hat{\mathbf{n}}_{i,j,k}^{S,2} \cdot \mathbf{C}_{i,j,k}^{P,2}$ and $\hat{\mathbf{n}}_{i,j,k}^{S,3} \cdot \mathbf{C}_{i,j,k}^{P,3}$ as well as for the $\hat{\mathbf{n}}_{i,j,k}^{P,m} \cdot \mathbf{C}_{i,j,k}^{S,m}$ elements.

Applying the previous nomenclature to Maxwell's laws, (2.1) and (2.2), in a lossless free-space medium, for each primary face, it can be derived that

$$\mu_0 \iint_{A_{i,j,k}^{P,m}} \frac{\partial \mathbf{H}}{\partial t} \cdot \hat{\mathbf{n}}_{i,j,k}^{P,m} \mathrm{d}A = - \oint_{\partial A_{i,j,k}^{P,m}} \mathbf{E} \cdot \mathrm{d}\mathbf{l}, \qquad (3.46)$$

which supposes the projection of $\mathbf{E}$ along the cell edges and simultaneously provides the projection of $\mathbf{H}$ toward the face unit vectors. Because (3.46), through the Stokes theorem, will eventually require the incorporation of $\mathbf{H} \cdot \mathbf{C}_{i,j,k}^{S,m}$ in Ampère's law for the update of $\mathbf{E}$, one must compute the consequent relation

$$\mathbf{C}_{i,j,k}^{S,m} = \left(\mathbf{C}_{i,j,k}^{S,m} \cdot \hat{\mathbf{n}}_{i,j,k}^{P,m} \right) \hat{\mathbf{n}}_{i,j,k}^{P,m} + \left(\mathbf{C}_{i,j,k}^{S,m} \cdot \hat{\mathbf{n}}_{i,j,k}^{P,m+1} \right) \hat{\mathbf{n}}_{i,j,k}^{P,m+1} + \left(\mathbf{C}_{i,j,k}^{S,m} \cdot \hat{\mathbf{n}}_{i,j,k}^{P,m+2} \right) \hat{\mathbf{n}}_{i,j,k}^{P,m+2}, \quad (3.47)$$

with the $m+1$ and $m+2$ standing for the cyclic permutation of indices 1, 2, 3. In this context, (3.47) leads to

$$\mu_0 A_{i,j,k}^{P,m} \frac{\partial \left(\mathbf{H} \cdot \mathbf{C}_{i,j,k}^{S,m} \right)}{\partial t} = \mu_0 A_{i,j,k}^{P,m} \left[\frac{\partial \left(\mathbf{H} \cdot \hat{\mathbf{n}}_{i,j,k}^{P,m} \right)}{\partial t} \left(\hat{\mathbf{n}}_{i,j,k}^{P,m} \cdot \mathbf{C}_{i,j,k}^{S,m} \right) \right.$$

$$+ \frac{\partial \left(\mathbf{H} \cdot \hat{\mathbf{n}}_{i,j,k}^{P,m+1} \right)}{\partial t} \left(\hat{\mathbf{n}}_{i,j,k}^{P,m+1} \cdot \mathbf{C}_{i,j,k}^{S,m} \right) + \frac{\partial \left(\mathbf{H} \cdot \hat{\mathbf{n}}_{i,j,k}^{P,m+2} \right)}{\partial t} \left. \left(\hat{\mathbf{n}}_{i,j,k}^{P,m+2} \cdot \mathbf{C}_{i,j,k}^{S,m} \right) \right] \quad (3.48)$$

At this stage, it is judged beneficial to distinguish the impact of the approximation order in (3.48). Both first- and second-order configurations may be used [17]. Consistent with the former, the last two terms in the right-hand side of (3.48) are neglected, thus simplifying the resulting expressions. Despite the fact that they are proven to be less accurate in some EMC structures than their second-order counterparts, the complexity of the latter is sometimes prohibitive. Therefore, the topological analysis, without loss of generality, will next deal with higher-order arrangements only in the 2-D case.

The first-order approximation of (3.48) reads

<u>Faraday's law:</u> $\mu_0 A_{i,j,k}^{P,m} \dfrac{\partial \left(\mathbf{H} \cdot \mathbf{C}_{i,j,k}^{S,m} \right)}{\partial t} = - \left(\hat{\mathbf{n}}_{i,j,k}^{P,m} \cdot \mathbf{C}_{i,j,k}^{S,m} \right) \displaystyle\oint_{\partial A_{i,j,k}^{P,m}} \left(\mathbf{E} \cdot \mathbf{C}_{i,j,k}^{P,m} \right) \mathrm{d}l,$ (3.49)

with the respective and fully dyadic relation for the magnetic field curl given by

$$(3.50)$$

<u>Ampère's law:</u> $\varepsilon_0 A_{i,j,k}^{S,m} \dfrac{\partial \left(\mathbf{E} \cdot \mathbf{C}_{i,j,k}^{P,m} \right)}{\partial t} = \left(\hat{\mathbf{n}}_{i,j,k}^{S,m} \cdot \mathbf{C}_{i,j,k}^{P,m} \right) \displaystyle\oint_{\partial A_{i,j,k}^{S,m}} \left(\mathbf{H} \cdot \mathbf{C}_{i,j,k}^{S,m} \right) \mathrm{d}l$

The discretization of (3.49), for $m = 1$, provides

$$\mu_0 A_{i,j,k}^{P,1} \frac{\mathbf{H}_{i,j,k}^{n+1/2} \cdot \mathbf{C}_{i,j,k}^{S,1} - \mathbf{H}_{i,j,k}^{n-1/2} \cdot \mathbf{C}_{i,j,k}^{S,1}}{\Delta t} = - \left(\hat{\mathbf{n}}_{i,j,k}^{P,1} \cdot \mathbf{C}_{i,j,k}^{S,1} \right) \left[\left(\mathbf{E}_{i,j+1,k}^{n} \cdot \mathbf{C}_{i,j+1,k}^{P,3} \right) L_{i,j+1,k}^{P,3} \right.$$

$$- \left(\mathbf{E}_{i,j,k}^{n} \cdot \mathbf{C}_{i,j,k}^{P,3} \right) L_{i,j,k}^{P,3} - \left(\mathbf{E}_{i,j,k+1}^{n} \cdot \mathbf{C}_{i,j,k+1}^{P,2} \right) L_{i,j,k+1}^{P,2}$$

$$\left. + \left(\mathbf{E}_{i,j,k}^{n} \cdot \mathbf{C}_{i,j,k}^{P,2} \right) L_{i,j,k}^{P,2} \right]$$

$$(3.51)$$

together with two supplementary equations for $\mathbf{H} \cdot \mathbf{C}_{i,j,k}^{S,2}$ and $\mathbf{H} \cdot \mathbf{C}_{i,j,k}^{S,3}$. Similar 3-D FVTD forms are promptly derived for the temporal update of $\mathbf{E} \cdot \mathbf{C}_{i,j,k}^{P,1}$, $\mathbf{E} \cdot \mathbf{C}_{i,j,k}^{P,2}$, and $\mathbf{E} \cdot \mathbf{C}_{i,j,k}^{P,3}$ from (3.50).

The modeling competences of the FVTD algorithm are verified by means of a 3-D T-junction, whose cross-section is considered to be uniform along the y-axis. The particular EMC structure is discretized into $200 \times 26 \times 120$ cells, and the excitation is inserted at the $z = 0$ plane. Additionally, the open ends of the waveguide are truncated by the suitable absorbing boundary conditions (see Chapter 4). Figure 3.5 shows the magnitude of the E_y component at the $j = 10$ plane at four distinct time instants and unveils the ability of the scheme to simulate propagating waves in a rapid and precise manner.

The reduced 2-D case. In two dimensions, (3.48) and its dual counterpart for $\mathbf{E} \cdot \mathbf{C}_{i,j}^{P,m}$ become

$$\mu_0 A_{i,j}^{P,m} \frac{\partial \left(\mathbf{H} \cdot \mathbf{C}_{i,j}^{S,m} \right)}{\partial t} = \mu_0 A_{i,j}^{P,m} \left[\frac{\partial \left(\mathbf{H} \cdot \hat{\mathbf{n}}_{i,j}^{P,m} \right)}{\partial t} \left(\hat{\mathbf{n}}_{i,j}^{P,m} \cdot \mathbf{C}_{i,j}^{S,m} \right) + \frac{\partial \left(\mathbf{H} \cdot \overline{\mathbf{n}}_{i,j}^{P,m} \right)}{\partial t} \left(\overline{\mathbf{n}}_{i,j}^{P,m} \cdot \mathbf{C}_{i,j}^{S,m} \right) \right],$$

$$(3.52)$$

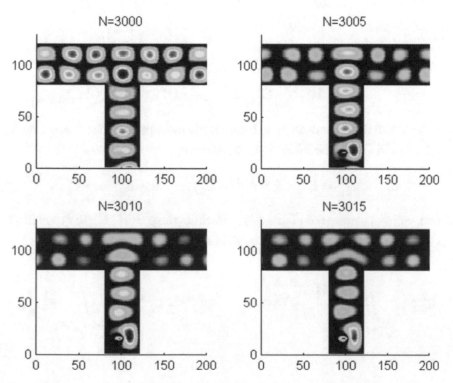

FIGURE 3.5: Magnitude of the E_y field component at different time steps. The numbers on the axes indicate the respective cell.

$$\varepsilon_0 A_{i,j}^{S,m} \frac{\partial \left(\mathbf{E} \cdot \mathbf{C}_{i,j}^{P,m}\right)}{\partial t} = \varepsilon_0 A_{i,j}^{S,m} \left[\frac{\partial \left(\mathbf{E} \cdot \hat{\mathbf{n}}_{i,j}^{S,m}\right)}{\partial t} \left(\hat{\mathbf{n}}_{i,j}^{S,m} \cdot \mathbf{C}_{i,j}^{P,m}\right) + \frac{\partial \left(\mathbf{E} \cdot \overline{\mathbf{n}}_{i,j}^{S,m}\right)}{\partial t} \left(\overline{\mathbf{n}}_{i,j}^{S,m+1} \cdot \mathbf{C}_{i,j}^{P,m}\right) \right]$$

$$(3.53)$$

with $\overline{\mathbf{n}}_{i,j}^{P,m} \overline{\mathbf{n}}_{i,j}^{S,m}$ orthogonal vectors $\hat{\mathbf{n}}_{i,j}^{P,m} \hat{\mathbf{n}}_{i,j}^{S,m}$. Moreover, in the above equation

$$\mathbf{C}_{i,j}^{S,m} = \left(\mathbf{C}_{i,j}^{S,m} \cdot \hat{\mathbf{n}}_{i,j}^{P,m}\right) \hat{\mathbf{n}}_{i,j}^{P,m} + \left(\mathbf{C}_{i,j}^{S,m} \cdot \overline{\mathbf{n}}_{i,j}^{P,m}\right) \overline{\mathbf{n}}_{i,j}^{P,m} \qquad (3.54)$$

Retaining the first-order approximation logic as in the 3-D case, the last terms on the right-hand side of (3.52) and (3.53) are omitted, and in this way, the final FVTD expressions are

$$\mu_0 \frac{\mathbf{H}_{i,j}^{n+1/2} \cdot \mathbf{C}_{i,j}^{S,1} - \mathbf{H}_{i,j}^{n-1/2} \cdot \mathbf{C}_{i,j}^{S,1}}{\Delta t} = -\left(\hat{\mathbf{n}}_{i,j}^{P,1} \cdot \mathbf{C}_{i,j}^{S,1}\right) \frac{E_z|_{i,j+1}^{n} - E_z|_{i,j}^{n}}{L_{i,j}^{P,2}}, \qquad (3.55)$$

$$\mu_0 \frac{\mathbf{H}_{i,j}^{n+1/2} \cdot \mathbf{C}_{i,j}^{S,2} - \mathbf{H}_{i,j}^{n-1/2} \cdot \mathbf{C}_{i,j}^{S,2}}{\Delta t} = -\left(\hat{\mathbf{n}}_{i,j}^{P,2} \cdot \mathbf{C}_{i,j}^{S,2}\right) \frac{E_z|_{i+1,j}^{n} - E_z|_{i,j}^{n}}{L_{i,j}^{P,1}}, \qquad (3.56)$$

$$\varepsilon_0 \frac{E_z|_{i,j}^{n+1} - E_z|_{i,j}^{n}}{\Delta t} = \frac{1}{A_{i-1,j-1}^{S,3}} \left[\left(\mathbf{H}_{i,j}^{n+1/2} \cdot \mathbf{C}_{i,j}^{S,2}\right) L_{i,j-1}^{S,2} - \left(\mathbf{H}_{i-1,j}^{n+1/2} \cdot \mathbf{C}_{i-1,j}^{S,2}\right) L_{i-1,j-1}^{S,2} \right.$$
$$\left. - \left(\mathbf{H}_{i,j}^{n+1/2} \cdot \mathbf{C}_{i,j}^{S,1}\right) L_{i-1,j}^{S,1} + \left(\mathbf{H}_{i,j-1}^{n+1/2} \cdot \mathbf{C}_{i,j-1}^{S,1}\right) L_{i-1,j-1}^{S,1} \right] \qquad (3.57)$$

To proceed with the second-order and sufficiently more precise schemes, taking into account all terms in (3.53) and (3.54), the subsequent conventions

$$\hat{\mathbf{n}}_{i,j}^{P,1} \times \overline{\mathbf{n}}_{i,j}^{P,1} = \hat{\mathbf{z}}, \qquad \hat{\mathbf{z}} \times \hat{\mathbf{n}}_{i,j}^{P,1} = \mathbf{C}_{i,j}^{P,2}, \qquad \hat{\mathbf{z}} \times \hat{\mathbf{n}}_{i,j}^{P,2} = -\mathbf{C}_{i,j}^{P,1}, \qquad (3.58)$$

are established, which, by means of Figure 3.5a, substitute the $\mu_0 \partial (\mathbf{H} \cdot \mathbf{C}_{i,j}^{S,m})/\partial t$ term in (3.52) by the corrected $\mu_0 (\hat{\mathbf{n}}_{i,j}^{P,m+1} \cdot \mathbf{C}_{i,j}^{S,m+1}) \partial (\mathbf{H} \cdot \hat{\mathbf{n}}_{i,j}^{P,m+1})/\partial t$, so yielding

$$\mu_0 \left(\hat{\mathbf{n}}_{i,j}^{P,m} \cdot \mathbf{C}_{i-1,j}^{S,m}\right) \iint_{A_{i,j}^{P,L}+A_{i,j}^{P,R}} \frac{\partial \left(\mathbf{H} \cdot \overline{\mathbf{n}}_{i,j}^{P,m}\right)}{\partial t} dA = -\left(\mathbf{C}_{i,j}^{P,m+1} \cdot \mathbf{C}_{i-1,j}^{S,m}\right) \oint_{\partial \left(A_{i,j}^{P,L}+A_{i,j}^{P,R}\right)} \hat{\mathbf{n}}_0 \cdot \left(\mathbf{E} \times \mathbf{C}_{i,j}^{P,m+1}\right)$$

$$(3.59)$$

where $A_{i,j}^{P,L}$, $A_{i,j}^{P,L}$ are shown in Figure 3.6b and $\hat{\mathbf{n}}_0$ is the outward unit vector normal to $\partial (A_{i,j}^{P,L}, A_{i,j}^{P,R})$ contour. The argument of the circulation in (3.59) is computed via

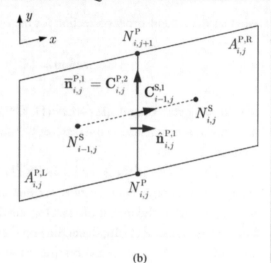

FIGURE 3.6: Topology of primary and secondary with an offset of +1 (a) edge direction cosines as well as inward-pointing unit normals and (b) scalar areas.

$$\hat{\mathbf{n}}_0 \cdot \left(\mathbf{E} \times \mathbf{C}_{i,j}^{P,m+1} \right) = -E_z \mathbf{C}_{i,j}^{P,m+1} \cdot (\hat{\mathbf{x}}dx + \hat{\mathbf{y}}dy) = -E_z \left[C_{i,j}^{P,(m+1)x}dx + C_{i,j}^{P,(m+1)y}dy \right],$$

(3.60)

which through the substitution of $A_{i,j}^P = A_{i,j}^{P,L} + A_{i,j}^{P,R}$ gives

$$\mu_0 \left(A_{i,j}^{P,L} + A_{i,j}^{P,R} \right) \left(\hat{\mathbf{n}}_{i,j}^{P,m} \cdot \mathbf{C}_{i-1,j}^{S,m} \right) \iint\limits_{A_{i,j}^{P,L} + A_{i,j}^{P,R}} \frac{\partial \left(\mathbf{H} \cdot \overline{\mathbf{n}}_{i,j}^{P,m} \right)}{\partial t} dA =$$

$$- \left(\mathbf{C}_{i,j}^{P,m+1} \cdot \mathbf{C}_{i-1,j}^{S,m} \right) \left(B_{i,j}^{P,L} + B_{i,j}^{P,R} \right)$$

(3.61)

with $B_{i,j}^{P,L}$ and $B_{i,j}^{P,R}$ representing the contour integrals around $A_{i,j}^{P,L}$ and $A_{i,j}^{P,R}$. For example, if $m = 2$,

$$B_{i,j}^{P,L} = \frac{1}{2} \left[C_{i,j}^{P,2x} \left(N_{i,j}^{P,x} - N_{i-1,j}^{P,x} \right) + C_{i,j}^{P,2y} \left(N_{i,j}^{P,y} - N_{i-1,j}^{P,y} \right) \right] \left(E_x|_{i,j}^n + E_x|_{i-1,j}^n \right)$$

$$+ \frac{1}{2} \left[C_{i,j}^{P,2x} \left(N_{i,j+1}^{P,x} - N_{i,j}^{P,x} \right) + C_{i,j}^{P,2y} \left(N_{i,j+1}^{P,y} - N_{i,j}^{P,y} \right) \right] \left(E_x|_{i,j+1}^n + E_x|_{i,j}^n \right)$$

$$+ \frac{1}{2} \left[C_{i,j}^{P,2x} \left(N_{i-1,j+1}^{P,x} - N_{i,j+1}^{P,x} \right) + C_{i,j}^{P,2y} \left(N_{i-1,j+1}^{P,y} - N_{i,j+1}^{P,y} \right) \right] \left(E_x|_{i-1,j+1}^n + E_x|_{i,j+1}^n \right)$$

$$+ \frac{1}{2} \left[C_{i,j}^{P,2x} \left(N_{i-1,j}^{P,x} - N_{i-1,j+1}^{P,x} \right) + C_{i,j}^{P,2y} \left(N_{i-1,j}^{P,y} - N_{i-1,j+1}^{P,y} \right) \right] \left(E_x|_{i-1,j}^n + E_x|_{i-1,j+1}^n \right),$$

whereas the second-order correction for the $\partial(\mathbf{H} \cdot \mathbf{C}_{i,j}^{\mathrm{S},1})/\partial t$ derivative is

$$\text{Correction} = \left(\mathbf{C}_{i,j}^{\mathrm{P},2} \cdot \mathbf{C}_{i,j}^{\mathrm{P},2} \right) \frac{B^{\mathrm{P,L}} + B^{\mathrm{P,R}}}{\mu_0 \left(A_{i,j}^{\mathrm{P,L}} + A_{i,j}^{\mathrm{P,R}} \right)} , \tag{3.62}$$

with an analogous treatment for the $\partial(\mathbf{H} \cdot \mathbf{C}_{i,j}^{\mathrm{S},2})/\partial t$ term. It should be stressed that because in the 2-D case, z-axis is orthogonal to the plane of $\mathbf{C}_{i,j}^{\mathrm{P},1}$ and $\mathbf{C}_{i,j}^{\mathrm{P},2}$, no correction like (3.62), is required for the E_z component.

The stability of the 2- and 3-D FVTD algorithms is determined through the von Neumann method, which, after some calculus, certifies that the well-known Courant condition of the FDTD technique is also valid for the distinct formulations, if the smallest wavelength in the domain is carefully incorporated. Actually, depending on the geometry of the EMC problem under investigation, this limit may accept some modifications toward the direction of stricter, with respect to temporal increment, numerical implementations. In conclusion, the second-order approximation enhances notably the performance of the FVTD method, at the expense of additional complexity, without affecting its overall stability and convergence.

3.3.2   The Conservation Law Formulation

Apart from the involvement of different cell topologies in its extraction, the FVTD method can be equivalently derived by the conservation law theory, which expresses all field quantities as components of a solution vector $\mathbf{F} = [\mathbf{E}\mathbf{H}]^{\mathrm{T}}$ and casts Maxwell's equations in a fairly exploitable form. Thus, commencing from (2.1) and (2.2) for a homogeneous, isotropic, and lossy ($\mathbf{J}_c \neq \mathbf{0}$, $\mathbf{M}_c = \mathbf{0}$) medium, the curl of any vector $\mathbf{f}$ may be written in terms of the divergence of a matrix operator $\mathcal{P}$ acting on $\mathbf{f}$ as

$$\nabla \times \mathbf{f} = [\mathrm{div}\{\mathcal{P}(\mathbf{f})\}]^{T} = \begin{bmatrix} \partial_v f_w - \partial_w f_v \\ \partial_w f_u - \partial_u f_w \\ \partial_u f_v - \partial_v f_u \end{bmatrix}, \text{ where } \mathcal{P}(\mathbf{f}) = \begin{bmatrix} 0 & -f_w & f_v \\ f_w & 0 & -f_u \\ -f_v & f_v & 0 \end{bmatrix},$$

defined in a generalized coordinate system (u, v, w). In this manner, Maxwell's equations become

$$\varepsilon \frac{\partial \mathbf{E}}{\partial t} - [\mathrm{div}\{\mathcal{P}(\mathbf{H})\}]^{\mathrm{T}} + \sigma \mathbf{E} = -\mathbf{J}_s, \tag{3.63}$$

$$\mu \frac{\partial \mathbf{H}}{\partial t} + [\mathrm{div}\{\mathcal{P}(\mathbf{E})\}]^{\mathrm{T}} = \mathbf{0}, \tag{3.64}$$

or in a more compact form,

$$\frac{\partial \mathbf{F}}{\partial t} + \mathbf{b}^{-1}\mathbf{QF} = \mathbf{b}^{-1}(\mathbf{L} + \mathbf{RF}), \tag{3.65}$$

for

$$\mathbf{b} = \begin{bmatrix} \varepsilon\mathbf{I} & 0 \\ 0 & \mu\mathbf{I} \end{bmatrix}, \quad \mathbf{QF} = \begin{bmatrix} -\nabla \times \mathbf{H} \\ \nabla \times \mathbf{E} \end{bmatrix}, \quad \mathbf{L} = \begin{bmatrix} -\mathbf{J}_s \\ 0 \end{bmatrix}, \quad \mathbf{R} = \begin{bmatrix} -\sigma\mathbf{I} & 0 \\ 0 & 0 \end{bmatrix}$$

where $\mathbf{J}_s$ is the vector of external (imposed) sources and $\mathbf{I}$ is the identity matrix. If the computational domain Ω is divided into a preselected number of elements Ω_i with a boundary of $\partial\Omega_i$, then (3.65) can be integrated over each Ω_i and, through the divergence theorem, yields

$$\iiint_{V_i} \partial_t \mathbf{F} dV + \iint_{\partial\Omega_i} \mathbf{b}^{-1}\mathbf{\Lambda}(\hat{\mathbf{n}})\mathbf{F} ds = \iiint_{V_i} \mathbf{b}^{-1}\mathbf{L} d\mathbf{r} + \iiint_{V_i} \mathbf{b}^{-1}\mathbf{RF} d\mathbf{r} , \tag{3.66}$$

with $\mathbf{\Lambda}(\hat{\mathbf{n}}) = \begin{bmatrix} 0 & -\mathcal{P}(\hat{\mathbf{n}}) \\ \mathcal{P}(\hat{\mathbf{n}}) & 0 \end{bmatrix}$, V_i as the volume of Ω_i, and $\hat{\mathbf{n}}$ representing the outward unit vector perpendicular to surface $\partial\Omega_i$. Note that quantity $\mathbf{\Lambda}(\hat{\mathbf{n}})\mathbf{F}$ is actually the flux through surface $\partial\Omega_i$. Next, the concept of the cell average—i.e., the average of $\mathbf{F}$ located at the barycenter of every element Ω_i—is introduced as

$$F_i = \frac{1}{V_i} \iiint_{V_i} \mathbf{F}(\mathbf{r}) dV , \tag{3.67}$$

where $\mathbf{r} = u\hat{\mathbf{u}} + u\hat{\mathbf{v}} + w\hat{\mathbf{w}}$ is a position vector. To conduct the required flux splitting, according to the conservation-law formulation, matrix $\bar{\mathbf{\Lambda}}(\hat{\mathbf{n}}) = \mathbf{b}^{-1}\mathbf{\Lambda}(\hat{\mathbf{n}})$, which is decomposable to a sum of matrices with positive and negative eigenvalues, is additionally launched [22]. Hence, and bearing in mind that the six eigenvalues of $\bar{\mathbf{\Lambda}}(\hat{\mathbf{n}})$ are given by diag$\{0,0,v,v,-v,-v\}$ for $v = 1\sqrt{\mu\varepsilon}$, one arrives at

$$\bar{\mathbf{\Lambda}}(\hat{\mathbf{n}}) = \bar{\mathbf{\Lambda}}(\hat{\mathbf{n}})^+ + \bar{\mathbf{\Lambda}}(\hat{\mathbf{n}})^- = \frac{1}{2}\begin{bmatrix} -v[\mathcal{P}(\hat{\mathbf{n}})]^2 & -\varepsilon^{-1}\mathcal{P}(\hat{\mathbf{n}}) \\ \mu^{-1}\mathcal{P}(\hat{\mathbf{n}}) & -v[\mathcal{P}(\hat{\mathbf{n}})]^2 \end{bmatrix} + \frac{1}{2}\begin{bmatrix} v[\mathcal{P}(\hat{\mathbf{n}})]^2 & -\varepsilon^{-1}\mathcal{P}(\hat{\mathbf{n}}) \\ \mu^{-1}\mathcal{P}(\hat{\mathbf{n}}) & v[\mathcal{P}(\hat{\mathbf{n}})]^2 \end{bmatrix}$$

$$\tag{3.68}$$

The surface integral on the left-hand section of (3.66) can be calculated if the $\bar{\mathbf{\Lambda}}(\hat{\mathbf{n}})$ $\mathbf{F}$ flux is already known on all faces of boundary $\partial\Omega_i$. For this purpose, let $\mathbf{F}^{\text{IN}}$ denote the required solution vector on the internal side of every face belonging to $\partial\Omega_i$ and $\mathbf{F}^{\text{OUT}}$ its counterpart located at the outside. Both quantities are acquired by an interpolation procedure centered on the respective face side. Amid them, the following approach

$$\bar{\Lambda}(\hat{\mathbf{n}})\mathbf{F} = \mathbf{b}^{-1}\left(\boldsymbol{\Gamma}_p\bar{\Lambda}(\hat{\mathbf{n}})^+\mathbf{F}^{\text{IN}} + \boldsymbol{\Gamma}_q\bar{\Lambda}(\hat{\mathbf{n}})^-\mathbf{F}^{\text{OUT}}\right), \tag{3.69}$$

which refers to faces separating two different volumes V_p, V_q, has been proven very efficient. In (3.69),

$$\boldsymbol{\Gamma}_p = 2\begin{bmatrix} Y_q(Y_p + Y_q)^{-1}\mathbf{I} & \mathbf{0} \\ \mathbf{0} & Z_q(Z_p + Z_q)^{-1}\mathbf{I} \end{bmatrix}, \quad \boldsymbol{\Gamma}_q = 2\begin{bmatrix} Y_p(Y_p + Y_q)^{-1}\mathbf{I} & \mathbf{0} \\ \mathbf{0} & Z_p(Z_p + Z_q)^{-1}\mathbf{I} \end{bmatrix},$$

$Z = Y^{-1}$ is the impedance, $\mathbf{I}$ is the 3×3 identity matrix, and $+,-$ superscripts are the inward or outward direction of $\hat{\mathbf{n}}$. Hence, for a lattice consisting of M elements, the flux integration at each element Ω_i for $i = 1, 2, \ldots, M$, is written in vector notation as

$$\left[\frac{1}{V_1}\left(\int_{\partial\Omega_1} b^{-1}\Lambda(\hat{\mathbf{n}})\mathbf{F}d\mathbf{s}\right)^{\text{T}} \frac{1}{V_2}\left(\int_{\partial\Omega_2} b^{-1}\Lambda(\hat{\mathbf{n}})\mathbf{F}d\mathbf{s}\right)^{\text{T}} \cdots \frac{1}{V_M}\left(\int_{\partial\Omega_M} b^{-1}\Lambda(\hat{\mathbf{n}})\mathbf{F}d\mathbf{s}\right)^{\text{T}}\right]^{\text{T}} \simeq \mathbf{AS}, \tag{3.70}$$

where $\mathbf{A}$ is the matrix of media structural properties and $\mathbf{S} = [\mathbf{F}_1^{\text{T}}\ \mathbf{F}_2^{\text{T}} \ldots \mathbf{F}_M^{\text{T}}]^{\text{T}}$ is the vector of unknowns within every element. Plugging (3.70) in the volume integration of (3.65) gives the result of

$$\frac{\partial\mathbf{S}}{\partial t} + \mathbf{AS} = \mathbf{B}, \tag{3.71}$$

with $\mathbf{B}$ as a source term incorporating the contributions of all elements Ω_i. For the FVTD method to be complete, the temporal derivative of (3.71) must be discretely approximated. Thus, via an ordinary second-order finite-difference scheme, acting in the sense of a predictor–corrector process, one receives

$$\mathbf{S}^{n+1/2} = \mathbf{S}^n - 0.5\Delta t\mathbf{AS}^n \quad \text{and} \quad \mathbf{S}^{n+1} = \mathbf{S}^n - 0.5\Delta t\mathbf{AS}^{n+1/2}, \tag{3.72}$$

that is fully explicit and subject to the stability conditions of Section 3.3.1.

3.4   TIME-DOMAIN SCHEMES FOUNDED ON MULTIRESOLUTION ANALYSIS

The incorporation of multiresolution abstractions to Maxwell's curl equations leads to the so called multiresolution time-domain (MRTD) method with a set of key intrinsic properties [23–30]. Actually, the development of these schemes is based on the systematic use of scaling and wavelet functions, which yield variable lattice gradings. For this purpose, electromagnetic fields are expressed by means of twofold expansions with regard to space. In particular, the expansion in terms of scaling functions only is proven ideal for the precise modeling of smoothly fluctuating fields (homogeneous media distributions), whereas the combined implementation of both scaling and wavelet counterparts is very instructive for regions of intense singularities or abrupt variations (high spectral frequencies). Both realizations are extracted via cubic Battle–Lemarie or Haar scaling and wavelet orthonormal expansions, which, owing to the lack of compact support, are spatially truncated. Although at the outset, this issue can be thought of as a shortcoming, the low- and band-pass attributes of the MRTD forms in the spectral domain permit the a priori estimation of the necessary resolution for a successful simulation, thus seriously contributing to the overall efficiency. In this way, the resultant algorithm exhibits highly linear dispersion traits, which guarantee notable accuracy levels for a discretization close to the Nyquist sampling condition. So, the minimum resolution for reliable MRTD outcomes is around two points per wavelength, unlike the usual choice of 10 points required by the FDTD approach. Under these perspectives, the present section provides the basic theoretical background for the establishment of the MRTD method and the extraction of its time-advancing equations through a method-of-moments formulation.

3.4.1   Derivation Through Scaling Functions

The extraction of the specific method, frequently designated as S-MRTD algorithm, is based on the exclusive usage of scaling functions in space and pulse functions in time for the representation of electric and magnetic vectors [23, 24]. In fact, the construction procedure is similar to the TLM technique's one,[3] described in Section 2.4.2, with the noteworthy exception that in the latter, pulse functions are used solely for spatial, while scaling functions for both spatial and temporal field expansions. Preserving the notation of Chapter 2 and starting from Maxwell's laws (2.63), $\mathbf{E}$ and $\mathbf{H}$ components are given by

$$E_x(x,y,z,t) = \sum_{i,j,k,n=-\infty}^{+\infty} E_x\big|_{i+1/2,j,k}^{n} S_{i+1/2,j,k}^{n}, \qquad (3.73a)$$

$$E_y(x,y,z,t) = \sum_{i,j,k,n=-\infty}^{+\infty} E_y\big|_{i,j+1/2,k}^{n} S_{i,j+1/2,k}^{n}, \qquad (3.73b)$$

$$E_z(x,y,z,t) = \sum_{i,j,k,n=-\infty}^{+\infty} E_z\big|_{i,j,k+1/2}^{n} S_{i,j,k+1/2}^{n}, \qquad (3.73c)$$

---

[3]It is interesting to mention that this convenient method-of-moments formulation may be successfully applied to the derivation of the FDTD update expressions, as well. In such a case, all field representations in space and time are performed via pulse functions only. Thus, through this universal approach, one can easily distinguish the structural and temporal differences among the three methods.

$$H_x(x,y,z,t) = \sum_{i,j,k,n=-\infty}^{+\infty} H_x\big|_{i,j+1/2,k+1/2}^{n+1/2} \, S_{i,j+1/2,k+1/2}^{n+1/2}, \tag{3.74a}$$

$$H_y(x,y,z,t) = \sum_{i,j,k,n=-\infty}^{+\infty} H_y\big|_{i+1/2,j,k+1/2}^{n+1/2} \, S_{i+1/2,j,k+1/2}^{n+1/2}, \tag{3.74b}$$

$$H_z(x,y,z,t) = \sum_{i,j,k,n=-\infty}^{+\infty} H_z\big|_{i+1/2,j+1/2,k}^{n+1/2} \, S_{i+1/2,j+1/2,k}^{n+1/2}, \tag{3.74c}$$

with $S_{i,j,k}^n = q_i(x)q_j(y)q_k(z)p_n(t)$ and $q_m(u)$, for $m = i, j, k$, and Δu as the spatial increment expressed as

$$q_m(u) = Q\left(\frac{u}{\Delta u} - m\right), \tag{3.75}$$

where $u = x, y, z$ and $Q(u)$ is the cubic spline Battle–Lemarie scaling function. Also, $p_n(t)$ are picked to be

$$p_n(t) = P\left(\frac{t}{\Delta t} - n\right), \quad \text{in which} \quad P(t) = \begin{cases} 1 & \text{for} \quad |t| \quad 1/2 \\ 1/2 & \text{for} \quad |t| = 1/2 \\ 0 & \text{for} \quad |t| \quad 1/2 \end{cases} \tag{3.76}$$

Subsequently, (3.73) and (3.74) are substituted into (2.63), and the resulting equations are sampled by means of pulse functions in time and scaling functions in space. For this procedure, the generalized integrals

$$\int_{-\infty}^{+\infty} p_n(t)p_m(t)\mathrm{d}t = \delta_{n,m}\Delta t, \tag{3.77}$$

$$\int_{-\infty}^{+\infty} q_l(u)q_m(u)\mathrm{d}u = \delta_{l,m}\Delta u, \tag{3.78}$$

$$\int_{-\infty}^{+\infty} q_{l(u)}\frac{\partial q_{m+1/2}(u)}{\partial u}\mathrm{d}u = \delta_{l,m} - \delta_{l,m+1}, \tag{3.79}$$

should be evaluated, where $\delta_{l,m}$ ($l, m = i, j, k$) is the Kronecker delta. More specifically for (3.79), the closed form expression of the scaling function is required. Toward this aim, the Galerkin method provides adequate treatment, if one opts for the complex conjugant of the basis function in the role of test functions. This implies that the Fourier transform of (3.75) has to be computed. The most practical way for this action is the numerical calculation, which results in the infinite sum of

$$\int_{-\infty}^{+\infty} q_l(u) \frac{\partial q_{m+1/2}(u)}{\partial u} du = \sum_{s=-\infty}^{+\infty} \xi(s)\delta_{l+s,m} ,$$

with coefficients $\xi(s)$, for $0 \leq s \leq 8$, given in Table 3.1 under the convention of $\xi(-s-1) = -\xi(s)$. It is emphasized, herein, that whereas Battle–Lemarie functions have an exponential decay for $s > 8$ and $\xi(s)$ are not exactly zero, their contribution has been found to be trivial in the majority of EMC problems. Therefore, for the development of the S-MRTD schemes and without loss of precision, one can safely write

$$\int_{-\infty}^{+\infty} q_l(u) \frac{\partial q_{m+1/2}(u)}{\partial u} du \simeq \sum_{\tau=-9}^{8} \xi(\tau)\delta_{l+\tau,m} \qquad (3.80)$$

In this context and after sampling the x, y, z quantities of (2.63) in space and time according to (3.75)–(3.80), a system of difference equations is extracted. For example, the x-component of Ampère's law yields

$$\frac{\varepsilon}{\Delta t} \left(E_x|_{i+1/2,j,k}^{n+1} - E_x|_{i+1/2,j,k}^{n} \right) = \sum_{\tau=-9}^{8} \xi(\tau) \left[\frac{1}{\Delta y} H_z|_{i+1/2,j+\tau+1/2,k}^{n+1/2} - \frac{1}{\Delta z} H_y|_{i+1/2,j,k+\tau+1/2}^{n+1/2} \right],$$

$$(3.81)$$

which can be written in a more compact form via the state-space representation of Section 2.4.2. Therefore, a 3-D product space is constructed to express electromagnetic fields as

TABLE 3.1: Coefficients $\xi(\tau)$ for the computation of generalized integrals			
τ	$\xi(\tau)$	τ	$\xi(\tau)$
0	1.291846	5	-0.008189
1	-0.156076	6	0.004378
2	0.059639	7	-0.002343
3	-0.029309	8	0.001254
4	0.015371		

$$|\mathcal{E}\rangle = \sum_{i,j,k,n=-\infty}^{+\infty} \mathbf{E}|_{i,j,k}^{n}|n,i,j,k\rangle \quad \text{and} \quad |\mathcal{H}\rangle = \sum_{i,j,k,n=-\infty}^{+\infty} \mathbf{H}_d|_{i,j,k}^{n}|n,i,j,k\rangle, \qquad (3.82)$$

namely, the already described ket-vector discipline. Such a notation assigns a basis vector $|i,j,k\rangle$ to every node with discrete coordinates (i,j,k) and a basis vector $|n\rangle$ to each discrete time step n. Owing to the summation in (3.81), $|\mathcal{E}\rangle$ and $|\mathcal{H}\rangle$ incorporate all electric and magnetic quantities in the entire grid at all temporal instants. Furthermore, through the shift operators of (2.82) and (2.83), (3.81) becomes

$$\varepsilon \mathbf{X}^{\dagger}\mathbf{T}^{\dagger}\mathbf{D}_{\mathrm{T}}|\mathcal{E}_x\rangle = \mathbf{X}^{\dagger}\mathbf{T}^{\dagger}\left(\mathbf{D}_{\mathrm{Y}}|\mathcal{H}_z\rangle - \mathbf{D}_{\mathrm{Z}}|\mathcal{H}_y\rangle\right), \qquad (3.83)$$

where

$$\mathbf{D}_{\mathrm{T}} = \frac{1}{\Delta t}\left(\mathbf{T}^{\dagger} - \mathbf{T}\right), \quad \mathbf{D}_{\mathrm{Y}} = \frac{1}{\Delta y}\mathbf{Y}^{\dagger}\sum_{\tau=-9}^{8}\xi(\tau)\mathbf{Y}^{-\tau} \quad \text{and} \quad \mathbf{D}_{\mathrm{Z}} = \frac{1}{\Delta z}\mathbf{Z}^{\dagger}\sum_{\tau=-9}^{8}\xi(\tau)\mathbf{Z}^{-\tau},$$

where $\mathbf{Y}, \mathbf{Z}$ are the corresponding shift operators along spatial coordinates j, k, respectively. In a comparable fashion, one can easily obtain the remaining five Cartesian Maxwell's expressions and thus conclude to

$$\Theta|\mathcal{F}\rangle = \mathbf{0}, \qquad (3.84)$$

with $|\mathcal{F}\rangle = [|\mathcal{E}_x\rangle \, |\mathcal{E}_y\rangle \, |\mathcal{E}_z\rangle \, |\mathcal{H}_x\rangle \, |\mathcal{H}_y\rangle \, |\mathcal{H}_z\rangle]^{\mathrm{T}}$,

$$\Theta = \begin{bmatrix} \varepsilon\mathbf{X}^{\dagger}\mathbf{T}^{\dagger}\mathbf{D}_{\mathrm{T}} & 0 & 0 & 0 & \mathbf{T}^{\dagger}\mathbf{X}^{\dagger}\mathbf{D}_{\mathrm{Z}} & -\mathbf{T}^{\dagger}\mathbf{X}^{\dagger}\mathbf{D}_{\mathrm{Y}} \\ 0 & \varepsilon\mathbf{Y}^{\dagger}\mathbf{T}^{\dagger}\mathbf{D}_{\mathrm{T}} & 0 & -\mathbf{T}^{\dagger}\mathbf{Y}^{\dagger}\mathbf{D}_{\mathrm{Z}} & 0 & \mathbf{T}^{\dagger}\mathbf{Y}^{\dagger}\mathbf{D}_{\mathrm{X}} \\ 0 & 0 & \varepsilon\mathbf{Z}^{\dagger}\mathbf{T}^{\dagger}\mathbf{D}_{\mathrm{T}} & \mathbf{T}^{\dagger}\mathbf{Z}^{\dagger}\mathbf{D}_{\mathrm{Y}} & -\mathbf{T}^{\dagger}\mathbf{Z}^{\dagger}\mathbf{D}_{\mathrm{X}} & 0 \\ 0 & -\mathbf{Y}^{\dagger}\mathbf{Z}^{\dagger}\mathbf{D}_{\mathrm{Z}} & \mathbf{Y}^{\dagger}\mathbf{Z}^{\dagger}\mathbf{D}_{\mathrm{Y}} & \mu\mathbf{Y}^{\dagger}\mathbf{Z}^{\dagger}\mathbf{D}_{\mathrm{T}} & 0 & 0 \\ \mathbf{X}^{\dagger}\mathbf{Z}^{\dagger}\mathbf{D}_{\mathrm{Z}} & 0 & -\mathbf{X}^{\dagger}\mathbf{Z}^{\dagger}\mathbf{D}_{\mathrm{X}} & 0 & \mu\mathbf{X}^{\dagger}\mathbf{Z}^{\dagger}\mathbf{D}_{\mathrm{T}} & 0 \\ -\mathbf{X}^{\dagger}\mathbf{Y}^{\dagger}\mathbf{D}_{\mathrm{Y}} & \mathbf{X}^{\dagger}\mathbf{Y}^{\dagger}\mathbf{D}_{\mathrm{X}} & 0 & 0 & 0 & \mu\mathbf{X}^{\dagger}\mathbf{Y}^{\dagger}\mathbf{D}_{\mathrm{T}} \end{bmatrix},$$

and $\mathbf{D}_{\mathrm{X}}$ denoted in an analogous manner as $\mathbf{D}_{\mathrm{Y}}, \mathbf{D}_{\mathrm{Z}}$. Proceeding to the stability of (3.84) via the well-known von Neumann analysis for a uniform 3-D mesh ($\Delta = \Delta x = \Delta y = \Delta z$) and v the speed of light, it is acquired that $v\Delta t \leq 0.368112\Delta$, unlike the looser FDTD stability limit $v\Delta t \leq 0.57735\Delta$ [23].

To conclude the presentation of the S-MRTD schemes, it would be very instructive to observe the location of field components with respect to the nodes of the unit cell. Shown in Figure

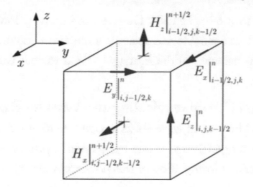

FIGURE 3.7: The elementary cell of the S-MRTD method.

3.7, its topology demonstrates an unarguable resemblance to that of the FDTD cell in Figure 2.1. Nevertheless, because of the diverse field expansions, the quantities at the two structures are not exactly the same. Although the components of the FDTD technique signify the total field at a certain point, their counterparts in the S-MRTD approach refer to only a portion of the total field. Actually, for the complete calculation of the latter at a prefixed node of the domain, one must use all field expansions of (3.73) and (3.74) and sample them in space and time.

The proficiency of the S-MRTD technique is validated through the EMC structure of Figure 3.8, which constitutes of a dielectric bridge of length d interconnecting two different GaAs substrates of a microstrip line. Such devices are typically encountered during the development of microwave monolithic integrated circuits (MMIC) and require a very systematic treatment to capture the abrupt variation of the propagating fields and avoid unwanted reflections from the bridge. This can be attained by selecting the relative permittivity of the bridge material to be near the value of its respective effective dielectric constant. In this study, dimension $a = 1$ mm and $d = 3.2a$, and

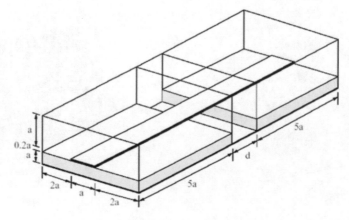

FIGURE 3.8: Substrate microstrip interconnection via a dielectric bridge of length d.

the overall mesh contains $26 \times 68 \times 12$ cells. Despite the relatively low resolution, the S-MRTD algorithm manages to conduct a reliable and accurate simulation, as concluded from the snapshots of Figure 3.9, displaying the behavior of the E_z magnitude at various time steps.

3.4.2   Development via Combined Scaling and Wavelet Functions

Overview of wavelet bases. The idea of wavelets as a robust mathematical tool for function hierarchical decomposition in orthogonal expansions has gained a prominent recognition in scientific research. Essentially, the general formulation of this powerful means is based on

$$g(s) = \sum_l a_l \phi(s-l) + \sum_m \sum_l b_{ml} \psi(2^m s - l), \qquad (3.85)$$

with $g(s)$ a real function belonging to the Hilbert space of twice integrable functions and a_l, b_{ml} weighting parameters. In the above, the first sum signifies the projection of $g(s)$ onto a subspace

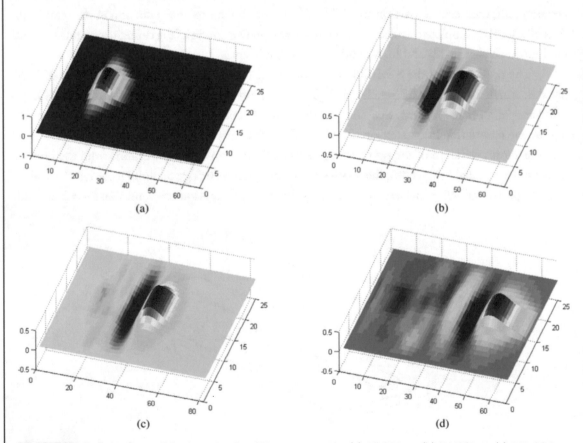

FIGURE 3.9: Snapshots of the magnitude of E_z component at (a) 35.891 ps, (b) 74.552 ps, (c) 163.906 ps, and (d) 202.751 ps.

G_0 referring to an approximation of $g(s)$ at a rough resolution level. The basis of G_0 is created via orthogonal translations of $\phi(s)$, designated as the *scaling function*. The resolution of G_0 is consecutively refined by the second sum of (3.85), which comprises projections of $g(s)$ onto the subspaces L_m, each one being defined by a *wavelet basis* $\{\psi(2^m s - l)\}$. Function $\psi_{m,l}(s)$ is called the *mother wavelet*, as all other wavelet members are produced by recursive dilations (i.e., enhanced resolutions) and translations of $\psi(s)$. Differently speaking, the addition of one wavelet improves the resolution of the overall approximation by a factor of 2. Thus, preferred levels of accuracy may be attained by adding the necessary amount of wavelet stages. In fact, the strength of such a process stems from the dependence of the wavelet coefficients magnitude on the local regularity of $g(s)$. This is exactly why these coefficients take large values near geometric discontinuities as soon as the EMC setup of interest is analyzed through the expansion of the unknown electromagnetic quantities.

Because their initial advent wavelets keep on evolving with new orthogonal systems; yet the earliest and easier-to-implement one is the *Haar basis*, which is literally the most appropriate for the interpretation of multiresolution mechanisms and the treatment of several real-world applications [25–28]. Defined as unitary pulses

$$\phi(s) = \begin{cases} 1, & 0 \leq s < 1 \\ 0, & \text{otherwise} \end{cases} \tag{3.86}$$

Haar scaling functions produce their basis via orthogonal translations of (3.86) in the form of $\phi_l(s) = \phi(s - l)$, with the mother wavelet function, for normalized s variables, given by

$$\psi(s) = \begin{cases} 1, & 0 \leq s < 1/2 \\ -1, & 1/2 \leq s < 1 \\ 0, & \text{otherwise} \end{cases} \tag{3.87}$$

Consequently, the respective wavelet basis is created in terms of translations and dilations, as

$$\psi_{m,l}(s) = \sqrt{2^m}\,\psi(2^m s - l), \tag{3.88}$$

a procedure that is actually proven to be universal for any dyadic set of functions. It is noteworthy to mention that the Haar basis exhibits a *remarkable* localization in the space or time domain; however, unlike its other counterparts, it is rather problematic in the respective Fourier domain. Sufficient alleviation of this nontrivial drawback can be trailed in the Battle–Lemarie basis, whose mother and member schemes are truncated with regard to space for the increase of their localization in both domains and the achievement of reliable estimations concerning the required resolution levels for accurate field computations.

Numerical implementation. The process of extracting the specific schemes entails an extra term in electromagnetic field expansions, for space representation, weighted by means of a predetermined wavelet set. Bearing in mind that higher dimensionalities—i.e., resolution levels—are also easily supported, analysis concentrates on the y-directed case. Therefore, if one substitutes the expansions in (3.73) and (3.74) with twofold equivalents in scaling and wavelet functions, it is acquired

$$E_x(x,y,z,t) = \sum_{i,j,k,n=-\infty}^{+\infty} \left[E_x|_{i+1/2,j,k}^{n} q_j(y) + \bar{E}_x|_{i+1/2,j+1/2,k}^{n} \psi_{j+1/2}(y) \right] Y_{i+1/2,k}^{n}, \quad (3.89a)$$

$$E_y(x,y,z,t) = \sum_{i,j,k,n=-\infty}^{+\infty} \left[E_y|_{i,j+1/2,k}^{n} q_{j+1/2}(y) + \bar{E}_y|_{i,j,k}^{n} \psi_j(y) \right] Y_{i,k}^{n}, \quad (3.89b)$$

$$E_z(x,y,z,t) = \sum_{i,j,k,n=-\infty}^{+\infty} \left[E_z|_{i,j,k+1/2}^{n} q_j(y) + \bar{E}_z|_{i,j+1/2,k+1/2}^{n} \psi_{j+1/2}(y) \right] Y_{i,k+1/2}^{n}, \quad (3.89c)$$

$$H_x(x,y,z,t) = \sum_{i,j,k,n=-\infty}^{+\infty} \left[H_x|_{i,j+1/2,k+1/2}^{n+1/2} q_{j+1/2}(y) + \bar{H}_x|_{i,j,k+1/2}^{n+1/2} \psi_j(y) \right] Y_{i,k+1/2}^{n+1/2}, \quad (3.90a)$$

$$H_y(x,y,z,t) = \sum_{i,j,k,n=-\infty}^{+\infty} \left[H_y|_{i+1/2,j,k+1/2}^{n+1/2} q_j(y) + \bar{H}_y|_{i+1/2,j+1/2,k+1/2}^{n+1/2} \psi_{j+1/2}(y) \right] Y_{i+1/2,k+1/2}^{n+1/2},$$
$$(3.90b)$$

$$H_z(x,y,z,t) = \sum_{i,j,k,n=-\infty}^{+\infty} \left[H_z|_{i+1/2,j+1/2,k}^{n+1/2} q_{j+1/2}(y) + \bar{H}_z|_{i+1/2,j,k}^{n+1/2} \psi_j(y) \right] Y_{i+1/2,k}^{n+1/2}, \quad (3.90c)$$

where the bar over the E, H components denotes the additional expansion coefficients of the technique and $Y_{j,k}^{n} = q_i(x)q_k(z)p_n(t)$, with $q_m(u)$—for $m = i, j, k$ and $u = x, y, z$—as well as $p_n(t)$ described in (3.75, and 3.76), respectively.[4] Moreover, $\psi_{m+1/2}(y)$ is given by

$$\psi_{m+1/2}(y) = \psi\left(\frac{y}{\Delta y} - m \right), \quad (3.91)$$

with $\psi(y)$ the Battle–Lemarie wavelet function. It must be stated, herein, that although $q_m(u)$ and $p_n(t)$ are evenly symmetrical with regard to $x = 0$, $\psi(y)$ has an even symmetry with reference to $x = 1/2$. This issue justifies the 1/2 spatial offset in the indices of $\psi(y)$, appearing in (3.89) and (3.90). Should the temporal resolution level be incorporated in (3.91), wavelet function

$$\psi_{n,m+1/2}(y) = \sqrt{2^n}\,\psi\left(2^n \frac{y}{\Delta y} - m \right), \quad (3.92)$$

according to (3.88) is deemed applicable. In this way, the resulting algorithm yields accurate approximations of electromagnetic fields with sufficient dispersion error characteristics. Observe that

---

[4]Similar definitions $X_{j,k}^{n} = q_j(y)q_k(z)p_n(t)$ and $Z_{i,j}^{n} = q_i(x)q_j(y)p_n(t)$ hold if the 1-D wavelet class for the requested field representation is selected along x- and z-coordinates, correspondingly.

for large frequency spectra, the latter errors are virtually trivial without any deviations from the anticipated behavior. Nonetheless, in realistic EMC problems, where higher modes coexist, the final MRTD implementation opts for supplementary wavelets toward the remaining directions in the grid.

Retaining the process of Section 3.4.1 and substituting (3.89) and (3.90) into Maxwell's laws, one has to take into account certain orthogonality relations such as the ones of (3.77)–(3.79). Such a prerequisite, after the necessary algebraic manipulations, leads to the numerical computation of

$$\int_{-\infty}^{+\infty} \psi_l(y) \frac{\partial \phi_m(y)}{\partial y} dy \simeq \sum_{\tau=-9}^{9} \vartheta(\tau)\delta_{l+\tau,m}, \tag{3.93}$$

$$\int_{-\infty}^{+\infty} \phi_l(y) \frac{\partial \psi_{m+1/2}(y)}{\partial y} dy \simeq \sum_{\tau=-9}^{9} \vartheta(\tau)\delta_{l+\tau,m+1}, \tag{3.94}$$

$$\int_{-\infty}^{+\infty} \psi_l(y) \frac{\partial \psi_{m+1/2}(y)}{\partial y} dy \simeq \sum_{\tau=-9}^{8} \zeta(\tau)\delta_{l+\tau,m}, \tag{3.95}$$

with coefficients $\vartheta(\tau)$ and $\zeta(\tau)$ summarized in Table 3.2, for $\vartheta(-\tau) = -\vartheta(\tau)$ and $\zeta(-\tau-1) = -\zeta(\tau)$. The truncation in the sums of (3.93)–(3.95) is attributed to the exponentially decaying content of both $q_m(u)$ and $\psi_{m+1/2}(y)$ that render the $\tau > 9$ (for $\vartheta(\tau)$) and the $\tau > 8$ (for $\zeta(\tau)$) terms of minor contribution.

Evidently, the next step is the sampling of the x, y, z components of (2.63) in space and time. As an illustration and for the sake of comparison with (3.81), the x-quantity of Ampère's law provides

TABLE 3.2: Coefficients $\vartheta(\tau)$ and $\zeta(\tau)$ for the calculation of generalized integrals

τ	$\vartheta(\tau)$	$\zeta(\tau)$	τ	$\vartheta(\tau)$	$\zeta(\tau)$
0	0.0	2.472538	5	−0.011153	0.013493
1	−0.046597	0.0956228	6	0.005976	−0.002858
2	0.054539	0.166058	7	−0.003202	0.002778
3	−0.036999	0.093924	8	0.001714	−0.001129
4	0.020574	0.003141	9	−0.000917	

$$\frac{\varepsilon}{\Delta t}\left(E_x|_{i+1/2,j,k}^{n+1} - E_x|_{i+1/2,j,k}^{n}\right) = \sum_{\tau=-9}^{8} \xi(\tau)\left[\frac{1}{\Delta y}H_z|_{i+1/2,j+\tau+1/2,k}^{n+1/2} - \frac{1}{\Delta z}H_y|_{i+1/2,j,k+\tau+1/2}^{n+1/2}\right]$$

$$+ \frac{1}{\Delta y}\sum_{\tau=-9}^{9}\vartheta(\tau)\bar{H}_z|_{i+1/2,j+\tau,k}^{n+1/2} \qquad (3.96a)$$

$$\frac{\varepsilon}{\Delta t}\left(\bar{E}_x|_{i+1/2,j+1/2,k}^{n+1} - E_x|_{i+1/2,j+1/2,k}^{n}\right) = \sum_{\tau=-9}^{8}\left[\frac{\zeta(\tau)}{\Delta y}\bar{H}_z|_{i+1/2,j+\tau+1,k}^{n+1/2} - \frac{\xi(\tau)}{\Delta z}\bar{H}_y|_{i+1/2,j+1/2,k+\tau+1/2}^{n+1/2}\right]$$

$$+ \frac{1}{\Delta y}\sum_{\tau=-9}^{9}\vartheta(\tau)H_z|_{i+1/2,j+\tau+1/2,k}^{n+1/2} \qquad (3.96b)$$

which via the ket-vector notation of (3.82) and operator definitions (2.82) and (2.83), become

$$\varepsilon\mathbf{X}^{\dagger}\mathbf{T}^{\dagger}\mathbf{D}_T |\mathcal{E}_x\rangle = \mathbf{X}^{\dagger}\mathbf{T}^{\dagger}\left(\mathbf{D}_Y|\mathcal{H}_z\rangle - \mathbf{D}_Z|\mathcal{H}_y\rangle + \bar{\mathbf{D}}_Y|\bar{\mathcal{H}}_z\rangle\right), \qquad (3.97a)$$

$$\varepsilon\mathbf{X}^{\dagger}\mathbf{Y}^{\dagger}\mathbf{T}^{\dagger}\mathbf{D}_T |\bar{\mathcal{E}}_x\rangle = \mathbf{X}^{\dagger}\mathbf{Y}^{\dagger}\mathbf{T}^{\dagger}\left(\bar{\mathbf{D}}_Y|\mathcal{H}_z\rangle + \bar{\bar{\mathbf{D}}}_Y|\bar{\mathcal{H}}_z\rangle + \mathbf{D}_Z|\bar{\mathcal{H}}_y\rangle\right), \qquad (3.97b)$$

for

$$\bar{\mathbf{D}}_Y = \frac{1}{\Delta y}\sum_{\tau=-9}^{9}\zeta(\tau)\mathbf{Y}^{-\tau}, \qquad \bar{\bar{\mathbf{D}}}_Y = \frac{1}{\Delta y}\mathbf{Y}^{\dagger}\sum_{\tau=-9}^{8}\vartheta(\tau)\mathbf{Y}^{-\tau},$$

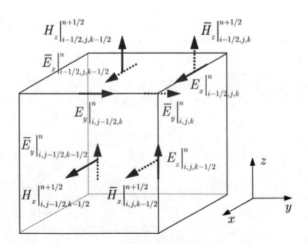

FIGURE 3.10: The elementary cell of the combined MRTD method.

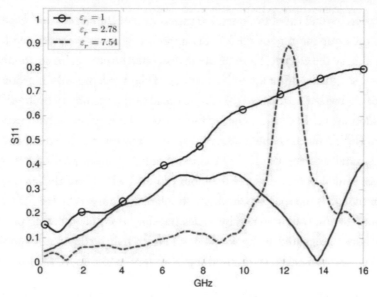

FIGURE 3.11: S_{11}-parameter calculation of the structure of Figure 3.8 for three different relative permittivities of the bridge material.

and $\mathbf{D}_T$ as signified in (3.83). Similar expressions are derived for the other five Cartesian Maxwell's constituents in an effort to construct the complete matrix system for MRTD method. Concerning the stability of this wavelet-oriented algorithm, the von Neumann analysis concludes to the even stricter, compared with the S-MRTD one, limit of $v\Delta t \leq 0.253064\Delta$ for a 3-D uniform lattice ($\Delta = \Delta x = \Delta y = \Delta z$) [25, 26]. Finally, the element cell of the method is depicted in Figure 3.10. As observed, the number of independent field variables is 12, namely, the double amount of the FDTD and S-MRTD approaches. This simply implies that the particular technique, although more robust and precise than the others in several cases, is computationally more expensive because it demands additional memory storage and CPU simulation time.

As a simple example, let us return to the structure of Figure 3.8, where the influence of the bridge's relative dielectric permittivity on the S_{11}-parameter of the entire structure is analyzed, with the rest of the implementation aspects maintained the same. The results, depicted in Figure 3.11, verify the significant changes on the parameter's profile within a wide frequency spectrum.

3.5   THE PSTD METHOD

The primary motive for the development of the PSTD technique has been the rather large number of nodes (i.e., a spatial sampling of at least 10–20 cells per minimum wavelength) required by the FDTD algorithm to achieve a reasonably acceptable precision. However, when real-world EMC applications with elongated simulation times must be modeled, it is usually necessary to increase grid resolution beyond the above range for the control of the cumulative dispersion and aniso-tropy errors. So, the sampling rate of the FDTD approach is not merely defined by the minimum wavelength, but also by the electrical size of the investigated structure. To surmount this inevitable drawback, without sacrificing other attractive features of the fundamental time-domain philosophy, the PSTD algorithm has been proposed [31, 32]. Instead of the customary finite-difference approxi-mators, this method uses either trigonometric functions or Chebyshev polynomials, through a fast Fourier transform (FFT), for the representation of spatial derivatives in Maxwell's equations. Because of its enhanced spectral accuracy, the PSTD technique entails only two cells per wavelength accord-ing to the Nyquist sampling theorem and therefore can deal with more challenging problems. One critical limitation though is the spatial periodicity stipulation addressed by the FFT usage. Late-time solutions are, consequently, contaminated by waves traveling from neighboring periods and become misleading. This issue, designated as the wraparound effect, may be drastically suppressed through a suitably modified perfectly matched layer absorbing boundary condition, whose extraction is discussed in Chapter 4.

Derivation through trigonometric functions. For the purpose of the analysis, let us first express Maxwell's curl laws, (2.1) and (2.2), in terms of a general coordinate-stretching framework, which introduces the corresponding variables of

$$\xi_u = b_u + j\,\frac{\omega_u}{\omega}, \qquad \text{where} \qquad u = x, y, z \tag{3.98}$$

and b_u an independent scaling factor that accelerates the annihilation of strenuous wave fronts, like evanescent ones, in lossy media. Substituting the curl operator in (2.1) and (2.2) with

$$\nabla_\xi \times [.] \equiv \sum_{u=x,y,z} \frac{1}{\xi_u}\frac{\partial}{\partial u}[.]\,\hat{\mathbf{u}},$$

it can be readily derived that

$$\varepsilon b_u \frac{\partial \mathbf{E}}{\partial t} + (b_u\sigma + \omega_u\varepsilon)\mathbf{E} = \frac{\partial}{\partial t}(\hat{\mathbf{u}}\times\mathbf{H}) - \omega_u\sigma\int_{-\infty}^{t}\mathbf{E}\,dt, \tag{3.99}$$

$$\mu\left(b_u\frac{\partial \mathbf{H}}{\partial t} + \omega_u\mathbf{H}\right) = -\frac{\partial}{\partial u}(\hat{\mathbf{u}}\times\mathbf{E}) - \mathbf{M}_c \tag{3.100}$$

In opposition to the FDTD method, which implements a staggered mesh, the PSTD scheme employs a lattice where all $\mathbf{E}$ and $\mathbf{H}$ field components are positioned at the cell centers. Such a mesh offers the unique benefits of specifying media properties and removing field singularities in curvilinear coordinates. If the computational domain is divided into $M_x \times M_y \times M_z$ cells, then spatial derivatives are approximated by

$$\frac{\partial f(\mathbf{r})}{\partial u} = \frac{2\pi}{M_u \Delta u} \mathcal{F}_u^{-1}\{jk_u \mathcal{F}_u\{f(\mathbf{r})\}\}, \tag{3.101}$$

with

$$\mathcal{F}_u\{f(\mathbf{r})\} \equiv \int_{-\infty}^{+\infty} f(\mathbf{r})e^{-jk_u u}du$$

as the Fourier transform definition of any field component $f(\mathbf{r})$, $\mathcal{F}_u^{-1}\{.\}$ the respective inverse transform, Δu the spatial increment toward u, and k_u the index of FFT which takes places for all u coordinates along a straight line cut through the plane specified by the other two coordinates. It is noteworthy to underline that because of the Nyquist theorem, the derivative in (3.101) is exact as long as $\Delta u \le \lambda/2$ (λ is the wavelength) [31]. So, even by means of two cells per wavelength, PSTD schemes exhibit no dispersion error attributed to spatial discretization. For example, in the case of trigonometric functions, spatial derivatives are given by

$$\left.\frac{\partial f(\mathbf{r})}{\partial u}\right|_{s\Delta u} = \frac{1}{M_u \Delta u} \sum_{m=-M_u/2}^{M_u/2-1} jk_{u,m}\tilde{f}(m)e^{jk_{u,m}s\Delta u}, \tag{3.102}$$

where $k_{u,m} = 2\pi m / (M_u \Delta u)$ and $\tilde{f}(m)$ is the Fourier series

$$\tilde{f}(m) = \Delta u \sum_{s=0}^{M_u-1} f(s\Delta u)e^{-jk_{u,m}s\Delta u}$$

In this manner, (3.99) and (3.100) are further expressed as

$$\varepsilon b_u \frac{\partial \mathbf{E}}{\partial t} + (b_u \sigma + \omega_u \varepsilon)\mathbf{E} = -\hat{\mathbf{u}} \times \mathcal{F}_u^{-1}\{jk_u \mathcal{F}_u\{\mathbf{H}\}\} - \omega_u \sigma \int_{-\infty}^{t} \mathbf{E}dt, \tag{3.103}$$

$$\mu\left(b_u \frac{\partial \mathbf{H}}{\partial t} + \omega_u \mathbf{H}\right) = \hat{\mathbf{u}} \times \mathcal{F}_u^{-1}\{jk_u \mathcal{F}_u\{\mathbf{E}\}\} - \mathbf{M}_c \tag{3.104}$$

To complete the updating procedure, it should be stressed that for temporal integration the well-known central finite differencing is used. Analytically, $\mathbf{E}$ components are defined at $t = n\Delta t$, whereas $\mathbf{H}$ components at $t = (n + 1/2)\Delta t$. For the rest of the terms in (3.103) and (3.104) not directly available at the required time steps, a conventional averaging scheme is devised, which retains second-order accuracy and does not degrade the efficiency of the discretization. The previous formulation stipulates only 1-D FFTs, thus permitting rapid calculations with an order of $O(M_u \log_2$

M_u) operations. Regarding the stability of the PSTD technique, the necessary criterion in a lossless, homogeneous domain has the form of

$$v\Delta t \leq \frac{2\Delta u}{\pi\sqrt{\text{dim}}} ,\qquad(3.105)$$

with v as the propagation velocity and dim as the dimensionality of the application. Notice that (3.105) is somewhat stricter than the FDTD Courant limit because of the $2/\pi$ term. Such a restriction is not meant to be a problem because for structures of considerable electrical size, Δt is dominated by accuracy aspects rather than the numerical stability itself. Hence, notwithstanding the potential requirement for small Δt, the PSTD approach can attain notable reductions in CPU and memory needs—as compared with the FDTD algorithm—for EMC geometries devoid of oddities or discontinuities smaller than half of the minimum wavelength [33, 34]. These remarks can be substantiated through the surface and contour plots of Figure 3.12, which provides the evolution of a Gaussian pulse, launched by means of the H_z field component, in a 3-D $50 \times 50 \times 60$ space.

The above attainments are not applicable to piecewise-smooth media with strong internal inhomogeneities, which degenerate the overall accuracy because of the Gibbs phenomenon. Nevertheless, because (3.103) and (3.104) assure the tangential field continuity across material interfaces, the resultant degree of smoothness can lead to rigorous spatial-derivative evaluations, even for slightly dissimilar media. Practically, for very large material contrasts, a finer mesh resolution should be established at the expense of an intensively heavy burden and a seriously confined time step. Otherwise, alternative solutions have to be trailed with the most convenient one being the nonuniform cosine-transform process, embedded in the PSTD method, which uses Chebyshev polynomials as a substitute to trigonometric functions. These polynomials in conjunction with the aforesaid perfectly matched layer absorber are also a complementary remedy to the spatial periodicity constraint (i.e., the wraparound effect) because they can successfully treat nonperiodic structures.

The Chebyshev PSTD method. Avoiding fixed global distributions of lattice nodes, the Chebyshev PSTD technique creates versatile handlings of arbitrary arrangements filled with inhomogeneous materials by decomposing the entire domain into a number of curved hexahedral subdomains. Then, to launch the particular polynomials within each region, a coordinate transformation that maps curvilinear boundaries in (x, y, z) into a unit cube in $(\zeta_1, \zeta_2, \zeta_3)$ coordinates is introduced under the convention of $\zeta_s = \zeta_s(x, y, z)$ for $s = 1, 2, 3$. If the content of every subdomain is presumed homogeneous, (2.1) and (2.2) become

$$\frac{\partial \mathbf{F}}{\partial t} = -\frac{1}{\sqrt{\mu\varepsilon}}\left(\mathbf{K}\frac{\partial \mathbf{F}}{\partial \zeta_1} + \Lambda\frac{\partial \mathbf{F}}{\partial \zeta_2} + \mathbf{M}\frac{\partial \mathbf{F}}{\partial \zeta_3} + \mathbf{NF}\right) = \mathbf{U}(t, \mathbf{F}),\qquad(3.106)$$

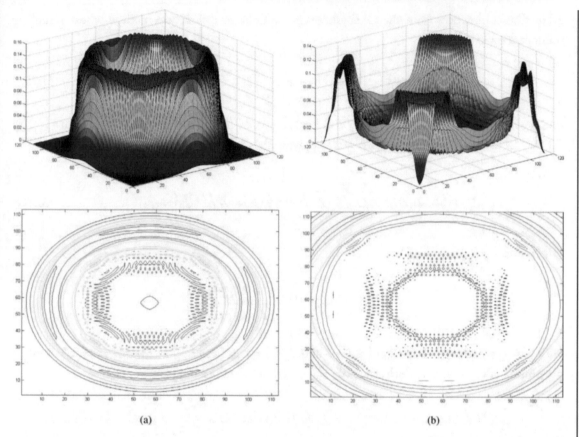

FIGURE 3.12: Temporal evolution of the H_z magnetic field component at (a) 5.378 ps and (b) 138.251 ps.

where $\mathbf{F} = [\mathbf{H}/\varepsilon^{1/2}, \mathbf{E}/\mu^{1/2}]^{\mathrm{T}}$ and matrices $\mathbf{K}$, $\mathbf{\Lambda}$, $\mathbf{M}$ contain the metrics of each coordinate ζ_1, ζ_2, ζ_3 as

$$
\left.\begin{array}{l}
\mathbf{K} : \text{for } s = 1 \\
\mathbf{\Lambda} : \text{for } s = 2 \\
\mathbf{M} : \text{for } s = 3
\end{array}\right\} =
\begin{bmatrix}
 & \mathbf{0} & & \begin{matrix} 0 & -\zeta_s^z & \zeta_s^y \\ \zeta_s^z & 0 & -\zeta_s^x \\ -\zeta_s^y & \zeta_s^x & 0 \end{matrix} \\
\begin{matrix} 0 & \zeta_s^z & -\zeta_s^y \\ -\zeta_s^z & 0 & \zeta_s^x \\ \zeta_s^y & -\zeta_s^x & 0 \end{matrix} & & \mathbf{0} &
\end{bmatrix}
, \text{ while } \mathbf{N} = \sigma\sqrt{\frac{\mu}{\varepsilon}}
\begin{bmatrix}
\mathbf{0} & & \mathbf{0} \\
& \begin{matrix} 1 & 0 & 0 \\ 0 & 1 & 0 \\ 0 & 0 & 1 \end{matrix}
\end{bmatrix}
$$

and $\mathbf{0}$ denoting a 3×3 zero matrix. Essentially, the main idea is to obtain field vector $\mathbf{F}(x, y, z)$ inside every subdomain by means of its mapped constituent $\mathbf{F}(\zeta_1, \zeta_2, \zeta_3)$. Thus, spatial derivative evaluation commences from the Chebyshev–Gauss–Lobatto collocation method whose sampling points are set at

$$\zeta_{1,i} = -\cos\left(\frac{i\pi}{I}\right), \quad \zeta_{2,j} = -\cos\left(\frac{j\pi}{J}\right), \quad \zeta_{3,k} = -\cos\left(\frac{k\pi}{K}\right), \tag{3.107}$$

nodes with $i = 1, 2, \ldots, I, j = 1, 2, \ldots, J$ and $k = 1, 2, \ldots, K$. In this context, any component $f(\zeta_1, \zeta_2, \zeta_3)$ may be interpolated through Chebyshev polynomials $p_i(\zeta_1), p_j(\zeta_2), p_k(\zeta_3)$, via

$$f(\zeta_1, \zeta_2, \zeta_3) = \sum_{i=1}^{I} \sum_{j=1}^{J} \sum_{k=1}^{K} f(\zeta_{1,i}, \zeta_{2,j}, \zeta_{3,k}) p_i(\zeta_1) p_j(\zeta_2) p_k(\zeta_3) \tag{3.108}$$

For instance, $p_i(\zeta_1)$ are given by

$$p_i(\zeta_1) = (-1)^{i+1+I} \frac{(1 - \zeta_1^2) T_I'(\zeta_1)}{g_i I^2 (\zeta_1 - \zeta_{1,i})},$$

with $g_1 = g_I = 2$ and $g_i = 1$ for $1 \le i \le I - 1$ and $T_I(\zeta_1) = \cos(I \cos^{-1} \zeta_1)$ as the Ith-order Chebyshev polynomial. So, the spatial derivatives of $f(\zeta_1, \zeta_2, \zeta_3)$ are acquired in terms of

$$\frac{\partial f}{\partial \zeta_1} = \sum_{l=0}^{I} d_{il}^{\zeta_1} f(\zeta_{1,l}, \zeta_{2,j}, \zeta_{3,k}), \quad \frac{\partial f}{\partial \zeta_2} = \sum_{l=0}^{J} d_{jl}^{\zeta_2} f(\zeta_{1,i}, \zeta_{2,l}, \zeta_{3,k}), \quad \frac{\partial f}{\partial \zeta_3} = \sum_{l=0}^{K} d_{kl}^{\zeta_3} f(\zeta_{1,i}, \zeta_{2,j}, \zeta_{3,l})$$

where the differential matrix element $d_{il}^{\zeta_s}$ (similarly for the other two) is denoted as

$$d_{il}^{\zeta_s} = p_l'(\zeta_s) = \begin{cases} \dfrac{(-1)^{i+l} g_i}{g_l \zeta_{s,i} - \zeta_{s,l}}, & \text{for} \quad i \neq l \\[2mm] -\dfrac{\zeta_{s,l}}{2(1 - \zeta_{s,l}^2)}, & \text{for} \quad 1 \le i = l \le I - 1 \\[2mm] -\dfrac{(2I^2 + 1)}{6}, & \text{for} \quad i = l = 0 \\[2mm] \dfrac{(2I^2 + 1)}{6}, & \text{for} \quad i = l = I \end{cases}$$

indicating that the Chebyshev PSTD algorithm requires a fairly restricted number of points per wavelength while concurrently accomplishing high levels of accuracy [32].

On the other hand, temporal update is executed by an M-stage, Nth-order Runge–Kutta integration,

$$\tau_0 = \mathbf{F}^n, \ \forall q \in [1, M] : \begin{cases} \beta_q = c_q\beta_{q-1} + \Delta t\,\mathbf{U}\left[(n + \gamma_q)\Delta t, \tau_{q-1}\right] \\ \tau_q = \tau_{q-1} + \kappa_q\beta_q \end{cases}, \ \mathbf{F}^{n+1} = \tau_M \quad (3.109)$$

applied within each subdomain, with β_q, c_q, γ_q, κ_q computable coefficients. As a matter of fact, for some problems where Δt is small, the two-stage second-order scheme is proven to be a reasonable configuration, owing to its comparatively reduced error, although the five-stage fourth-order analogue could be a satisfactory option as well. Finally, a comment on the fulfillment of physical continuity conditions at subdomain boundaries must be made. Expressly, on the interface of two such regions A and B, the tangential components of electric and magnetic fields are enforced to be continuous, namely, $\mathbf{E}_A^{\tan} = \mathbf{E}_B^{\tan}$ and $\mathbf{H}_A^{\tan} = \mathbf{H}_B^{\tan}$. In this way, the PSTD method extends its efficacy to inhomogeneous and irregular structures, without the need of unduly system expenditures because its enhanced sampling density decreases the simulation time and storage necessities.

3.6   ADI CONCEPTS FOR TIME-DOMAIN ALGORITHMS

The Courant–Friedrich–Levy criterion that controls the stability of most explicit time-domain methods—and intensely that of the FDTD technique—has been, despite its worth, one of the most limiting factors for the modeling of electrically large EMC structures related to shielding, immunity, signal integrity, or packaging. Recent progress in integrated circuit technology has lead to the fabrication of complex high-speed interconnect devices with demanding clock rates and low-pass signal bandwidths that extend well into the microwave frequency spectrum. At these speeds, electromagnetic wave propagation, radiation effects, and parasitic wave phenomena become rather involved and hard-to-handle. For instance, a long interconnect operating at the gigahertz range is anticipated to behave like an antenna because its length corresponds to a serious wavelength fraction. Although numerous enhancements have been so far presented, there remains a family of modern applications for which the FDTD algorithm requires heavy overheads. Indeed, to model tiny-scale features, the cell size should be on the order of microns (or less), thus enforcing an upper threshold on the time step at 10^{-12} or even 10^{-15} seconds, which in turn leads to thousands of time iterations for the steady state. These observations indicate the far stricter action of Δt bound than the one really needed for EMC analysis. Obviously, such a hindrance would be mitigated if the time-step choice was not connected to spatial increment via any stability condition. Toward this purpose, a semi-explicit technique that circumvents the Courant limit and yields an unconditionally stable scheme regardless of spatial and temporal intervals has been developed [35, 36]. By introducing the ADI time marching in the time-domain process, the only restriction for selecting a time step—at least theoretically—seems to be the numerical accuracy. However, effective studies revealed that the dispersion errors of the resulting method depend closely on grid resolution, whereas their values are seriously increased as time steps become larger. Thus, to attain a high precision level, the technique's maximum time increment should be eventually restricted. This defect is drastically limited through several approaches that do not augment memory storage or algorithmic complexity [37–45], even for other time-domain schemes [46–49].

In the present section, the general ADI-FDTD method is elaborately discussed for the case of physical media. According to the main concept, the discretized field components are staggered in space like in an ordinary grid but collocated rather than staggered in time. Hence, for the solution evolving from time step n to $n + 1$, the single FDTD iteration is divided into two distinct subiterations: the first to advance from n to $n + 1/2$ and the second to advance from $n + 1/2$ to $n + 1$. Clearly, one may proceed as follows:

First subiteration: Update from time step n to $n + 1/2$. Let us focus on (2.8a). For the first half-step, every quantity in this relation is discretized using finite differences centered at $n + 1/4$ to obtain

$$\varepsilon \left.\frac{\partial E_x}{\partial t}\right|^{n+1/4} + \sigma E_x\big|^{n+1/4} = \left.\frac{\partial H_z}{\partial y}\right|^{n+1/2} - \left.\frac{\partial H_y}{\partial z}\right|^{n} \qquad (3.110)$$

Note that spatial derivatives involving magnetic fields are computed at two different time steps, giving an overall center point of $n + 1/4$. Differently speaking, partial derivative $\partial H_z/\partial y$ is replaced with a semi-implicit difference approximation of its as-yet unknown pivotal values at $n + 1/2$, whereas the other term, $\partial H_y/\partial z$, is evaluated explicitly from known field data at n. Through a fixed cell origin, (3.110) provides

$$\frac{2\varepsilon_{i,j,k}}{\Delta t}\left[E_x\big|_{i,j,k}^{n+1/2} - E_x\big|_{i,j,k}^{n}\right] + \frac{\sigma_{i,j,k}}{2}\left[E_x\big|_{i,j,k}^{n+1/2} + E_x\big|_{i,j,k}^{n}\right]$$
$$= \frac{1}{\Delta y}\left[H_z\big|_{i,j+1/2,k}^{n+1/2} - H_z\big|_{i,j-1/2,k}^{n+1/2}\right] - \frac{1}{\Delta z}\left[H_y\big|_{i,j,k+1/2}^{n} - H_y\big|_{i,j,k-1/2}^{n}\right], \quad (3.111a)$$

where the unknown E_x quantity at $n + 1/4$, corresponding to the losses term, has been substituted by the temporal average of its already calculated constituents at n and $n + 1/2$ time steps. Moreover, the expressions for the E_y and E_z components in (2.8b) and (2.8c) at this subiteration turn into

$$\frac{2\varepsilon_{i,j,k}}{\Delta t}\left[E_y\big|_{i,j,k}^{n+1/2} - E_y\big|_{i,j,k}^{n}\right] + \frac{\sigma_{i,j,k}}{2}\left[E_y\big|_{i,j,k}^{n+1/2} + E_y\big|_{i,j,k}^{n}\right]$$
$$= \frac{1}{\Delta z}\left[H_x\big|_{i,j,k+1/2}^{n+1/2} - H_x\big|_{i,j,k-1/2}^{n+1/2}\right] - \frac{1}{\Delta x}\left[H_z\big|_{i+1/2,j,k}^{n} - H_z\big|_{i-1/2,j,k}^{n}\right], \quad (3.111b)$$

$$\frac{2\varepsilon_{i,j,k}}{\Delta t}\left[E_z|_{i,j,k}^{n+1/2} - E_z|_{i,j,k}^n\right] + \frac{\sigma_{i,j,k}}{2}\left[E_z|_{i,j,k}^{n+1/2} + E_z|_{i,j,k}^n\right]$$

$$= \frac{1}{\Delta x}\left[H_y|_{i+1/2,j,k}^{n+1/2} - H_y|_{i-1/2,j,k}^{n+1/2}\right] - \frac{1}{\Delta y}\left[H_x|_{i,j+1/2,k}^n - H_x|_{i,j-1/2,k}^n\right] \qquad (3.111c)$$

Our intention is to find the unknown electric components at time step $n + 1/2$. However, there are also unknown magnetic terms on the right-hand side of (3.111). To eliminate these quantities, which may be defined as synchronous variables, a set of finite-difference schemes centered at $n + 1/4$ is applied to (2.7) and

$$\frac{2\mu_{i,j,k}}{\Delta t}\left[H_x|_{i,j,k}^{n+1/2} - H_x|_{i,j,k}^n\right] + \frac{\rho'_{i,j,k}}{\Delta t}\left[H_x|_{i,j,k}^{n+1/2} + H_x|_{i,j,k}^n\right]$$

$$= \frac{1}{\Delta z}\left[E_y|_{i,j,k+1/2}^{n+1/2} - E_y|_{i,j,k-1/2}^{n+1/2}\right] - \frac{1}{\Delta y}\left[E_z|_{i,j+1/2,k}^n - E_z|_{i,j-1/2,k}^n\right], \qquad (3.112a)$$

$$\frac{2\mu_{i,j,k}}{\Delta t}\left[H_y|_{i,j,k}^{n+1/2} - H_y|_{i,j,k}^n\right] + \frac{\rho'_{i,j,k}}{2}\left[H_y|_{i,j,k}^{n+1/2} + H_y|_{i,j,k}^n\right]$$

$$= \frac{1}{\Delta x}\left[E_z|_{i+1/2,j,k}^{n+1/2} - E_z|_{i-1/2,j,k}^{n+1/2}\right] - \frac{1}{\Delta z}\left[E_x|_{i,j,k+1/2}^n - E_x|_{i,j,k-1/2}^n\right], \qquad (3.112b)$$

$$\frac{2\mu_{i,j,k}}{\Delta t}\left[H_z|_{i,j,k}^{n+1/2} - H_z|_{i,j,k}^n\right] + \frac{\rho'_{i,j,k}}{2}\left[H_z|_{i,j,k}^{n+1/2} + H_z|_{i,j,k}^n\right]$$

$$= \frac{1}{\Delta y}\left[E_x|_{i,j+1/2,k}^{n+1/2} - E_x|_{i,j-1/2,k}^{n+1/2}\right] - \frac{1}{\Delta x}\left[E_y|_{i+1/2,j,k}^n - E_y|_{i-1/2,j,k}^n\right] \qquad (3.112c)$$

Consider again the update of E_x in (3.111a) and use (3.112c) to eliminate the unknown H_z terms at $n + 1/2$. In this manner, the following equation is extracted

$$\kappa_1 E_x|_{i,j,k}^{n+1/2} - \kappa_2 E_x|_{i,j+1,k}^{n+1/2} - \kappa_3 E_x|_{i,j-1,k}^{n+1/2} = \kappa_4 E_x|_{i,j,k}^n + \kappa_5 H_z|_{i,j+1/2,k}^n - \kappa_6 H_z|_{i,j-1/2,k}^n$$

$$- \kappa_7\left[H_y|_{i,j,k+1/2}^n - H_y|_{i,j,k-1/2}^n\right] - \kappa_8\left[E_y|_{i+1/2,j+1/2,k}^n - E_y|_{i-1/2,j+1/2,k}^n\right]$$

$$+ \kappa_9\left[E_y|_{i+1/2,j-1/2,k}^n - E_y|_{i-1/2,j-1/2,k}^n\right] \qquad (3.113)$$

where

$$\kappa_1 = 1 + \left(\frac{2\Delta t}{\Delta y}\right)^2 p_{i,j,k}\left(q_{i,j+1/2,k} - q_{i,j-1/2,k}\right),$$

$$\kappa_2 = \left(\frac{2\Delta t}{\Delta y}\right)^2 p_{i,j,k} q_{i,j+1/2,k}, \quad \kappa_3 = \left(\frac{2\Delta t}{\Delta y}\right)^2 p_{i,j,k} q_{i,j-1/2,k}$$

$$\kappa_4 = p_{i,j,k}\left(4\varepsilon_{i,j,k} - \sigma_{i,j,k}\Delta t\right), \qquad \kappa_5 = \frac{2\Delta t}{\Delta y} p_{i,j,k} q_{i,j+1/2,k} r_{i,j+1/2,k},$$

$$\kappa_6 = \frac{2\Delta t}{\Delta y} p_{i,j,k} q_{i,j-1/2,k} r_{i,j-1/2,k}, \qquad \kappa_7 = \frac{2\Delta t}{\left(4\varepsilon_{i,j,k} + \sigma_{i,j,k}\Delta t\right)\Delta z} p_{i,j,k},$$

$$\kappa_8 = \frac{(2\Delta t)^2}{\Delta x \Delta y} p_{i,j,k} q_{i,j+1/2,k}, \qquad \kappa_9 = \frac{(2\Delta t)^2}{\Delta x \Delta y} p_{i,j,k} q_{i,j-1/2,k},$$

with

$$p_{i,j,k} = \left(4\varepsilon_{i,j,k} + \sigma_{i,j,k}\Delta t\right)^{-1}, \quad q_{i,j,k} = \left(4\mu_{i,j,k} + \rho'_{i,j,k}\Delta t\right)^{-1}, \quad r_{i,j,k} = 4\mu_{i,j,k} - \rho'_{i,j,k}\Delta t$$

It becomes apparent that (3.113) concludes to a system of simultaneous equations for E_x when repeatedly applied to every j-coordinate along a y-directed line through the grid. The matrix of this system is tridiagonal and easily solved with available numerical packages. Similarly, application of the preceding process to (3.111b) and (3.112a) yields a tridiagonal matrix system for each z-cut through the lattice to obtain E_y at $n + 1/2$, whereas the same idea holds for E_z at each x-cut with (3.111c, and 3.112b). If all electric field components are determined in the mesh, the first subiteration is completed with the computation of $H_x H_y$, and H_z components at $n + 1/2$ by means of (3.112), which is direct and fully explicit.

Second subiteration: Update from time step $n + 1/2$ to $n + 1$. Recalling (2.8a), the ADI-FDTD method for the second half step requires that finite-difference approximations centered at $n + 3/4$, to get

$$\varepsilon \left.\frac{\partial E_x}{\partial t}\right|^{n+3/4} + \sigma E_x|^{n+3/4} = \left.\frac{\partial H_z}{\partial y}\right|^{n+1/2} - \left.\frac{\partial H_y}{\partial z}\right|^{n+1}, \qquad (3.114)$$

where magnetic quantities are discretized at time steps $n + 1/2$ and $n + 1$ to supply an overall center point of $n + 3/4$. At this stage, all fields at $n + 1/2$ are already known, whereas their counterparts at $n + 1$ are the unknowns. Now the $\partial H_z/\partial y$ term is treated explicitly from known value, and the $\partial H_y/\partial z$ derivative accepts the semi-implicit evaluation via the so far unknown pivotal field data at $n + 1$. Moreover, an analogous time-averaging around $n + 3/4$ holds for E_x, associated with the losses, on the left-hand side of (3.114). In this framework, electric field components are given by

$$\frac{2\varepsilon_{i,j,k}}{\Delta t}\left[E_x\big|_{i,j,k}^{n+1} - E_x\big|_{i,j,k}^{n+1/2}\right] + \frac{\sigma_{i,j,k}}{2}\left[E_x\big|_{i,j,k}^{n+1/2} + E_x\big|_{i,j,k}^{n+1}\right]$$

$$= \frac{1}{\Delta y}\left[H_z\big|_{i,j+1/2,k}^{n+1/2} - H_z\big|_{i,j-1/2,k}^{n+1/2}\right] - \frac{1}{\Delta z}\left[H_y\big|_{i,j,k+1/2}^{n+1} - H_y\big|_{i,j,k-1/2}^{n+1}\right], \quad (3.115a)$$

$$\frac{2\varepsilon_{i,j,k}}{\Delta t}\left[E_y\big|_{i,j,k}^{n+1} - E_y\big|_{i,j,k}^{n+1/2}\right] + \frac{\sigma_{i,j,k}}{2}\left[E_y\big|_{i,j,k}^{n+1/2} + E_y\big|_{i,j,k}^{n+1}\right]$$

$$= \frac{1}{\Delta z}\left[H_x\big|_{i,j,k+1/2}^{n+1/2} - H_x\big|_{i,j,k-1/2}^{n+1/2}\right] - \frac{1}{\Delta x}\left[H_z\big|_{i+1/2,j,k}^{n+1} - H_z\big|_{i-1/2,j,k}^{n+1}\right], \quad (3.115b)$$

$$\frac{2\varepsilon_{i,j,k}}{\Delta t}\left[E_z\big|_{i,j,k}^{n+1} - E_z\big|_{i,j,k}^{n+1/2}\right] + \frac{\sigma_{i,j,k}}{2}\left[E_z\big|_{i,j,k}^{n+1/2} + E_z\big|_{i,j,k}^{n+1}\right]$$

$$= \frac{1}{\Delta x}\left[H_y\big|_{i+1/2,j,k}^{n+1/2} - H_y\big|_{i-1/2,j,k}^{n+1/2}\right] - \frac{1}{\Delta y}\left[H_x\big|_{i,j+1/2,k}^{n+1} - H_x\big|_{i,j-1/2,k}^{n+1}\right] \quad (3.115c)$$

Due to the unknown (synchronous) terms encountered on the right-hand side of (3.115), relations (2.7) are approximated by finite-difference schemes centered at $n + 3/4$. This leads to

$$\frac{2\mu_{i,j,k}}{\Delta t}\left[H_x\big|_{i,j,k}^{n+1} - H_x\big|_{i,j,k}^{n+1/2}\right] + \frac{\rho'_{i,j,k}}{2}\left[H_x\big|_{i,j,k}^{n+1/2} + H_x\big|_{i,j,k}^{n+1}\right]$$

$$= \frac{1}{\Delta z}\left[E_y\big|_{i,j,k+1/2}^{n+1/2} - E_y\big|_{i,j,k-1/2}^{n+1/2}\right] - \frac{1}{\Delta y}\left[E_z\big|_{i,j+1/2,k}^{n+1} - E_z\big|_{i,j-1/2,k}^{n+1}\right], \quad (3.116a)$$

$$\frac{2\mu_{i,j,k}}{\Delta t}\left[H_y\big|_{i,j,k}^{n+1} - H_y\big|_{i,j,k}^{n+1/2}\right] + \frac{\rho'_{i,j,k}}{2}\left[H_y\big|_{i,j,k}^{n+1/2} + H_y\big|_{i,j,k}^{n+1}\right]$$

$$= \frac{1}{\Delta x}\left[E_z\big|_{i+1/2,j,k}^{n+1/2} - E_z\big|_{i-1/2,j,k}^{n+1/2}\right] - \frac{1}{\Delta z}\left[E_x\big|_{i,j,k+1/2}^{n+1} - E_x\big|_{i,j,k-1/2}^{n+1}\right], \quad (3.116b)$$

$$\frac{2\mu_{i,j,k}}{\Delta t}\left[H_z\big|_{i,j,k}^{n+1} - H_z\big|_{i,j,k}^{n+1/2}\right] + \frac{\rho'_{i,j,k}}{2}\left[H_z\big|_{i,j,k}^{n+1/2} + H_z\big|_{i,j,k}^{n+1}\right]$$

$$= \frac{1}{\Delta y}\left[E_x\big|_{i,j+1/2,k}^{n+1/2} - E_x\big|_{i,j-1/2,k}^{n+1/2}\right] - \frac{1}{\Delta x}\left[E_y\big|_{i+1/2,j,k}^{n+1} - E_y\big|_{i-1/2,j,k}^{n+1}\right] \quad (3.116c)$$

Contrary to the first subiteration, E_x in (3.115a) is updated if we eliminate the unknown H_y terms at time step $n + 1$ via (3.116b) and acquire

$$v_1 E_x|_{i,j,k}^{n+1} - v_2 E_x|_{i,j,k+1}^{n+1} - v_3 E_x|_{i,j,k-1}^{n+1} = v_4 E_x|_{i,j,k}^{n+1/2} - v_5 H_y|_{i,j,k+1/2}^{n+1/2} + v_6 H_y|_{i,j,k-1/2}^{n+1/2}$$

$$+ v_7 \left[H_z|_{i,j+1/2,k}^{n+1/2} - H_z|_{i,j-1/2,k}^{n+1/2} \right] - v_8 \left[E_z|_{i+1/2,j,k+1/2}^{n+1/2} - E_z|_{i-1/2,j,k+1/2}^{n+1/2} \right]$$

$$+ v_9 \left[E_z|_{i+1/2,j,k-1/2}^{n+1/2} - E_z|_{i-1/2,j,k-1/2}^{n+1/2} \right] \tag{3.117}$$

where

$$v_1 = 1 + \left(\frac{2\Delta t}{\Delta z} \right)^2 p_{i,j,k} \left(q_{i,j,k+1/2} - q_{i,j,k-1/2} \right), \quad v_2 = \left(\frac{2\Delta t}{\Delta z} \right)^2 p_{i,j,k} q_{i,j,k+1/2},$$

$$v_3 = \left(\frac{2\Delta t}{\Delta z} \right)^2 p_{i,j,k} q_{i,j,k-1/2},$$

$$v_4 = p_{i,j,k} \left(4\varepsilon_{i,j,k} - \sigma_{i,j,k} \Delta t \right), \quad v_5 = \frac{2\Delta t}{\Delta z} p_{i,j,k} q_{i,j,k+1/2} r_{i,j,k+1/2},$$

$$v_6 = \frac{2\Delta t}{\Delta z} p_{i,j,k} q_{i,j,k-1/2} r_{i,j,k-1/2}, \quad v_7 = \frac{2\Delta t}{\left(4\varepsilon_{i,j,k} + \sigma_{i,j,k} \Delta t \right) \Delta y} \alpha_{i,j,k},$$

$$v_8 = \frac{(2\Delta t)^2}{\Delta x \Delta z} p_{i,j,k} q_{i,j,k+1/2}, \quad v_9 = \frac{(2\Delta t)^2}{\Delta x \Delta z} p_{i,j,k} q_{i,j,k-1/2},$$

which, in repeated fashion, create a tridiagonal matrix system for every z-cut. The other two systems of equations are formed in a similar way; for E_y (and every x-cut though the mesh), one must combine (3.115b) and (3.116c), whereas for E_z (and every y-cut though the grid), expressions (3.115c) and (3.116a) are taken into account. Finally, upon solving the above systems and assessing electric field components everywhere in the mesh, magnetic quantities H_x, H_y, and H_z at time step $n + 1$ are explicitly updated by (3.116).

It has to be stated that the particular ADI formulation, incorporated in the FDTD method, is somewhat different from the typical one. Thus, instead of performing the alternations toward the computation directions with regard to spatial coordinates, this modification is applied directly to each of the two spatial derivatives of (2.7) and (2.8) during the subiterations. The advantage of this process can be easily esteemed in 3-D problems, where only two alternations are required, and furthermore, calculations are conducted along all three coordinate directions. For example, in (3.111) and (3.112), one deduces that H_z, H_x, and H_y, are implicitly evaluated along x, y, and z axes, whereas in (3.115) and (3.116), they receive an explicit treatment.

To get a more concrete idea on the capabilities of the ADI-FDTD method, presume a simple rectangular waveguide, whose cutoff frequency is set at 7.5 GHz. The structure is discretized in terms of $40 \times 20 \times 160$ cells and maintains a spatial uniformity along y-axis. Figure 3.13 gives the dominant TE_{10} mode inside the waveguide for different time steps. As can be seen, the propagation is very smooth, and the profile of the mode is clearly visible. However, the interest issue, herein, lies on the total simulation time: the ADI-FDTD technique allowed the use of a time increment almost 1.7 times larger than the traditional Courant limit, thus reducing the total CPU time. Similar deductions are obtained from Figure 3.14, where the anticipated lack of propagation for frequencies smaller than the cutoff one is also depicted.

In general, (3.113) and (3.117) along with the similarly derived equations for the other electric and magnetic components can be summarized in matrix notation as

$$\text{First subiteration:} \quad \mathbf{Z}_1 \mathbf{W}^{n+1/2} = \mathbf{Y}_1 \mathbf{W}^n \tag{3.118a}$$

$$\text{Second subiteration:} \quad \mathbf{Z}_2 \mathbf{W}^{n+1} = \mathbf{Y}_2 \mathbf{W}^{n+1/2} \tag{3.118b}$$

Here, $\mathbf{W}^n$ is a column vector comprising all field quantities at time-step n. Likewise, $\mathbf{Z}_1, \mathbf{Z}_2, \mathbf{Y}_1$ and $\mathbf{Y}_2$ are coefficient matrices whose elements are related to the constitutive parameters and FDTD

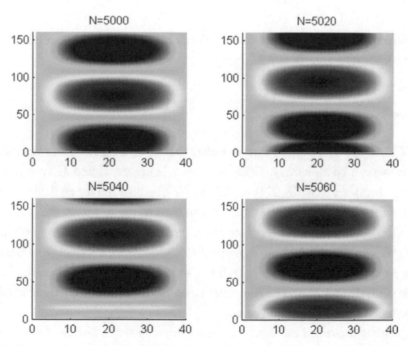

FIGURE 3.13: Propagation of the dominant TE_{10} mode inside a rectangular waveguide.

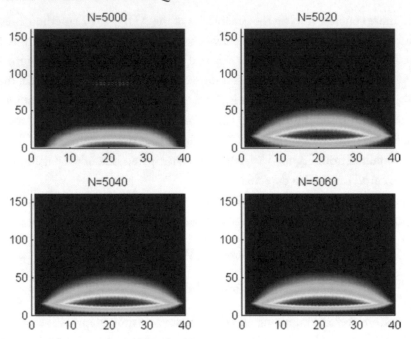

FIGURE 3.14: Snapshots of the field inside a rectangular waveguide for frequencies below the cutoff one.

simulation characteristics. The solution of (3.118) can be conducted either implicitly or explicitly. Isolating the unknown vectors $\mathbf{W}$ and merging the results, one has

$$\left.\begin{array}{l} \mathbf{W}^{n+1/2} = \mathbf{Z}_1^{-1}\mathbf{Y}_1\mathbf{W}^n \\ \mathbf{W}^{n+1} = \mathbf{Z}_2^{-1}\mathbf{Y}_2\mathbf{W}^{n+1/2} \end{array}\right\} \rightarrow \mathbf{W}^{n+1} = \mathbf{Z}_2^{-1}\mathbf{Y}_2\mathbf{Z}_1^{-1}\mathbf{Y}_1\mathbf{W}^n \rightarrow \mathbf{W}^{n+1} = \Xi\mathbf{W}^n , \quad (3.119)$$

with $\Xi = \mathbf{Z}_2^{-1}\mathbf{Y}_2\mathbf{Z}_1^{-1}\mathbf{Y}_1$. It is straightforward to realize that $\mathbf{Z}_1$ and $\mathbf{Z}_2$ are sparse matrices, and therefore, their inverses can be effectively obtained. Once this occurs, $\mathbf{E}$ and $\mathbf{H}$ vectors are recursively updated at each time step. Actually, $\mathbf{Z}_1^{-1}$ and $\mathbf{Z}_2^{-1}$ need not be resolved for all field values but only for the two left-most lattice points, enabling the computation of the rest from their adjacent cells. For example, in (3.113), matrix $\mathbf{Z}_1^{-1}$ is used only to assess the leftmost E_x at $j = 0$ and $j = 1$ during $n + 1/2$. The remaining E_x can then be computed by applying (3.113) with a sequence of ascending j along the transverse y-cut. Repetition of the process for E_y and E_z accelerates the algorithm and avoids unnecessary matrix inversions. Magnetic components, on the other hand, are directly calculated after their electric counterparts have been updated.

Investigation of numerical stability. The unconditionally stable profile of the ADI-FDTD algorithm is validated by means of the von Neumann method. For the sake of simplicity, presume that the medium does not contain any losses. Introducing instantaneous electric/magnetic field values distributed in space across the grid and taking their Fourier transform, the suitable propagating sinusoidal modes in the spatial spectral domain are extracted. Substitution of these waves into (3.111), (3.112), and (3.115), (3.116), leads to the recursive system of $\mathbf{W}^{n+1} = \mathbf{GW}^n$—just like (3.119)—where $\mathbf{G}$ is the 6×6 amplification matrix. The next step is to find the six eigenvalues of $\mathbf{G}$ and verify their location in the unit circle. On condition that their magnitudes are less than unity, the technique is considered stable. After the appropriate mathematical manipulations, the desired eigenvalues are found to be

$$\lambda_1 = \lambda_2 = 1, \ \lambda_3 = \lambda_5 = \frac{\sqrt{\Theta_1^2 - \Theta_2^2} + j\Theta_2}{\Theta_1}, \ \lambda_4 = \lambda_6 = \frac{\sqrt{\Theta_1^2 - \Theta_2^2} - j\Theta_2}{\Theta_1}, \quad (3.120)$$

with

$$\Theta_1 = \left(\mu\varepsilon + \vartheta_x^2\right)\left(\mu\varepsilon + \vartheta_y^2\right)\left(\mu\varepsilon + \vartheta_z^2\right),$$

$$\Theta_2 = 2\mu^2\varepsilon^2\vartheta_x\vartheta_y\vartheta_z\sqrt{\mu\varepsilon\vartheta_x^2 + \mu\varepsilon\vartheta_y^2 + \mu\varepsilon\vartheta_z^2 + \vartheta_z^2\vartheta_y^2 + \vartheta_y^2\vartheta_z^2 + \vartheta_z^2\vartheta_x^2},$$

and

$$\vartheta_u = \frac{\Delta t}{\Delta u}\sin\left(\frac{k_u\Delta u}{2}\right), \quad \text{for } u = x, y, z$$

Aside from λ_1 and λ_2, the magnitudes of the other four eigenvalues are also equal to one, Because $\Theta_1 \geq \Theta_2$ and the square roots in the numerators of $\lambda_3, \lambda_4, \lambda_5,$ and λ_6 become real numbers. Hence, it is concluded that the method is unconditionally stable, and the Courant limit may be safely surpassed. This is, in fact, the most significant merit of the ADI-FDTD method. Although it requires additional memory storage and is affected by the undesired lattice reflection errors, it allows larger time intervals, thus concluding to faster simulations.

3.7 THE NONSTANDARD FDTD TECHNIQUE

The nonstandard FDTD method constitutes a generalized perspective of traditional models for the approximation of differential [50–53]. Although the latter cannot always adequately treat involved EMC problems, the specific forms preserve the core properties of their continuous counterparts and avoid detrimental oscillations. The construction of such an algorithm stems from the idea of exact schemes that ensure the amelioration of consistency, stability, and convergence concerns. To underline this issue, it is stressed that a properly built method allows the unperturbed mapping of every physical quantity. Actually, this last assertion plays a serious role in the proficiency of a potential difference formula, as it enables the successful handling of strenuous cases without the need of obligatory simplifications. Hence, the extraction of the principal aspects regarding the derivation of nonstandard operators comprises a significantly enlightening guide.

Before the basic formulation, let us focus on the definition of the nonstandard operator. It is well-acknowledged that the conventional second-order finite-difference analogue is

$$\mathbf{d}_x^{\mathrm{conv}}\left[f|_x^t\right] = \frac{1}{\Delta x}\mathbf{d}_x\left[f|_x^t\right], \quad \text{with} \quad \mathbf{d}_x\left[f|_x^t\right] = f|_{x+\Delta x/2}^t - f|_{x-\Delta x/2}^t \tag{3.121}$$

Then, the nonstandard finite-difference operator is defined by

$$\mathbf{d}_x^{\mathrm{nst}}\left[f|_x^t\right] = \frac{1}{\Phi(k,\Delta x)}\mathbf{d}_x\left[f|_x^t\right], \tag{3.122}$$

where $\Phi(k, \Delta x)$ is a correction function picked to minimize the difference $|(\partial_x - \mathbf{d}_x^{\mathrm{nst}})f(x)|$ with regard to a basis functions set, typically containing sinusoidal and exponential terms with arguments toward each propagation direction in the grid. A normal choice for $\Phi(k, \Delta x)$ may be acquired in the subsequent manner: consider the evaluation of $\mathbf{d}_x[e^{jkx}]$, which due to (3.121) is found to be $\mathbf{d}_x[e^{jkx}]$ $= 2je^{jkx}\sin(k\Delta x/2)$. To satisfy the aforesaid minimization condition, $\Phi(k, \Delta x)$ is denoted as

$$\Phi(k,\Delta x) = 2\sin(k\Delta x/2)/k \tag{3.123}$$

So, for an $f \in \{\sin(kx), \cos(kx), e^{\pm jkx}\}$, function (3.123) yields $\partial_x f^{jkx} = \mathbf{d}_x^{\mathrm{nst}}[f^{jkx}]$ and leads to a zero-error finite-difference technique relative to plane waves that travel along the prefixed direction of wavevector $\mathbf{k}$. Unfortunately, this directional anisotropy inhibits the practicality of the resulting scheme, thus opting for a more drastic treatment of (3.122) provided in the next paragraphs of the present section.

3.7.1 Discrete Schemes for the Wave Equation

Suppose the general wave equation that governs the propagation of any electric or magnetic field quantity f. Its homogeneous version in Cartesian coordinates (x, y, z), becomes

$$\frac{\partial^2 f(\mathbf{r},t)}{\partial t^2} - v^2(\mathbf{r})\nabla^2 f(\mathbf{r},t) = 0, \quad \text{where} \quad \nabla^2 \equiv \frac{\partial}{\partial x^2} + \frac{\partial}{\partial y^2} + \frac{\partial}{\partial z^2}, \tag{3.124}$$

$\mathbf{r}$ is the position vector and $v(\mathbf{r},t)$ is the propagation velocity. For simplicity and notational convenience, the nonstandard finite-difference concepts are, at the outset, applied in a 2-D domain.

Derivation of 2-D operators. Herein and according to the background of [51], there exist two Laplacian difference operators for the discretization of (3.124). The first one is expressed as

$$\mathbf{P}_1^2\left[f|_{x,y}^t\right] = \mathbf{d}_x^2\left[f|_{x,y}^t\right] + \mathbf{d}_y^2\left[f|_{x,y}^t\right], \quad \text{with} \quad \begin{aligned} \mathbf{d}_x^2\left[f|_{x,y}^t\right] &= f|_{x+\Delta,y}^t - 2f|_{x,y}^t + f|_{x-\Delta,y}^t \\ \mathbf{d}_y^2\left[f|_{x,y}^t\right] &= f|_{x,y+\Delta}^t - 2f|_{x,y}^t + f|_{x,y-\Delta}^t \end{aligned} \quad (3.125)$$

and $\Delta = \Delta x = \Delta y$ as the spatial step that represents a uniform mesh. In contrast, the second operator is determined by the nodal pattern of

$$\mathbf{P}_2^2\left[f|_{x,y}^t\right] = \frac{1}{2}\left(f|_{x+\Delta,y+\Delta}^t + f|_{x-\Delta,y+\Delta}^t + f|_{x+\Delta,y-\Delta}^t + f|_{x-\Delta,y-\Delta}^t\right) - 2f|_{x,y}^t, \quad (3.126)$$

that offers an extra degree of freedom for the computation of $\nabla^2 f(x,y)$ through a weighted superposition of $\mathbf{P}_1^2[.]$ and $\mathbf{P}_2^2[.]$. This procedure concludes to

$$\nabla^2 f(x, y) \cong \frac{1}{[\Psi(k,\Delta)]^2}\mathbf{P}_0^2\left[f|_{x,y}^t\right] = \mathbf{P}_0^{2,\text{nst}}\left[f|_{x,y}^t\right],$$

with

$$\mathbf{P}_0^2\left[f|_{x,y}^t\right] = h\mathbf{P}_1^2\left[f|_{x,y}^t\right] + (1 - h)\mathbf{P}_2^2\left[f|_{x,y}^t\right], \quad (3.127)$$

for $0 \leq h \leq 1$, a coefficient depending on the location and the parameters of the partial differential equation, under study. In (3.127) and via (3.122), $\mathbf{P}_0^{2,\text{nst}}[.]$ is designated as the generalized nonstandard operator whose duty is to accomplish the most accurate spatial derivative approximations. Correction function $\Phi(k, \Delta)$ is selected to minimize the difference $|(\nabla^2 - \mathbf{P}_0^{2,\text{nst}})f(x,y)|$. Note that, because of its configuration, $\mathbf{P}_0^{2,\text{nst}}[.]$ can spread its performance in a more broadband rationale than (3.122) [51].

Having acquired the main characteristics of (3.127), the study will, now, proceed to the development of a representation for $\nabla^2 e^{j\mathbf{k}\cdot\mathbf{r}}$ with $\mathbf{r} = x\hat{\mathbf{x}} + y\hat{\mathbf{y}}$ and $\hat{\mathbf{x}},\hat{\mathbf{y}}$ as the respective unit vectors. Because in the 1-D case, the use of (3.125) gives $\mathbf{d}_x^2[e^{jkx}] = 2[\cos(k\Delta) - 1]e^{jkx}$, the aim is to get a value of h so that

$$\mathbf{P}_0^2\left[e^{j\mathbf{k}\cdot\mathbf{r}}\right] \cong 2\left[\cos(k\Delta) - 1\right]e^{j\mathbf{k}\cdot\mathbf{r}}, \quad (3.128)$$

for all $\mathbf{k}$ directions at a preset magnitude $k = |\mathbf{k}|$. Plugging (3.128) in (3.127), one derives

$$h\left(\mathbf{P}_1^2\left[e^{j\mathbf{k}\cdot\mathbf{r}}\right] - \mathbf{P}_2^2\left[e^{j\mathbf{k}\cdot\mathbf{r}}\right]\right) + \mathbf{P}_2^2\left[e^{j\mathbf{k}\cdot\mathbf{r}}\right] = 2\left[\cos(k\Delta) - 1\right]e^{j\mathbf{k}\cdot\mathbf{r}}, \qquad (3.129)$$

which, if satisfied by a specific h with a tolerable precision, permits the extraction of an equally effective $\nabla^2 e^{j\mathbf{k}\cdot\mathbf{r}}$ calculation. In this framework, let us define $\mathbf{P}_i^2[e^{j\mathbf{k}\cdot\mathbf{r}}] = 2e^{j\mathbf{k}\cdot\mathbf{r}}\,\Pi_i(k\Delta,\theta)$ for $i = 1,2$ and after some calculus by means of $k_x = k\cos\theta$ and $k_y = k\sin\theta$ obtain

$$\Pi_1(k\Delta,\theta) = \cos(k_x\Delta) + \cos(k_y\Delta) - 2, \qquad (3.130)$$

$$\Pi_2(k\Delta,\theta) = \cos(k_x\Delta)\cos(k_y\Delta) - 1, \qquad (3.131)$$

where θ is the angle created by $\mathbf{k}$ and x-axis. So, from (3.129), $h = h(k\Delta,\theta)$ is given by

$$h(k\Delta,\theta) = \frac{\cos(k\Delta) - \Pi_2(k\Delta,\theta) - 1}{\Pi_1(k\Delta,\theta) - \Pi_2(k\Delta,\theta)} \qquad (3.132)$$

Scrutinizing (3.132), one comes up with the deduction that, regardless of the main objective, h does depend on θ. This relation is quite convenient and points to $h = h(k\Delta, \theta)$, which exactly fulfills (3.131) at a distinct $\theta = \theta_{ex}$ value. An indicative θ_{ex} for minimizing the $\mathbf{P}_0^2[e^{j\mathbf{k}\cdot\mathbf{r}}]/e^{j\mathbf{k}\cdot\mathbf{r}}$ deviation from $2[\cos(k\Delta) - 1]$ can be derived by $\cos^4\theta_{ex} = 1/2$ [50]. Therefore with the aid of (3.130) and (3.131), (3.132) becomes

$$h(k\Delta) = h(k\Delta,\theta_{ex}) = \frac{\cos(k_x^{ex}\Delta)\cos(k_y^{ex}\Delta) - \cos(k\Delta)}{1 + \cos(k_x^{ex}\Delta)\cos(k_y^{ex}\Delta) - \cos(k_x^{ex}\Delta) - \cos(k_y^{ex}\Delta)}, \qquad (3.133)$$

where $k_x^{ex} = k\cos\theta_{ex}$ and $k_y^{ex} = k\sin\theta_{ex}$. Next, if one denotes as $\mathbf{d}_t^2[.]$ the temporal counterpart of $\mathbf{d}_x^2[.]$ or $\mathbf{d}_y^2[.]$, wave (3.124) receives the nonstandard FDTD form

$$\mathbf{d}_t^2\left[f\big|_{x,y}^t\right] - \upsilon^2(\mathbf{r})\frac{1}{[\Phi(\Delta)]^2}\mathbf{P}_0^2\left[f\big|_{x,y}^t\right] = 0 \Rightarrow \mathbf{d}_t^2\left[f\big|_{x,y}^t\right] - \upsilon^2(\mathbf{r})\mathbf{P}_0^{2,\mathrm{nst}}\left[f\big|_{x,y}^t\right] = 0, \;(3.134)$$

which, after the appropriate substitutions and the involvement of time-step Δt, yields

$$f\big|_{x,y}^{t+\Delta t} = 2f\big|_{x,y}^t - f\big|_{x,y}^{t-\Delta t} + \upsilon^2(\mathbf{r})\mathbf{P}_0^{2,\mathrm{nst}}\left[f\big|_{x,y}^t\right] \qquad (3.135)$$

The solution of (3.134) in a 3-D domain is equivalently prompt because the formulation maintains the previous notions, and only $\mathbf{P}_0^2[.]$ operator has to be defined under a different viewpoint.

Derivation of 3-D operators. As can be easily observed, the extension of nonstandard finite-differencing in 3-D EMC structures needs three Laplacian operators for spatial derivative approximation. Retaining the notation of the 2-D case and working on a uniform grid, $\mathbf{P}_1^2[.]$ is, now, described by

$$\mathbf{P}_1^2 \left[f|_{x,y,z}^t \right] = \mathbf{d}_x^2 \left[f|_{x,y,z}^t \right] + \mathbf{d}_y^2 \left[f|_{x,y,z}^t \right] + \mathbf{d}_z^2 \left[f|_{x,y,z}^t \right],$$

$$\mathbf{d}_x^2 \left[f|_{x,y,z}^t \right] = f|_{x+\Delta,y,z}^t - 2f|_{x,y,z}^t + f|_{x-\Delta,y,z}^t$$

$$\text{for} \quad \mathbf{d}_y^2 \left[f|_{x,y,z}^t \right] = f|_{x,y+\Delta,z}^t - 2f|_{x,y,z}^t + f|_{x,y-\Delta,z}^t$$

$$\mathbf{d}_z^2 \left[f|_{x,y,z}^t \right] = f|_{x,y,z+\Delta}^t - 2f|_{x,y,z}^t + f|_{x,y,z-\Delta}^t \qquad (3.136)$$

Furthermore, the second operator, $\mathbf{P}_2^2[.]$, extends its nodal spectrum at two diverse z-planes, as

$$\mathbf{P}_2^2 \left[f|_{x,y,z}^t \right] = \frac{1}{4} \left(f|_{x+\Delta,y+\Delta,z+\Delta}^t + f|_{x+\Delta,y-\Delta,z+\Delta}^t + f|_{x+\Delta,y+\Delta,z-\Delta}^t + f|_{x+\Delta,y-\Delta,z-\Delta}^t \right.$$

$$\left. + f|_{x-\Delta,y+\Delta,z+\Delta}^t + f|_{x-\Delta,y-\Delta,z+\Delta}^t + f|_{x-\Delta,y+\Delta,z-\Delta}^t + f|_{x-\Delta,y-\Delta,z-\Delta}^t \right) - 2f|_{x,y,z}^t$$

$$(3.137)$$

Conversely, for the supplementary operator, $\mathbf{P}_3^2[.]$, three 2-D Laplacians with the properties of (3.136) should be combined for each of the three coordinate planes. This concept provides

$$\mathbf{P}_3^2 \left[f|_{x,y,z}^t \right] = \frac{1}{2} \left(\mathbf{P}_2^2 \left[f|_{x,y,z}^t \right]_{x=0} + \mathbf{P}_2^2 \left[f|_{x,y,z}^t \right]_{y=0} + \mathbf{P}_2^2 \left[f|_{x,y,z}^t \right]_{z=0} \right) = \frac{1}{4} \left[f|_{x+\Delta,y-\Delta,z}^t \right.$$

$$\left. + f|_{x+\Delta,y,z-\Delta}^t + f|_{x,y+\Delta,z-\Delta}^t + f|_{x,y-\Delta,z+\Delta}^t + f|_{x-\Delta,y+\Delta,z}^t + f|_{x-\Delta,y,z+\Delta}^t \right] - 3f|_{x,y,z}^t$$

$$(3.138)$$

If $\mathbf{P}_i^2[.]$ $(i = 1, 2, 3)$ is applied to $e^{j\mathbf{k}\cdot\mathbf{r}}$ and $\mathbf{P}_i^2[e^{j\mathbf{k}\cdot\mathbf{r}}] = 2e^{j\mathbf{k}\cdot\mathbf{r}}\Pi_i(k\Delta,\theta,\varphi)$, functions Π_i are acquired via

$$\Pi_1(k\Delta, \theta, \varphi) = \cos(k_x\Delta) + \cos(k_y\Delta) + \cos(k_z\Delta) - 3, \qquad (3.139)$$

$$\Pi_2(k\Delta, \theta, \varphi) = \cos(k_x\Delta)\cos(k_y\Delta)\cos(k_z\Delta) - 1, \qquad (3.140)$$

$$2\Pi_3(k\Delta, \theta, \varphi) = \cos(k_x\Delta)\cos(k_y\Delta) + \cos(k_x\Delta)\cos(k_z\Delta) + \cos(k_y\Delta)\cos(k_z\Delta) - 3, \quad (3.141)$$

with $k_x = k \sin\theta \cos\varphi$, $k_y = k \sin\theta \sin\varphi$, and $k_z = k \cos\theta$ at spherical coordinates (r, θ, φ).

For the construction of the 3-D $\mathbf{P}_0^2[.]$ analogue of (3.127), the formulation requires a combination like

$$\mathbf{P}_0^2\left[f|_{x,y,z}^t\right] = \sum_{i=1}^{3} q_i \mathbf{P}_i^2\left[f|_{x,y,z}^t\right], \quad \text{where} \quad q_1 + q_2 + q_3 = 1, \tag{3.142}$$

such that

$$\Pi_0 = \sum_{i=1}^{3} q_i \Pi_i \cong \cos(k\Delta) - 1 \tag{3.143}$$

is purely isotropic for every θ and φ. To this goal, one at first defines $\mathbf{P}_{12}^2[.] = h_{12}\mathbf{P}_1^2[.] + (1 - h_{12})\mathbf{P}_2^2[.]$ and $\mathbf{P}_{13}^2[.] = h_{13}\mathbf{P}_1^2[.] + (1 - h_{13})\mathbf{P}_3^2[.]$ and tries to minimize the fluctuation of linear relations

$$\Pi_{12} = h_{12}\Pi_1 + (1 - h_{12})\Pi_2 \quad \text{and} \quad \Pi_{13} = h_{13}\Pi_1 + (1 - h_{13})\Pi_3, \tag{3.144}$$

with respect to θ angle at $\varphi = 0°$. Substitution of (3.139)–(3.141) into (3.144) gives

$$h_{12}(k\Delta,\theta,0) = \frac{\cos(k\Delta) - \Pi_2(k\Delta,\theta,0) - 1}{\Pi_1(k\Delta,\theta,0) - \Pi_2(k\Delta,\theta,0)}, \tag{3.145}$$

$$h_{13}(k\Delta,\theta,0) = \frac{\cos(k\Delta) - \Pi_3(k\Delta,\theta,0) - 1}{\Pi_1(k\Delta,\theta,0) - \Pi_3(k\Delta,\theta,0)} \tag{3.146}$$

Obviously, from (3.132), (3.133), (3.145), and (3.146) for the optimal $\theta = \theta_{ex}$ value and the relation of h to θ, it may be proven that $h_{12}(k\Delta, \theta) = h(k\Delta, \theta) = h(k\Delta)$ and $h_{13}(k\Delta, \theta) = 2h(k\Delta) - 1$. So,

$$\mathbf{P}_{12}^2\left[f|_{x,y,z}^t\right] = h\mathbf{P}_1^2\left[f|_{x,y,z}^t\right] + (1 - h)\mathbf{P}_2^2\left[f|_{x,y,z}^t\right], \tag{3.147}$$

$$\mathbf{P}_{13}^2\left[f|_{x,y,z}^t\right] = (2h - 1)\mathbf{P}_1^2\left[f|_{x,y,z}^t\right] + 2(1 - h)\mathbf{P}_3^2\left[f|_{x,y,z}^t\right], \tag{3.148}$$

can be regarded practically isotropic at $\varphi = 0°$. Subsequently, (3.147) is superposed to (3.148) as

$$\mathbf{P}_0^2\left[f|_{x,y,z}^t\right] = q\mathbf{P}_{12}^2\left[f|_{x,y,z}^t\right] + (1 - q)\mathbf{P}_{13}^2\left[f|_{x,y,z}^t\right], \tag{3.149}$$

whereas its φ dependence is suppressed through $\Pi_{12} = h\Pi_1 + (1 - h)\Pi_2$, $\Pi_{13} = (2h - 1)\Pi_1 + 2(1 - h)\Pi_3$, and $\theta = \pi/4$. Repeating the idea of (3.142)–(3.145) for (3.149), one derives

$$q(k\Delta) = q(k\Delta, \varphi_{ex}) = \frac{\Pi_{13}(k\Delta, \pi/4, \varphi_{ex}) - \cos(k\Delta) + 1}{\Pi_{13}(k\Delta, \pi/4, \varphi_{ex}) - \Pi_{12}(k\Delta, \pi/4, \varphi_{ex})} \qquad (3.150)$$

Notice that φ_{ex} imposes a weak dependence of s on φ, i.e., it minimizes φ variation. After the pertinent computations, $\varphi \cong 0.11811\pi$, and finally, $\mathbf{P}_0^{2,\mathrm{nst}}[.]$ becomes

$$\mathbf{P}_0^{2,\mathrm{nst}}\left[f|_{x,y,z}^t\right] = \frac{1}{\Phi(\Delta)}\mathbf{P}_0^2\left[f|_{x,y,z}^t\right] = \frac{1}{\Phi(\Delta)}\left(q_1\mathbf{P}_1^2\left[f|_{x,y,z}^t\right] + q_2\mathbf{P}_2^2\left[f|_{x,y,z}^t\right] + q_3\mathbf{P}_3^2\left[f|_{x,y,z}^t\right]\right),$$

$$(3.151)$$

with

$$q_1 = q(1-h) + (2h-1), \quad q_2 = q(1-h), \quad q_3 = 1 - q_1 - q_2, \qquad (3.152)$$

again constituting (3.135) as the nonstandard solution of the 3-D wave equation.

3.7.2 Discretizing the System of Maxwell's Equations

When Maxwell's curl laws are to be discretized, the profile of the resulting difference operators stipulates the decomposition of the potential $\mathbf{P}_0^2[.]$ to assure a reliable consistency. Therefore,

$$\mathbf{P}_0^2\left[f(\mathbf{r},t)\right] = \sum_{\xi=1}^{\dim} \mathbf{d}_\xi^A\left[f(\mathbf{r},t)\right]\mathbf{d}_\xi^B\left[f(\mathbf{r},t)\right], \qquad (3.153)$$

where $\mathbf{d}_\xi^B[.]$ represents central differencing, $\mathbf{d}_\xi^A[.]$ is a new operator that must be specified, and dim is the dimensionality of vector $\mathbf{r}$, associated with the problem (i.e., $2 \to (x, y)$ or $3 \to (x, y, z)$).

Next in terms of (3.122), $\mathbf{d}_t^{\mathrm{nst}}[f(\mathbf{r}, t)] = \mathbf{d}_t[f(\mathbf{r}, t)]/\Phi(\omega, \Delta t)$ is defined with $\Phi(\omega, \Delta t) = 2\omega^{-1} \sin(\omega\Delta t/2)$ as the correction function for time differentiation and $\mathbf{d}_t[f(\mathbf{r}, t)] = f(\mathbf{r}, t + \Delta t/2) - f(\mathbf{r}, t - \Delta t/2)$. In this sense, the nonstandard expression of Maxwell's equations is written as

$$\mu(\mathbf{r}_f)\frac{\partial\mathbf{H}(\mathbf{r}_f,t)}{\partial t} = -\nabla \times \mathbf{E}(\mathbf{r}_e,t) \Rightarrow \mu(\mathbf{r}_f)\mathbf{d}_t^{\mathrm{nst}}\left[\mathbf{H}(\mathbf{r}_f,t)\right] = -\bar{\mathbf{P}}_A^{\mathrm{nst}}\left[\mathbf{E}(\mathbf{r}_e,t)\right], \qquad (3.154)$$

$$\varepsilon(\mathbf{r}_e)\frac{\partial\mathbf{E}(\mathbf{r}_e,t + \Delta t/2)}{\partial t} = \nabla \times \mathbf{H}(\mathbf{r}_f,t + \Delta t/2)$$

$$\Rightarrow \varepsilon(\mathbf{r}_e)\mathbf{d}_t^{\mathrm{nst}}\left[\mathbf{E}(\mathbf{r}_e,t + \Delta t/2)\right] = \bar{\mathbf{P}}_B^{\mathrm{nst}}\left[\mathbf{H}(\mathbf{r}_f,t + \Delta t/2)\right], \qquad (3.155)$$

with $\mathbf{r}_f$ and $\mathbf{r}_e$ signifying that $\mathbf{H}$ and $\mathbf{E}$ components are positioned at the faces and the edges of the elementary cell, respectively [52]. Furthermore, vector operators $\bar{\mathbf{P}}_A^{\mathrm{nst}}[.]$, $\bar{\mathbf{P}}_B^{\mathrm{nst}}[.]$ are prescribed by

$$\bar{\mathbf{P}}_i^{\mathrm{nst}}\left[f(\mathbf{r},t)\right] = \frac{1}{\Phi(k,\Delta)}\bar{\mathbf{P}}_i\left[f(\mathbf{r},t)\right] = \frac{1}{\Phi(k,\Delta)}\sum_{\xi=1}^{\dim(\mathbf{r})}\mathbf{d}_\xi^i\left[f(\mathbf{r},t)\right]\hat{\xi}, \text{ for } i = \mathrm{A}, \mathrm{B}, \qquad (3.156)$$

$\Phi(k, \Delta) = 2k^{-1} \sin(k\Delta/2)$ and $\hat{\xi}$ as the suitable unit vector depending on dim($\mathbf{r}$). Through the expansion of $\mathbf{d}_t^{\mathrm{nst}}[.]$ and $\bar{\mathbf{P}}_i^{\mathrm{nst}}[.]$, the nonstandard FDTD version of (3.154) and (3.155) is

$$\mathbf{H}(\mathbf{r}_f, t + \Delta t/2) = \mathbf{H}(\mathbf{r}_f, t - \Delta t/2) - \frac{\Phi(\omega, \Delta t)}{\mu(\mathbf{r}_f)\Phi(k, \Delta)} \bar{\mathbf{P}}_A\left[\mathbf{E}(\mathbf{r}_e, t)\right], \qquad (3.157)$$

$$\mathbf{E}(\mathbf{r}_e, t + \Delta t) = \mathbf{E}(\mathbf{r}_e, t) + \frac{\Phi(\omega, \Delta t)}{\varepsilon(\mathbf{r}_e)\Phi(k, \Delta)} \bar{\mathbf{P}}_B\left[\mathbf{H}(\mathbf{r}_f, t + \Delta t/2)\right] \qquad (3.158)$$

Let us now think of the extraction of $\mathbf{d}_\xi^A[.]$. For the 2-D case, an auxiliary operator $\mathbf{d}_\xi^C[.]$ is launched as

$$\mathbf{d}_x^C\left[f|_{x,y}^t\right] = \frac{1}{2}\left(f|_{x+\Delta/2,y+\Delta}^t - f|_{x-\Delta/2,y+\Delta}^t + f|_{x+\Delta/2,y-\Delta}^t - f|_{x-\Delta/2,y-\Delta}^t\right), \qquad (3.159)$$

$$\mathbf{d}_y^C\left[f|_{x,y}^t\right] = \frac{1}{2}\left(f|_{x+\Delta,y+\Delta/2}^t - f|_{x-\Delta,y+\Delta/2}^t + f|_{x+\Delta,y-\Delta/2}^t - f|_{x-\Delta,y-\Delta/2}^t\right) \qquad (3.160)$$

Assuming the weighted relation of

$$\mathbf{d}_\xi^A\left[f|_{x,y}^t\right] = s\mathbf{d}_\xi^B\left[f|_{x,y}^t\right] + (1-s)\mathbf{d}_\xi^C\left[f|_{x,y}^t\right], \quad \text{for} \quad \xi \in (x,y), \qquad (3.161)$$

it can be certified that

$$\sum_{\xi=x,y} \mathbf{d}_\xi^B\left[f|_{x,y}^t\right] \mathbf{d}_\xi^C\left[f|_{x,y}^t\right] = 2\mathbf{P}_2^2\left[f|_{x,y}^t\right] - \mathbf{P}_1^2\left[f|_{x,y}^t\right], \qquad (3.162)$$

where $\mathbf{P}_1^2[.]$ and $\mathbf{P}_2^2[.]$ are described by (3.155) and (3.156). Coefficient s is determined so as (3.153) is completely fulfilled. Substituting (3.161) into (3.153) and employing (3.162), one gets $s = (1 + h)/2$, with h obtained from (3.133).

For the 3-D case, (3.133) reveals that the weighted relation of $\mathbf{d}_\xi^A[.]$, for $\xi \in (x, y, z)$, needs three independent difference operators, i.e., $\mathbf{d}_\xi^B[.]$—given by (3.121)—as well as the two auxiliary ones $\mathbf{d}_\xi^C[.]$ and $\mathbf{d}_\xi^D[.]$. Thus, opting for any partial derivative, e.g., ∂_z, the unknown operators $\mathbf{d}_z^C[.]$ and $\mathbf{d}_z^D[.]$ read

$$\mathbf{d}_z^C\left[f|_{x,y,z}^t\right] = \frac{1}{4}\left(f|_{x+\Delta,y+\Delta,z+\Delta/2}^t + f|_{x-\Delta,y+\Delta,z+\Delta/2}^t + f|_{x+\Delta,y-\Delta,z+\Delta/2}^t + f|_{x-\Delta,y-\Delta,z+\Delta/2}^t\right.$$
$$\left. - f|_{x+\Delta,y+\Delta,z-\Delta/2}^t - f|_{x-\Delta,y+\Delta,z-\Delta/2}^t - f|_{x+\Delta,y-\Delta,z-\Delta/2}^t - f|_{x-\Delta,y-\Delta,z-\Delta/2}^t\right),$$
$$(3.163)$$

$$\mathbf{d}_z^D\left[f|_{x,y,z}^t\right] = \frac{1}{4}\left(f|_{x,y+\Delta,z+\Delta/2}^t + f|_{x,y-\Delta,z+\Delta/2}^t + f|_{x+\Delta,y,z+\Delta/2}^t + f|_{x-\Delta,y,z+\Delta/2}^t\right.$$
$$\left. - f|_{x,y+\Delta,z-\Delta/2}^t - f|_{x,y-\Delta,z-\Delta/2}^t - f|_{x+\Delta,y,z-\Delta/2}^t - f|_{x-\Delta,y,z-\Delta/2}^t\right) \qquad (3.164)$$

Finally, the superposition for $\mathbf{d}_z^A[.]$ can be expressed as

$$\mathbf{d}_z^A\left[f|_{x,y,z}^t\right] = \sum_{i=B,C,D} s_i\mathbf{d}_z^i\left[f|_{x,y,z}^t\right] \quad \text{with} \quad \sum_{i=B,C,D} s_i = 1 \qquad (3.165)$$

with a similar procedure holding for ∂_x and ∂_y. To evaluate s_i, it is proven that

$$\sum_{\xi=x,y,z} \mathbf{d}_\xi^B\left[f|_{x,y,z}^t\right]\mathbf{d}_\xi^C\left[f|_{x,y,z}^t\right] = 3\mathbf{P}_2^2\left[f|_{x,y,z}^t\right] - \mathbf{P}_3^2\left[f|_{x,y,z}^t\right], \qquad (3.166)$$

$$\sum_{\xi=x,y,z} \mathbf{d}_\xi^B\left[f|_{x,y,z}^t\right]\mathbf{d}_\xi^D\left[f|_{x,y,z}^t\right] = 2\mathbf{P}_3^2\left[f|_{x,y,z}^t\right] - \mathbf{P}_1^2\left[f|_{x,y,z}^t\right], \qquad (3.167)$$

where $\mathbf{P}_1^2[.]$, $\mathbf{P}_2^2[.]$, and $\mathbf{P}_3^2[.]$ are derived via (3.136)–(3.138). Replacement of (3.165)–(3.167) into the 3-D analogue of (3.153) leads to

$$s_B = q_1 + \frac{1}{3}q_2 + \frac{1}{2}q_3, \quad s_C = \frac{1}{3}q_2, \quad s_D = \frac{1}{3}q_2 + \frac{1}{2}q_3, \qquad (3.168)$$

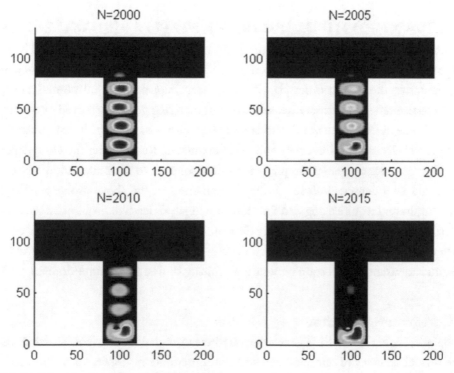

FIGURE 3.15: Magnitude of the E_y field component at various time instants.

with q_i ($i = 1,2,3$) retrieved from (3.152). In this way, $\mathbf{d}_\xi^A[.]$ and $\mathbf{d}_\xi^B[.]$ are fully specified, so permitting the systematic construction of (3.156) and the succeeding time update of 3-D Maxwell's laws. Overall, a notable asset of these operators is the involvement of extra nodal points for the approximation of partial derivatives. This means that, in contrast to the confined FDTD stencil, the nonstandard concepts offer an advanced management of lattice cells and, through extra degrees of freedom, allow the significant suppression of dispersion errors in EMC applications with curved details, arbitrary media discontinuities, and dissimilar interfaces.

To comprehend the action of the nonstandard FDTD method, let us reconsider the problem of the T junction analyzed in Section 3.3 via the FVTD approach. Nonetheless, herein, the problem becomes more complicated because the two arms of the junction are coupled only by a rectangular thin slot (iris). Such a device is rather difficult to handle by the simple time-domain algorithms because of the acute field fluctuations in the vicinity of the slot, especially from a resolution point of view. Actually, the slot and the PEC wall is expected to act as a high-pass filter. Indeed, the previous anticipations are validated in Figure 3.15, which presents the variation of the E_y magnitude inside the junction for a 55% coarser resolution than the FVTD one. Note that the outcomes are quite different from those of Figure 3.5 with the unobstructed propagation.

3.8 ADVANCED HIGHER-ORDER FDTD APPROXIMATORS

The accurate simulation of complex EMC problems by means of the FDTD method is strongly related to the efficiency of spatial/temporal derivative approximations in the discrete domain. Despite its great popularity, the second-order algorithm, as what happens with all numerical approaches, suffers from lattice reflection errors intensely when discretizing generally curved or electrically large applications. These defects influence the update of propagating wave fronts by imposing a nonphysical link of phase velocity with frequency and grid parameters. Recognizing the close dependence of dispersion discrepancies on spatial sampling, a possible solution to their alleviation, up to a satisfactory level, could be a denser mesh resolution. Nonetheless, in the above problems that expand in many wavelengths and require elongated time intervals, analysis leads to prohibitive burdens even for very powerful computing machines. To this difficulty, higher-order FDTD schemes are judged as an important healer for artificial error distortions [54–65]. Indeed, since their early advent, they have earned significant attention and soon became a promising tool in EMC time-domain modeling.

3.8.1 Cartesian Realizations

For the analysis, higher-order FDTD schemes are hereafter classified as (P,Q), with P and Q indicating the formal accuracy of temporal and spatial differentiation, respectively. Thus, let us express Faraday's and Ampère's laws in Cartesian coordinates as

$$\frac{\partial}{\partial t}\begin{bmatrix} \mathbf{E} \\ \mathbf{H} \end{bmatrix} = \begin{bmatrix} 0 & \varepsilon^{-1} \\ -\mu^{-1} & 0 \end{bmatrix}\begin{bmatrix} \nabla\times & 0 \\ 0 & \nabla\times \end{bmatrix}\begin{bmatrix} \mathbf{E} \\ \mathbf{H} \end{bmatrix} - \begin{bmatrix} \sigma\varepsilon^{-1} & 0 \\ 0 & \rho'\mu^{-1} \end{bmatrix}\begin{bmatrix} \mathbf{E} \\ \mathbf{H} \end{bmatrix} \Rightarrow \frac{\partial \mathbf{F}}{\partial t} = \mathbf{CDF} - \mathbf{LF},$$

$$(3.169)$$

where it has been assumed that $\mathbf{J}_c, \mathbf{M}_c$ are obtained through (2.6), $\mathbf{J}_s = \mathbf{M}_s = 0$, and $\mathbf{D}$ is the matrix curl operator acting on $\mathbf{F} = [\mathbf{E}\,\mathbf{H}]^T$. Expanding $\mathbf{F}$ in Taylor series with respect to time, one gets

$$\frac{\partial \mathbf{F}^n}{\partial t} = \frac{\mathbf{F}^{n+1/2} - \mathbf{F}^{n-1/2}}{\Delta t} - \frac{(\Delta t)^2}{24}\frac{\partial^3 \mathbf{F}^n}{\partial t^3} + O\left[(\Delta t)^4\right], \qquad (3.170)$$

which manages the precision of temporal differentiation. In fact, if the third-order differentiation terms in (3.170) are ignored, the desired $(2,Q)$ schemes with Qth-order spatial accuracy are extracted [54]. On the other hand, if such terms are considered, the resulting formulae are fourth-order accurate in time, namely, the $(4,Q)$ family is acquired. In the latter case, substitution of (3.170) into (3.169) leads to

$$\varepsilon\mathbf{E}^{n+1} = \varepsilon\mathbf{E}^n + \Delta t\nabla\times\mathbf{H}^{n+1/2} - \Delta t\sigma\mathbf{E}^{n+1/2} + \frac{\varepsilon(\Delta t)^2}{24}\frac{\partial^3\mathbf{E}^{n+1/2}}{\partial t^3}, \qquad (3.171)$$

$$\mu\mathbf{H}^{n+1/2} = \mu\mathbf{H}^{n-1/2} - \Delta t\nabla\times\mathbf{E}^n - \Delta t\rho'\mathbf{H}^n + \frac{\mu(\Delta t)^2}{24}\frac{\partial^3\mathbf{H}^n}{\partial t^3} \qquad (3.172)$$

As detected, the direct evaluation of the $\partial^3/\partial t^3$ derivatives is laborious because they involve additional time levels. To evade this shortcoming, they are transformed into spatial analogues via repeated differentiations as

$$\frac{\partial^3}{\partial t^3}\begin{bmatrix} \varepsilon\mathbf{E} \\ \mu\mathbf{H} \end{bmatrix} = \frac{1}{\mu\varepsilon}\nabla\times\nabla^2\begin{bmatrix} \mathbf{H} \\ -\mathbf{E} \end{bmatrix} - \nabla^2\begin{bmatrix} \dfrac{1}{\mu}\left(\dfrac{\rho'}{\mu} + 2\dfrac{\sigma}{\varepsilon}\right)\mathbf{E} \\ \dfrac{1}{\varepsilon}\left(\dfrac{\sigma}{\varepsilon} + 2\dfrac{\rho'}{\mu}\right)\mathbf{H} \end{bmatrix}$$

$$- \left(\frac{\sigma}{\varepsilon} + \frac{\rho'}{\mu}\right)\nabla\times\begin{bmatrix} -\dfrac{\rho'}{\mu}\mathbf{H} \\ \dfrac{\sigma}{\varepsilon}\mathbf{E} \end{bmatrix} + \frac{\partial}{\partial t}\begin{bmatrix} \dfrac{\sigma^2}{\varepsilon}\mathbf{E} \\ \dfrac{(\rho')^2}{\mu}\mathbf{H} \end{bmatrix} \qquad (3.173)$$

It should be stressed that (3.173) opts for a locally homogeneous medium, whereas the small-valued $(\Delta t)^2/24$ term in (3.171) and (3.172) allows further simplifications. Because all derivatives on the right-hand side of (3.173) are multiplied by it, simple second-order finite differences can be used for their approximation.

To accomplish the fourth-order spatial discretization of (3.172) and (3.173), a set of central finite-difference schemes, based on a staggered-mesh arrangement, is preferred. Therefore,

$$\left.\frac{\partial f}{\partial \xi}\right|_{pt}^{t} \simeq \left.\frac{\partial f}{\partial \xi}\right|_{s}^{n} = \frac{9}{8}\left(\frac{f|_{s+1/2}^{n} - f|_{s-1/2}^{n}}{\Delta\xi}\right) - \frac{1}{24}\left(\frac{f|_{s+3/2}^{n} - f|_{s-3/2}^{n}}{\Delta\xi}\right), \qquad (3.174)$$

which computes the ξ-derivative of f at point pt, with a stencil of $s\Delta\xi$ along the direction of differentiation, for $\xi \in (x, y, z)$ and $s \in (i, j, k)$. This procedure needs the field values at symmetrical nodes around pt on a straight line parallel to ξ-axis that crosses the point of interest. For larger orders of spatial approximation in the (P,Q) schemes, (3.74) yields the following operator, which spans in $Q - 1$ cells,

$$\left.\frac{\partial f}{\partial \xi}\right|_{pt}^{t} \simeq \left.\frac{\partial f}{\partial \xi}\right|_{s}^{n} = \frac{1}{\Delta\xi}\sum_{m=1}^{Q/2} K_m\left[f|_{s+(2m-1)/2}^{n} - f|_{s-(2m-1)/2}^{n}\right] \qquad (3.175)$$

Coefficients K_m are acquired via the Taylor expansion on the right-hand side of (3.175) and the prerequisite for minimum truncation error. Such a procedure gives the $Q/2 \times Q/2$ system

$$\begin{bmatrix} 1 & 3 & \cdots & Q-1 \\ 1 & 3^3 & \cdots & (Q-1)^3 \\ \vdots & \vdots & \ddots & \vdots \\ 1 & 3^{Q-1} & \cdots & (Q-1)^{Q-1} \end{bmatrix}\begin{bmatrix} K_1 \\ K_2 \\ \vdots \\ K_{Q/2} \end{bmatrix} = \begin{bmatrix} 1 \\ 0 \\ \vdots \\ 0 \end{bmatrix}, \qquad (3.176)$$

whose closed-form solution is expressed as

$$K_m = \frac{(-1)^{m+1}(Q-1)!!^2}{2^{Q-2}\left(\frac{1}{2}Q + m - 1\right)!\left(\frac{1}{2}Q - m\right)!(2m-1)^2}, \qquad (3.177)$$

with $M!! = M(M-2)(M-4)\ldots$ The order of truncation error

$$\frac{(\Delta\xi)^Q}{2^Q(Q+1)!}\sum_{m=1}^{Q/2} K_m(2m-1)^{Q+1}f^{(Q+1)}\Big|_{s}^{n}, \qquad (3.178)$$

identifies the order of spatial approximation. For comparison, Table 3.3 encapsulates the most common K_m values along with the corresponding truncation errors up to eighth order. An instructive

ORDER Q	K_1	K_2	K_3	K_4	K_5	TRUNCATION ERROR
2	1	0	0	0	0	$-\dfrac{(\Delta\xi)^2}{24}\, f^{(3)}\Big|_s^n$
4	$\dfrac{9}{8}$	$\dfrac{-1}{24}$	0	0	0	$\dfrac{3(\Delta\xi)^4}{640}\, f^{(5)}\Big|_s^n$
6	$\dfrac{75}{64}$	$-\dfrac{25}{384}$	$\dfrac{3}{640}$	0	0	$\dfrac{5(\Delta\xi)^6}{7168}\, f^{(7)}\Big|_s^n$
8	$\dfrac{1225}{1024}$	$-\dfrac{245}{3072}$	$\dfrac{49}{5120}$	$-\dfrac{5}{7168}$	0	$\dfrac{35(\Delta\xi)^8}{294912} f^{(9)}\Big|_s^n$

TABLE 3.3: Spatial central difference forms of diverse orders

indicator regarding the performance of (3.175) is the dispersion relation that displays the spectral wavenumber variation. So, for the $(2,Q)$ forms, by assuming plane-wave propagation in a 3-D Cartesian domain, one derives

$$\left[\frac{\sin(\omega\Delta t/2)}{v\Delta t}\right]^2 = \sum_{\xi=x,y,z} \frac{1}{(\Delta\xi)^2}\left[\sum_{m=1}^{Q/2} K_m\sin\left(\frac{2m-1}{2}k_\xi^{\text{num}}\Delta\xi\right)\right]^2, \qquad (3.179)$$

where $v = 1/\sqrt{\mu\varepsilon}$, and $\mathbf{k}^{\text{num}} = k_x^{\text{num}}\hat{\mathbf{x}} + k_y^{\text{num}}\hat{\mathbf{y}} + k_z^{\text{num}}\hat{\mathbf{z}}$ is the numerical wavenumber.

Additionally via the von Neumann method, the stability condition for any member of the $(2,Q)$ family can be readily extracted. For illustration, the $(2,4)$ case has

$$v\Delta t \leq \frac{6}{7}\left[\sqrt{\frac{1}{(\Delta x)^2} + \frac{1}{(\Delta y)^2} + \frac{1}{(\Delta z)^2}}\right]^{-1}, \qquad (3.180)$$

indicating that the solution achieved by this type of discretization will converge to the exact one as increments $\Delta\xi$ get smaller. Nonetheless, because the orders of spatial and temporal approximation are different, $(2,Q)$ schemes are not expected to present their optimal accuracy, when Δt is equal or very close to the maximum one. Moreover, even in the case of finer grids, their convergence rate does not surpass that of the usual FDTD technique, unless time steps become sufficiently small, so that the error of time advancing is not dominant. The aforementioned formulation reveals several motives for the successful application of the higher-order FDTD algorithm, but its most attractive feature is its increased precision. Moreover, the use of staggered meshes enhances the consistent treatment of diverse EMC geometries and simultaneously maintains the divergence of initial conditions in homogeneous areas. Actually, these benefits have activated a constant research on the development of various efficient renditions with improved nodal arrangements, lower dispersion or dissipation errors, and rigorous extensions to other coordinate systems. A pinpointing simple example for the (2,4) FDTD scheme is shown in Figure 3.16, which illustrates the propagation of H_z component in a 3-D space, terminated by the appropriate absorbing boundary conditions (see Chapter 4) in the presence of a rectangular and a cylindrical PEC shielding structure. Observe that the method attains a fairly even modeling of the propagating fields and—the most critical—with a significantly smaller resolution.

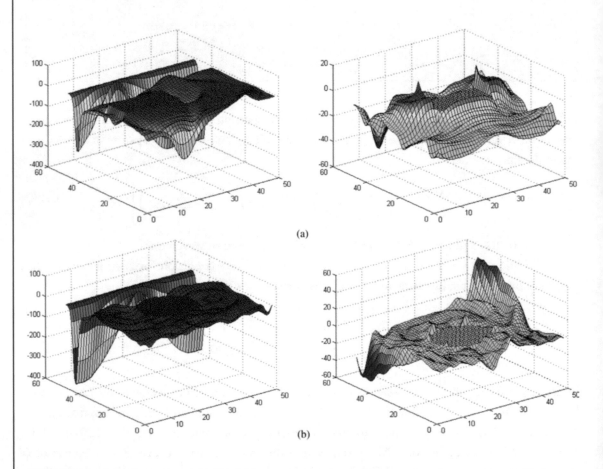

FIGURE 3.16: Magnitude of the H_z field component at two different time instants for a 3-D domain containing (a) a rectangular and (b) a cylindrical shielding structure.

3.8.2 Discretization of Arbitrarily Curved Models

An issue of major concern when implementing the FDTD method in curvilinear EMC applications is the inability of the staircase approximation process to offer the best possible level of accuracy. Having already discussed the particular topic in Section 2.2.3 through the integral-based extraction of the update equations, it is our purpose herein to devise the curvilinear counterparts of (3.175). As a matter of fact, a poor discrete model of a curved surface is likely to cancel the benefits stemming from their increased temporal and spatial accuracy. Hence, to present the key steps of the construction framework, a fourth-order FDTD method in an orthogonal curvilinear lattice, which conforms to metallic boundaries or dielectric surfaces, is described. The advantage from the use of such meshes is the preservation of simplicity in the extraction of Maxwell's equations and the manipulation of continuity conditions at the interfaces.

Assume the 2-D coordinate transformation $x = x(u, v)$, $y = y(u, v)$ that maps the non-Cartesian domain in the physical space, $\Omega = \{\Omega; u, v\}$, into a rectangular area in the transformed space, $\Gamma = \{\Gamma; u, v\}$. Along these lines, the *TM* mode of Maxwell's laws (2.1) and (2.2) in a homogeneous, isotropic, and lossy material with $\mathbf{J}_s = \mathbf{M}_s = \mathbf{0}$ is transformed to

$$\mu \frac{\partial H_u}{\partial t} = -\frac{1}{\bar{V}} \frac{\partial E_z}{\partial v} - \rho' H_u, \tag{3.181}$$

$$\mu \frac{\partial H_v}{\partial t} = \frac{1}{\bar{V}} \frac{\partial E_z}{\partial u} - \rho' H_v, \tag{3.182}$$

$$\varepsilon \frac{\partial E_z}{\partial t} = \frac{1}{\bar{V}} \left[\alpha_v \frac{\partial H_v}{\partial u} - \alpha_u \frac{\partial H_u}{\partial v} - \beta \left(\frac{\partial H_v}{\partial v} - \frac{\partial H_u}{\partial u} \right) \right] + \frac{1}{\bar{V}} (\Lambda H_u + \Psi H_v) - \sigma E_z, \tag{3.183}$$

for $\bar{V} = x_u y_v - x_v y_u$ the Jacobian, $\alpha_u = x_u^2 + y_u^2$, $\alpha_v = x_v^2 + y_v^2$ system metrics, $\beta = x_u x_v - y_u y_v$, and

$$\Lambda = y_{uu} y_v - y_{uv} y_u + x_{uu} x_v - x_{uv} x_u, \quad \Psi = y_{vu} y_v - y_{vv} y_u + x_{vu} x_v - x_{vv} x_u, \tag{3.184}$$

where the subscripts in x, y variables point out differentiation with respect to u and v.

If the previous transformation is also orthogonal, that is $\beta = 0$ and $\bar{V} \neq 0$, (3.183) becomes

$$\varepsilon \frac{\partial E_z}{\partial t} = \frac{1}{\bar{V}} \left(\alpha_v \frac{\partial H_v}{\partial u} - \alpha_u \frac{\partial H_u}{\partial v} \right) + \frac{1}{\bar{V}} (\Lambda H_u + \Psi H_v) - \sigma E_z \tag{3.185}$$

In this way, the explicit method takes avail of grid orthogonality properties, like the zero off-diagonal elements in the metric tensors, which reduce the computational burden [56, 65].

To develop the fourth-order FDTD scheme, let us apply the Taylor expansion series to H_u (similarly for the other components) and employ the central finite-differencing scheme of (3.170) as

$$\left.\frac{\partial H_u}{\partial t}\right|^n = \frac{H_u|^{n+1/2} - H_u|^{n-1/2}}{\Delta t} - \frac{(\Delta t)^2}{24} \left.\frac{\partial^3 H_u}{\partial t^3}\right|^n + O\left[(\Delta t)^4\right] \qquad (3.186)$$

The next stage involves the evaluation of $\partial^3/\partial t^3$ term in (3.186). Differentiating (3.181) for three consecutive times, one acquires

$$\left.\frac{\partial^3 H_u}{\partial t^3}\right|^n = -\left.\frac{\partial^2}{\partial t^2}\left(\frac{1}{\mu\bar{V}}\frac{\partial E_z}{\partial v} - \frac{\rho'}{\mu}H_u\right)\right|^n, \qquad (3.187)$$

where for the resulting $\partial^2 H_u/\partial t^2$, the corresponding second-order derivative of (3.182) leads to

$$\left.\frac{\partial^2 H_u}{\partial t^2}\right|^n = -\frac{1}{\mu\bar{V}}\left.\left(\frac{\partial}{\partial t} - \frac{\rho'}{\mu}\right)\frac{\partial E_z}{\partial v}\right|^n + \left(\frac{\rho'}{\mu}\right)^2 H_u|^n \qquad (3.188)$$

To calculate $H_u|^n$ in (3.188) through its already stored (known) values at time steps $n + 1/2$ and $n - 1/2$, the subsequent fourth-order time-averaging process is launched

$$H_u|^n = \frac{H_u|^{n+1/2} + H_u|^{n-1/2}}{2} - \frac{(\Delta t)^2}{8}\left.\frac{\partial^2 H_u}{\partial t^2}\right|^n + O\left[(\Delta t)^4\right] \qquad (3.189)$$

As a consequence, from (3.187) to (3.189) and a uniform grid $(u,v) = (i\Delta u, j\Delta v)$ with $\Delta = \Delta u = \Delta v$, the update FDTD formula of (3.181) reads

$$\left(1 + \rho'_{\mathrm{B}}\right) H_u|^{n+1/2}_{i,j+1/2} = \left(1 - \rho'_{\mathrm{B}}\right) H_u|^{n-1/2}_{i,j+1/2}$$

$$-\frac{\Delta t}{\mu\bar{V}|_{i,j+1/2}}\left\{1 + \frac{(\Delta t)^2}{24}\frac{\partial^2}{\partial t^2} + \frac{\rho'_{\mathrm{A}}}{12}\left[1 + \frac{(\rho'_{\mathrm{A}})^2}{8}\right]^{-1}\left(\Delta t\frac{\partial}{\partial t} - \rho'_{\mathrm{A}}\right)\right\}$$

$$\times D_v\left[E_z|^n_{i,j+1/2}\right], \qquad (3.190)$$

where

$$\rho'_{\mathrm{A}} = \frac{\rho'\Delta t}{\mu}, \quad \rho'_{\mathrm{B}} = \rho'_{\mathrm{A}}\frac{1 + (\rho'_{\mathrm{A}})^2/24}{2\left[1 + (\rho'_{\mathrm{A}})^2/8\right]}, \qquad (3.191)$$

and

$$D_v\left[f|^n_{i,j}\right] = \frac{9}{8}\left(\frac{f|^n_{i,j+1/2} - f|^t_{i,j-1/2}}{\Delta}\right) - \frac{1}{24}\left(\frac{f|^n_{i,j+3/2} - f|^n_{i,j-3/2}}{\Delta}\right), \qquad (3.192)$$

is the curvilinear analogue of the fourth-order spatial operator (3.174). Observe that, for briefness, the stencils of constitutive parameters, indicating the material positions in the interior of a cell, have been omitted.

Equivalently, (3.182) and (3.183) become

$$\left(1 + \rho_B'\right) H_v\big|_{i+1/2,j}^{n+1/2} = \left(1 - \rho_B'\right) H_v\big|_{i+1/2,j}^{n-1/2}$$

$$- \frac{\Delta t}{\mu \bar{V}\big|_{i+1/2,j}} \left\{ 1 + \frac{(\Delta t)^2}{24}\frac{\partial^2}{\partial t^2} + \frac{\rho_A'}{12}\left[1 + \frac{(\rho_A')^2}{8}\right]^{-1}\left(\Delta t\frac{\partial}{\partial t} - \rho_A'\right)\right\}$$

$$\times \; D_u\left[E_z\big|_{i+1/2,j}^n\right], \tag{3.193}$$

$$\left(1 + \sigma_B\right) E_z\big|_{i,j}^{n+1} = \left(1 - \sigma_B\right) E^z\big|_{i,j}^n - \frac{\Delta t}{\varepsilon \bar{V}\big|_{i,j}} \left\{ 1 + \frac{(\Delta t)^2}{24}\frac{\partial^2}{\partial t^2} + \frac{\sigma_A}{12}\left[1 + \frac{(\sigma_A)^2}{8}\right]^{-1}\left(\Delta t\frac{\partial}{\partial t} - \sigma_A\right)\right\}$$

$$\left(\alpha_v\big|_{i,j} D_u\left[H_v\big|_{i,j}^{n+1/2}\right] - \alpha_u\big|_{i,j} D_v\left[H_u\big|_{i,j}^{n+1/2}\right] + \Lambda\big|_{i,j} H_u\big|_{i,j}^{n+1/2} + \Psi\big|_{i,j} H_v\big|_{i,j}^{n+1/2}\right) \tag{3.194}$$

with σ_A and σ_B found by (3.191) under the substitution of μ with ε and ρ' with σ, whereas $D_u[.]$ is given by (3.192) for an interchange of the $\pm 1/2$ shift from j to i index. Furthermore, when H_u or H_v are located at integer mesh points, the necessary interpolations are performed by the fourth-order relations

$$H_u\big|_{i,j}^{n+1/2} = \frac{9}{16}\left(H_u\big|_{i,j-1/2}^{n+1/2} + H_u\big|_{i,j+1/2}^{n+1/2}\right) - \frac{1}{16}\left(H_u\big|_{i,j+3/2}^{n+1/2} + H_u\big|_{i,j-3/2}^{n+1/2}\right), \tag{3.195}$$

$$H_v\big|_{i,j}^{n+1/2} = \frac{9}{16}\left(H_v\big|_{i-1/2,j}^{n+1/2} + H_v\big|_{i+1/2,j}^{n+1/2}\right) - \frac{1}{16}\left(H_v\big|_{i+3/2,j}^{n+1/2} + H_v\big|_{i-3/2,j}^{n+1/2}\right) \tag{3.196}$$

Finally, the discrete values of coefficients Λ, Ψ, and metrics α_u, α_v need the knowledge of second-order derivatives with respect to u and v. These quantities, typically reduced to first-order ones, are obtained by the five-point schemes

$$x_u\big|_{i,j} = \frac{1}{12\Delta}\left(x\big|_{i-2,j} - x\big|_{i+2,j}\right) + \frac{2}{3\Delta}\left(x\big|_{i+1,j} - x\big|_{i-1,j}\right), \tag{3.197}$$

$$y_v\big|_{i,j} = \frac{1}{12\Delta}\left(y\big|_{i,j-2} - y\big|_{i,j+2}\right) + \frac{2}{3\Delta}\left(y\big|_{i,j+1} - y\big|_{i,j-1}\right) \tag{3.198}$$

The previous approach, despite its constraint for some extra memory, hardly affects the overall computational overhead. In contrast, its capability to successfully handle curved domains or structural details and relatively abrupt material interfaces is proven to be very helpful and rewarding.

3.9   WENO SCHEMES IN THE TIME DOMAIN

Designed to devise a cell-averaged reconstruction that determines field fluxes at cell boundaries with high precision, the WENO forms can significantly annihilate spurious oscillations near EMC discontinuities [66–69]. Thus, rather than calculating electric and magnetic components based solely on one preset stencil, these schemes encompass a convex combination of all candidate stencils in the elementary cell. To each of these stencils, a certain weight is appointed, which specifies their contribution to the overall approximation. The weights can be defined so that near structural oddities a satisfactory precision is achieved and wave propagation is properly resolved. In an attempt to avoid the original implicit WENO profile, a generalized 3-D time-domain method is presented in this section [69]. Being completely explicit, the algorithm estimates the abruptness of intricate regions and establishes an adaptive finite-difference procedure that constructs the pertinent polynomials for spatial derivatives. Additionally, via the systematic use of higher-order nodal arrangements in curvilinear coordinates, the nonsymmetric meshing of curl operators is discarded, and lattice dissipation is reduced. Hence, discretization preserves numerical energy, suppresses dispersion discrepancies, and imposes broadband continuity conditions for fields across interfaces.

Suppose the succeeding vector rendition of Maxwell's time-dependent equations

$$\frac{\partial \mathbf{W}_1(\mathbf{r},t)}{\partial t} + \nabla \times \mathbf{W}_2(\mathbf{r},t) = \mathbf{g}(\mathbf{r},t), \qquad (3.199)$$

with $\mathbf{W}_1 = [-\mathbf{D}\ \mathbf{B}]^{\mathrm{T}}$ and $\mathbf{W}_2 = [\mathbf{H}\ \mathbf{E}]^{\mathrm{T}}$ as the electric/magnetic fluxes and intensities in (u,v,w) coordinates, $\mathbf{r}$ as a position indicator, and $\mathbf{g}$ as the source term. For the approximation of spatial derivatives in (3.199), the cell average of any $f|_{i,j,k}$ is denoted (see also in (3.67) for a similar definition) as

$$\tilde{f}(t)\big|_{i,j,k} = \frac{1}{V_s} \iiint_{V_s} \psi(\varepsilon,\mu)\, f(\mathbf{r},t)\, \mathrm{d}V , \qquad (3.200)$$

for a fixed $(i\Delta u, j\Delta v, k\Delta w)$ resolution and $V_s = \Delta u \Delta v \Delta w$ as the volume of the $s_{i,j,k}$ cell. The integral in (3.200) can be effectively computed by a Gaussian quadrature along the edges and over the faces of $s_{i,j,k}$ on condition that functions $\psi(\varepsilon,\mu)$ are known for every medium that fills the domain. For example, if the cell is curvilinear and spans in

$$s_{i,j,k} = \left[u|_{i-1/2}, u|_{i+1/2}\right] \times \left[v|_{j-1/2}, v|_{j+1/2}\right] \times \left[w|_{k-1/2}, w|_{k+1/2}\right], \qquad (3.201)$$

the cell average at the $(i, j + 1/2, k)$ face is given by

$$\bar{f}(t)|_{i,j+1/2,k} = \frac{1}{\Delta u \Delta w} \sum_{l=1}^{m} \sum_{m=1}^{m} f\left(\mathbf{r}(u|_l, v|_{j+1/2}, w|_m)\right) \Gamma_m \Gamma_l, \qquad (3.202)$$

where l,m represents different integration points and coefficients Γ_l, Γ_m. Note that (3.202) stipulates point-wise f values, whereas the technique evolves its volumetric design. Consequently, spatial derivatives in (3.199) are rigorously obtained through

$$\frac{\partial f(\mathbf{r},t)}{\partial \zeta} = \mathcal{K}_\zeta(\mathbf{r},t) + O\left[(\Delta \zeta)^{q+1}\right], \qquad \text{for} \quad \zeta \in (u,v,w), \qquad (3.203)$$

together with piecewise polynomials $\mathcal{K}_\zeta(\mathbf{r},t)$. The main intention is the development of a weighted convex combination of all functional stencils L_s applicable to the $s_{i,j,k}$ cell and the reconstruction of $\mathcal{K}_\zeta(\mathbf{r},t)$. As a matter of fact, the proposed $(2q - 1)$ th-order schemes assign an optimal nonlinear weight to each stencil. So, the essentially nonoscillatory property is absolutely exploited contrary to traditional ideas with two-point approximations. Presuming v as the direction under investigation, the resulting stencil family is

$$\{L_s\}_v^{q-1} = \left\{f|_{i,j-d+q/2,k}^n, f|_{i,j-d+3q/2,k}^n, \cdots, f|_{i,j+d-q/2,k}^n\right\}, \qquad (3.204)$$

for $d = 0,1, \ldots, q - 1$. Should all nodal locations are adequately found, a number of q interpolating polynomials $C_{q,v}(\mathbf{r},t)$ is annotated to the stencils of (3.204) with the recursive representation of

$$C_{q-1,v}^{2q-1}\left(\bar{f}|_{i,j-d+1/2,k}^n, \cdots, \bar{f}|_{i,j+d-1/2,k}^n\right) = \tau_q \sum_{m=0}^{q-1} \chi_{m,v}^q C_{m,v}^q\left(\bar{f}|_{i,j-d+q/2,k}^n, \cdots, \bar{f}|_{i,j+d-q/2,k}^n\right), \quad (3.205)$$

at time step $n\Delta t$ and likewise for u and w. Parameters $\tau_q, \chi_{m,v}^q$ are found via

$$\tau_q = \frac{1}{2\Delta v} \left|\mathbf{r}(u,v,w) - \mathbf{r}(u, v - q\Delta v/2, w)\right| \quad \text{and} \quad \sum_{m=0}^{q-1} \chi_{m,v}^q = 1, \qquad (3.206)$$

with the values of the latter for $q = 2, \ldots, 5$ described in Table 3.4. Observe that in (3.204), the stencil family is derived by the original f quantities, unlike (3.207) where cell averages are incorporated. This formulation prepares the algorithm to treat all physical interactions, before the numerical integration of (3.203). Hence, for $q = 4$, the first member of the $C_{q,v}(\mathbf{r},t)$ set is

TABLE 3.4: Parameters $\chi_{m,v}^q$ for polynomials $C_{q,v}(\mathbf{r},t)$

$\chi_{m,v}^q$	$m=0$	$m=1$	$m=2$	$m=3$	$m=4$
$q=2$	$\frac{2}{3}$	$\frac{1}{3}$	0	0	0
$q=3$	$\frac{1}{10}$	$\frac{4}{10}$	$\frac{5}{10}$	0	0
$q=4$	$\frac{10}{15}$	$-\frac{7}{15}$	$\frac{8}{15}$	$\frac{4}{15}$	0
$q=5$	$-\frac{11}{24}$	$\frac{3}{24}$	$\frac{19}{24}$	$-\frac{4}{24}$	$-\frac{17}{24}$

$$C_{1,v}^7(\mathbf{r},t) = \frac{\tau_4}{3}\left[\bar{f}\big|_{i,j,k}^n - \bar{f}\big|_{i,j-2,k}^n + \bar{f}\big|_{i,j-4,k}^n + \tau_4\left(\bar{f}\big|_{i,j-1,k}^n - 3\bar{f}\big|_{i,j-3,k}^n + 2\bar{f}\big|_{i,j-5,k}^n\right)\right] \qquad (3.207)$$

In this way, $\mathcal{K}_u(\mathbf{r},t)$ are assessed by the sum of

$$K_v(\mathbf{r},t) = \sum_{l=0}^{q-1}\frac{s_l^v}{\sum_{m=0}^{q-1}b_m^v}C_{s,v}(\mathbf{r},t), \text{ with } s_l^v = \frac{r_l^v}{(\eta+P_v)^q}, \qquad (3.208)$$

and η varying below 10^{-6} to prevent the denominator from being zeroed. Coefficients r_l are real and representative of media constitutive parameters. Also, P_v is a smoothness gauge written as

$$P_v = \sum_{\alpha=1}^{q-1}\left\{\sum_{\beta=1}^{\alpha}\left(\delta^{q-\alpha}\left[\bar{f}\big|_{i,j-q+\beta,k}^n\right]\right)^2\right\}\alpha^{-1}, \qquad (3.209)$$

with $\delta^\kappa[.]$ as a 3-D difference operator defined according to a recursive rationale by

$$\delta^1\left[\bar{f}\big|_{i,j,k}^n\right] = \bar{f}\big|_{i,j+1,k}^n - \bar{f}\big|_{i,j,k}^n \text{ and } \delta^\kappa\left[\bar{f}\big|_{i,j,k}^n\right] = \delta^{\kappa-1}\left[\bar{f}\big|_{i,j+1,k}^n\right] - \delta^{\kappa-1}\left[\bar{f}\big|_{i,j,k}^n\right] \qquad (3.210)$$

This dispersion-reduction strategy, repeated toward each axis, is proven accurate and convergent, even for coarse lattices. So, spatial derivatives accept an enhanced treatment without the increase of complexity, because (3.205)–(3.209) are conducted only once, ahead of the basic simulation stage.

Regarding the explicit update of $\mathbf{W}_1(\mathbf{r},t)$ in (3.199), the ensuing adapted leapfrog scheme is used

$$\mathbf{W}_1^{n+1}(\mathbf{r},t) = \mathbf{W}_1^n(\mathbf{r},t) - \Delta t \left[\nabla \times \mathbf{W}_2^n(\mathbf{r},t) - \mathbf{g}^n(\mathbf{r},t)\right], \qquad (3.211)$$

the principal trait of which is the involvement of curvilinear WENO forms in the time advancing algorithm. Instead of (3.211), higher-order temporal accuracy may be accomplished through a Runge–Kutta process. By expressing as $\boldsymbol{\Omega}[.]$, the bracketed quantity on the right-hand side of (3.211), one gets

$$\mathbf{W}_1^{(1)}(\mathbf{r},t) = \mathbf{W}_1^n(\mathbf{r},t) - \Delta t \boldsymbol{\Omega}\left[\mathbf{W}_2^n(\mathbf{r},t), \mathbf{g}^n(\mathbf{r},t)\right], \qquad (3.212a)$$

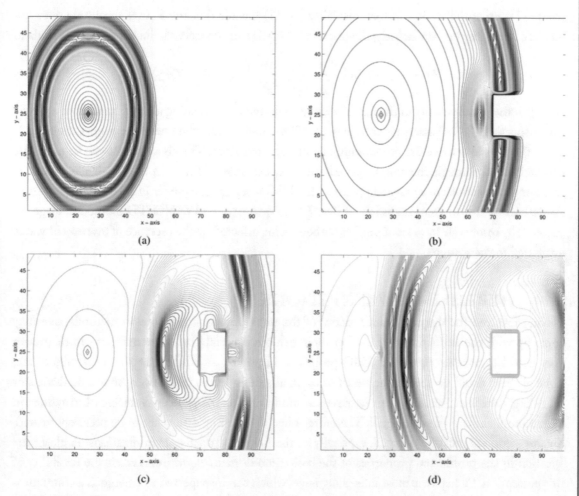

FIGURE 3.17: Snapshots of H_z-scattering interactions with a PEC rectangular cylinder at (a) 26.782 ms, (b) 68.314 ms, (c) 103.709 ms, and (d) 142.538 ms.

$$\mathbf{W}_1^{(2)}(\mathbf{r},t) = \frac{1}{4}\left\{ 3\mathbf{W}_1^n(\mathbf{r},t) + \mathbf{W}_1^{(1)}(\mathbf{r},t) - \Delta t \Omega\left[\mathbf{W}_1^{(1)}(\mathbf{r},t)\right]\right\}, \qquad (3.212b)$$

$$\mathbf{W}_1^{n+1}(\mathbf{r},t) = \frac{1}{3}\left\{ \mathbf{W}_1^n(\mathbf{r},t) + 2\mathbf{W}_1^{(2)}(\mathbf{r},t) - 2\Delta t \Omega\left[\mathbf{W}_1^{(2)}(\mathbf{r},t)\right]\right\} \qquad (3.212c)$$

The stability of (3.212) is explored via the von Neumann and the energy inequalities method. Focusing on the latter, the inner product on both sides of (3.199) by $\mathbf{W}_1^{n+1} + \mathbf{W}_1^n$ provides

$$\left\|\mathbf{W}_1^{n+1}(\mathbf{r},t)\right\| \leq \left\|\mathbf{W}_1^n(\mathbf{r},t)\right\| + \sum_{\gamma=0}^{n} G\left\|\Omega^{\gamma+1}\right\|, \qquad (3.213)$$

where G is an amplification factor. Inequality (3.213) unveils that the $\mathbf{F}_A(\mathbf{r},t)$ amplitude does not increase during its update, namely, the discrete energy is fully conserved. Such a deduction leads to

$$\Delta t \leq T_C \times \min_{ijk}\left(\frac{\Delta u}{b_{ijk}^{n,u}}, \frac{\Delta v}{b_{ijk}^{n,v}}, \frac{\Delta w}{b_{ijk}^{n,w}}\right), \text{ for } 0 < T_C \leq 1, \qquad (3.214)$$

and $b_{ijk}^{n,\zeta}$ is the phase velocity of the fastest wave at time step n propagating along ζ-axis. From a physical perspective, (3.214) assures that time steps will be smaller than the minimum propagation interval in the grid to achieve the proper sampling of any wave front. The above merits can be promptly comprehended through the simple, yet informative, example of Figure 3.17, which illustrates the scattering interactions of a Gaussian pulse with a PEC rectangular cylinder in a 3-D region of space. As derived by the different snapshots of the H_z field quantity, the WENO-TD method manages to successfully resolve all wave front profiles without being affected by the presence of evanescent waves attributed to the scatterer.

3.10   HYBRID IMPLEMENTATIONS

Having resolved the distinct characteristics of the most popular time-domain methods, one may come up with the deduction that each of them exhibits a certain set of merits that can be proven very useful in various classes of EMC problems. However, as efficient as these techniques are, none of them could be deemed universal or optimally designed to tackle every structure, device, or modeling situation. Actually, their weaknesses and limitations impose unavoidable obstructions in the accomplishment of such a goal. Therefore, when demanding setups are to be precisely treated, diverse solutions must be pursued, especially in the field of attractive hybridizations. A closer inspection in the profile and properties of the basic time-domain algorithms reveals the feasibility of this perception. Perhaps the most interesting issue, which can be expected to emerge as a competitor for specialized problems, is the relatively extended list of combinations with the FDTD method, although an assortment of other implementations has been presented in the recent archival literature.

For instance, in the case of extremely fine geometrical details, a prospective hybrid approach could harness the superior discretization competence of the FETD technique with the robustness and speed of the FDTD algorithm. Hence, the scope of applications is drastically widened, and the numerical outcomes are sufficiently more consistent, adhered to reality, and the most significant: they are delivered at a sensible computational cost. For the above reasons and because of the constantly augmenting research on the given topic, this section is dedicated to hybrid time-domain realizations and their principal attributes for several typical frameworks along with useful hints for the effective simulation of diverse difficult cases.

3.10.1 Efficient FETD/FDTD Algorithms

The mixed formulation described here attempts to accomplish a stable hybrid of a FDTD grid with a time-domain FE lattice comprising tetrahedral and pyramidal elements [70, 71]. Particularly, tetrahedra enable the accurate modeling of burdensome geometries, whereas the rectangular base of each pyramid allows the construction of robust interfaces with the FDTD cells. In essence, there are many points of equivalence between the two algorithms that favor this combination. First of all, if rectangular hexahedra are chosen and spatial integrations are computed via the trapezoidal rule, then the FDTD concepts are safely extracted from the FETD practice. Second, the use of edge and facet elements is analogous to the wave equation analysis on condition that the leapfrog time update is used for the coupled curl relations, and central time differencing is responsible for the wave equation. Therefore, returning to the aspects of Section 3.2.1 and as by now mentioned, the hybridization commences by denoting that edge elements are intended for the approximation of electric fields and Faedo–Galerkin method is next used to yield discrete systems of equations, characterized by symmetrical matrices. This interesting procedure can adequately represent vector wave equation $\nabla \times \mu^{-1} \nabla \times \mathbf{E} = \omega^2 \varepsilon \mathbf{E}$ by $\mathbf{SE} = \omega^2 \mathbf{ME}$, where $\mathbf{M}$ is the mass and $\mathbf{S}$ is the stiffness matrix. Because $\mathbf{M,S}$ are positive definite, numerical eigenvalues ω^2 are real and nonnegative, a fact that clearly gives the opportunity to develop a nonoscillatory and energy-conserving time integration scheme. A key profit of the discretization, so acquired, is the direct classification of the elements and not the edges themselves as in the pure finite-element or finite-difference arrangements. Albeit a probably more mature way of assembling the required spatial approximants would be the use of finite differences for edges that belong to the basic elements, such a policy leads to nonsymmetric matrices and, hence, stability may no longer be warranted.

After the necessary topological concerns, the time stepping of the hybrid FETD/FDTD algorithm is subsequently discussed. Evidently, on this type of mixed-element computational domain, it could be conducted directly from matrices $\mathbf{M,S}$; nevertheless, this idea is hardly economical because large sections of the mesh will comprise FDTD elements, thus increasing the total storage burden.

Mitigation of this difficulty comes from the advancement of fields in two consecutive stages as (A) every quantity handled by the FDTD method is temporally integrated in the usual explicit way and (B) the rest of the components are implicitly updated. According to this abstraction, the former stage must be able to treat certain tangential electric terms from the FETD grid (see Figure 3.18). In this case, all edges shared only by FDTD cells are manipulated explicitly, whereas all other edges—i.e., of a tetrahedron or a pyramid—implicitly, as elaborately reviewed in Table 3.5.

Actually, the update action of the second stage refers to the generalized Newmark scheme [69], which is proven very competent in tetrahedral and pyramidal grids. Regarding the stability of the hybrid formulation, it is easily proven—by the von Neumann analysis—that the Courant limit of the FDTD algorithm is sufficient for this objective, without any additional constraints. This fact is extremely convenient for realistic EMC configurations, where the FETD mesh may become rather involved because of the large number of detailed parts.

Returning to the spatial FETD/FDTD discretization, it is stressed that the development of the hybrid meshes is quite straightforward. The most immediate way is to build the FDTD lattice and use a triangular tessellation for complex shapes or discontinuities. This triangulated surface is then set in the FDTD grid, and all elements incorporating one or several triangles are marked to be ignored during the computations because the specific area is handled by the FETD technique. However, to elevate the quality of the entire mesh—issue of considerable importance in the FEM—a supplementary number of FDTD cells adjacent to the previous ones are also marked.

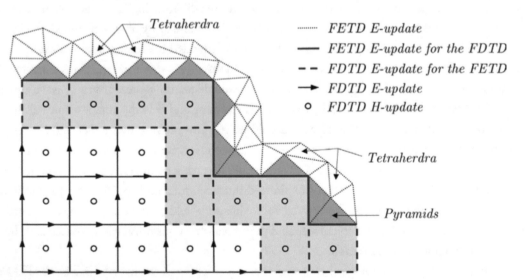

FIGURE 3.18: A 2-D transverse cut of the hybrid interface between the FETD and FDTD lattices. Thick lines signify FETD calculations required for the FDTD technique, whereas dashed lines indicate FDTD evaluations required for the FETD method.

TABLE 3.5: Basic time-integration actions of the hybrid FETD/FDTD technique

STAGE	ACTION TO BE PERFORMED
A	All components in the FDTD grid are updated through the leapfrog approach. Analytically, $\mathbf{H}$ and $\mathbf{E}$ quantities at time steps $n + 1/2$ and $n + 1$ are evaluated by their past FDTD $\mathbf{E}$ and $\mathbf{H}$ constituents at n and $n - 1/2$, respectively. Yet, the calculation of $\mathbf{H}$ components at $n + 1/2$ in the vicinity of the interface stipulates the FETD $\mathbf{E}$ values at time step n. These values are supplied from the cell edges with a thick continuous line of Figure 3.18.
B	The rest of the components situated in the FETD (dotted) lattice are implicitly advanced as follows: All $\mathbf{E}$ quantities at time step $n + 1$ are obtained by means of past FETD $\mathbf{E}$ components at n and $n - 1$ in conjunction with those already derived at $n + 1$ via the FDTD method and are located on the cell edges with dashed lines of Figure 3.18.

Successively, this limited volume of cells near the structural oddity is rediscretized by tetrahedra and pyramids, so that the base of the latter is always associated to the face of a usual FDTD cell. Of course, such a framework can be integrated into an automatic adaptive triangulation process with all the popular criteria for the refinement of the mesh [70]. In fact, nowadays, there exists a large number of very powerful—both from a topological and visualization viewpoint—commercial packages that offer a great deal of dynamic features for the rapid construction of optimal grids tailored to almost every application.

Conclusively, the hybridized method, thanks to its systematic formulation and high level of reliability, provides precise computational results, free from spectral contamination and nonphysical dissipation errors. Furthermore, it permits local spatial triangulations without the need to decrease the total time step to model field singularities near sharp corners or wedges. As a consequence, no restrictions on the materials existing at the interface of the two techniques should be posed. The only issue that must receive extra attention is the convergence rate of the iterative solvers engaged for the implicit time update. To this direction, the devise of enhanced matrix preconditioners remains a field of intensive investigation.

3.10.2 Conformal FVTD/FDTD Techniques

The notable flexibility in the development of robust and adaptive grids through the involvement of FE tessellations, described in Section 3.10.1, may be also effectively accomplished by the hybridization of the conformal FVTD method with the general family of FDTD solvers [72–74]. The presented methodology starts from the integral form of Maxwell's (2.1)–(2.4), which, for a lossless and isotropic medium, can be discretized with regard to time (leapfrog scheme), as

$$\frac{1}{\Delta t} \iint_S \left(\mathbf{D}^{n+1} - \mathbf{D}^n \right) \cdot \hat{\mathbf{n}} \mathrm{d}s = \underbrace{\oint_C \mathbf{H}^{n+1/2} \cdot \mathrm{d}\mathbf{l}}_{\text{magnetic circulation}} , \tag{3.215}$$

$$\frac{1}{\Delta t} \iint_S \left(\mathbf{B}^{n+1/2} - \mathbf{B}^{n-1/2} \right) \cdot \hat{\mathbf{n}} \mathrm{d}s = \underbrace{- \oint_C \mathbf{E}^n \cdot \mathrm{d}\mathbf{l}}_{\text{electric circulatic}} , \tag{3.216}$$

$$\frac{1}{\Delta t} \iiint_{V_m} \left(\mathbf{D}^{n+1} - \mathbf{D}^n \right) \mathrm{d}V = \underbrace{\iint_{\partial V_m} \hat{\mathbf{n}} \times \mathbf{H}^{n+1/2} \mathrm{d}s}_{\text{magnetic vorticity}} , \tag{3.217}$$

$$\frac{1}{\Delta t} \iiint_{V_e} \left(\mathbf{B}^{n+1/2} - \mathbf{B}^{n-1/2} \right) \mathrm{d}V = \underbrace{- \iint_{\partial V_e} \hat{\mathbf{n}} \times \mathbf{E}^n \mathrm{d}s}_{\text{electric vorticity}} \tag{3.218}$$

The above relations can be successfully used for the mathematical definition and topological formulation of both FVTD and FDTD methods. More specifically, one can readily point out that for the following: (a) <u>FVTD technique</u>: the electric (magnetic) vorticity should be computed over a surface to update the magnetic (electric) vector at the center of the volume enclosed by this surface; (b) <u>FDTD technique</u>: the electric (magnetic) circulation must be calculated around an electric (magnetic) contour to update the corresponding magnetic (electric) field component along the area vector of the surface enclosed by this contour.

The precision of the algorithm relies on the systematic construction of the grid to obtain the correct curves that enclose the object and the proper surfaces that surround its volume. Generally, the mesh, so developed, may comprise a variety of cells like hexahedra, tetrahedrons, or prisms. Retaining the notation of [72], this mesh will be designated as electric lattice and its vertices electric vertices, where $\mathbf{E}$ and $\mathbf{D}$ vectors are typically positioned. In a similar fashion, each face of an electric cell is called an electric face, whereas the center of the cell represents a magnetic vertex. Two magnetic vertices are connected if their two respective electric cells share a common face, hence forming a set of magnetic cells, namely the magnetic lattice, where $\mathbf{H}$ and $\mathbf{B}$ vectors are located. It is mentioned that volume V_m in (3.217) belongs to a magnetic cell and volume V_e in (3.218) to an electric one. For the special case of a nonplanar electric face, the concept of the barycenter (i.e., point whose coordinates are the average of those of the face vertices) is incorporated.

In this context, the resultant hybrid approach employs a composite grid that consists of a rectangular mesh intersecting a conformal one. The latter in turn is actually a combination of the aforesaid electric and magnetic lattices. Analytically, for smoothly varying objects, the electric lattice includes a small number of prisms perpendicularly erected along the objects' surfaces. It becomes apparent that the essential aim is to benefit from the triangular surface grid formed by this inter-section. As a consequence, the conformal electric lattice will comprise several layers of triangular prisms that will eventually overlap an outer rectangular mesh intended for the FDTD method. The height of the prisms is approximately the same as the FDTD spatial increment. In the internal electric or magnetic vertices, the FVTD algorithm is used for time integration, with the interior boundary of the rectangular mesh set sufficiently close to the object under study. On the other hand, the outer-boundary **E** components on the FVTD electric lattice are obtained via interpolation of their computed FDTD constituents, whereas **E** quantities on the FDTD mesh at the inner boundary of the rectangular grid are evaluated by interpolating the already calculated FVTD values. As deduced, each time step requires a double interpolation for the data from one grid to another; however, the whole procedure, because of its local application, hardly increases the memory or CPU burden to a noteworthy extent.

To complete the presentation of the combined method, a word on the critical issue of boundary conditions at the interface of the two schemes should be devoted. In fact, the strenuous wave interactions occurring at this region stipulate great attention. For this purpose, a general technique that introduces an intuitive interpolation/extrapolation concept with considerable accuracy levels is discussed. Focusing on the frequently encountered EMC arrangement of a smooth PEC boundary, the algorithm, at first place, flattens any corners or edges of the surface and then constructs the rect-angular FDTD ensemble that contains the volume surrounded by the structure. The outer-boundary vertices of this tessellation play the role of interior-boundary vertices of the overlapping FVTD mesh, whereas at each boundary vertex of the latter, a surface normal vector is assigned. The foot of this vector at the structure's surface provides two directions. Expressly, the tangential (to the surface) electric field components at a boundary vertex are determined by interpolation and the normal ones through extrapolation from outer-boundary values. Such a process is more rigorous than the stair-case approximation but less precise than the interpolation in a fine conformal grid.

Nonetheless, when the interface of two dissimilar media is to be modeled, the preceding approach must be properly modified. Let us inspect the two lossy dielectric regions of Figure 3.19, where across the interface both the tangential electric and magnetic field components are continuous. As a matter of fact, the policy that the FDTD method sets the variables makes it straightforward to simulate the boundary condition and concurrently provides the sufficient means to handle the hybrid FVTD/FDTD area in a numerical sense [73, 74]. In particular, region 1 is characterized by $\varepsilon_1, \mu_1, \sigma_1, \rho_1'$, whereas region 2 is characterized by $\varepsilon_2, \mu_2, \sigma_2, \rho_2'$ material parameters. Moreover, the

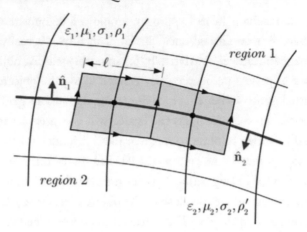

FIGURE 3.19: Integration procedure for a tangential electric field component at the interface of two dissimilar lossy materials by means of the conformal FVTD/FDTD method.

unit surface normal vectors pointing into the respective areas are depicted as $\hat{\mathbf{n}}_1$ and $\hat{\mathbf{n}}_2$. Setting $\hat{\mathbf{n}}_1 = -\hat{\mathbf{n}}_2 = \hat{\mathbf{n}}$, one obtains the $\hat{\mathbf{n}} \times \mathbf{E}_1 = \hat{\mathbf{n}} \times \mathbf{E}_2$ and $\hat{\mathbf{n}} \times \mathbf{H}_1 = \hat{\mathbf{n}} \times \mathbf{H}_2$ boundary conditions, which lead to

$$\hat{\mathbf{n}} \cdot \left(\mu_1 \frac{\partial \mathbf{H}_1}{\partial t} + \rho'_1 \mathbf{H}_1 \right) = \hat{\mathbf{n}} \cdot \left(\mu_2 \frac{\partial \mathbf{H}_2}{\partial t} + \rho'_2 \mathbf{H}_2 \right), \qquad (3.219a)$$

$$\hat{\mathbf{n}} \cdot \left(\varepsilon_1 \frac{\partial \mathbf{E}_1}{\partial t} + \sigma_1 \mathbf{E}_1 \right) = \hat{\mathbf{n}} \cdot \left(\varepsilon_2 \frac{\partial \mathbf{E}_2}{\partial t} + \sigma_2 \mathbf{E}_2 \right) \qquad (3.219b)$$

If $\hat{\mathbf{t}}$ indicates the tangential direction on the interface of the two media, (3.221) becomes

$$\oint_C \mathbf{H} \cdot d\mathbf{l} = A_1 \hat{\mathbf{t}} \cdot \left(\varepsilon_1 \frac{\partial \mathbf{E}_1}{\partial t} + \sigma_1 \mathbf{E}_1 \right) + A_2 \hat{\mathbf{t}} \cdot \left(\varepsilon_2 \frac{\partial \mathbf{E}_2}{\partial t} + \sigma_2 \mathbf{E}_2 \right)$$

$$= (A_1 \varepsilon_1 + A_2 \varepsilon_2) \hat{\mathbf{t}} \cdot \frac{\partial \mathbf{E}_1}{\partial t} + (A_1 \sigma_1 + A_2 \sigma_2) \hat{\mathbf{t}} \cdot \frac{\partial \mathbf{E}_2}{\partial t}, \qquad (3.220)$$

where $\mathbf{A}_1 = A_1 \hat{\mathbf{t}}$ and $\mathbf{A}_2 = A_2 \hat{\mathbf{t}}$ are specific area vectors characterizing magnetic contour C. Next, integration of (3.216) along a curve on the interface yields

$$-\oint_C \mathbf{E} \cdot d\mathbf{l} = A \left(\mu_1 \frac{\partial \mathbf{H}_1}{\partial t} + \rho'_1 \mathbf{H}_1 \right) \cdot \hat{\mathbf{n}} = A \left(\mu_2 \frac{\partial \mathbf{H}_2}{\partial t} + \rho'_2 \mathbf{H}_2 \right) \cdot \hat{\mathbf{n}}, \qquad (3.221)$$

with A as the total area enclosed. In this way, the adequately generalized (3.220) and (3.221) can supply the required field quantities at almost any interface, even at the demanding ones run into shielding, immunity, or printed circuit board problems whose correct modeling is strongly related to such calculations. As a final remark, it is emphasized that for the surface integral of (3.220), dot product $\hat{\mathbf{t}} \cdot \mathbf{E}$ has been assumed constant throughout the entire area of integration. Nevertheless, a more consistent and rigorous convention would dictate a linear variation of $\hat{\mathbf{t}} \cdot \mathbf{E}$ in $\mathbf{A}_1$ and $\mathbf{A}_2$ along the surface normal direction.

3.10.3 Accuracy-Optimized MRTD/FDTD Representations

Except for its continually growing usage as an individual tool in EMC analysis, the MRTD method, together with its wavelet–Galerkin counterpart, offers substantial cooperation opportunities with other time-domain techniques and principally with the FDTD one [75, 76]. This paragraph illustrates the theoretical framework of their basic hybrid rendition, along with some instructive connection conditions for the discretization of the numerically created interfaces that separate the computational domains of the two schemes.

Algorithmic formulation. As already discussed in Section 3.4, the Galerkin-weighted residuals procedure is substituted into Maxwell's curl laws, thus forming a system of six equations with scaling functions as well as their spatial and temporal derivatives. The expressions so acquired are multiplied by suitable basis functions, such as the Battle–Lemarie, the Haar, the Daubechies, or the Deslauriers–Dubuc ones with various numbers of vanishing moments N [25]. The final set of update formulae is fairly robust and versatile because it can be effectively applied to many electromagnetic problems characterized by abrupt wave fluctuations. Therefore, it can be deduced that the insertion of scaling functions preserves the simplicity in the algorithm's overall implementation. Furthermore, their orthogonality and compact support properties provide remarkable interpolation competences that allow the—otherwise not so "physical"—fulfillment of continuity conditions along boundaries or media interfaces. Owing to these inherent assets, the wavelet–Galerkin technique turns out to be a higher-order approximation strategy, far more efficient than those extracted from other polynomials.

For the hybridization with the FDTD scheme, let us concentrate on a waveguide structure, where conductive and arbitrarily placed wedges protrude from its walls. The computational domain is divided into regions of sharp (near wedges) and smooth field transition, respectively [75]. Each electric or magnetic field component in the former area is expanded in a set of scaling functions, like (3.81) or (3.89) and (3.90), hence taking avail of their ability to precisely simulate highly varying phenomena. In this manner, a significant enhancement of accuracy is attained because spatial derivatives in the MRTD or wavelet–Galerkin time-advancing equations are evaluated via more than two samples, as opposed to other realizations. Contrarily, smooth field transition regions, where singularities are absent and linear interactions occur, are treated by the simpler FDTD algorithm. The coexistence of the two time-domain tools at the interface of the previous areas is easy to configure, on condition that spatial and temporal increments are selected to be equal in the two areas. Consequently, each field quantity of the MRTD region is always calculated by $4N - 2$ samples, without allowing for the presence of the interface. For instance, the update of E_z component in the vicinity of the interface needs samples from both FDTD and MRTD areas, as described in Figure 3.20. The number of these samples depends on the distance from the interface and the chosen scaling functions, whereas no other topological assumptions should be devised. So, the treatment does not cause any late-time instabilities, evades unnecessary complications, and accomplishes satisfactory convergence rates.

However, the areas of abrupt transition must be systematically ascertained to represent the transient phenomenon and minimize the total overhead. Although the calculation of fields at a cubic cell around a singularity corner by the MRTD expressions is adequate to give trustworthy outcomes for coarse meshes, more components need to be accordingly updated when finer lattices are taken into account. This deduction can be primarily attributed to the fact that the cell volume decreases as the grid becomes denser, and therefore, the sharp transition region demands a larger number of cells for its consistent modeling. Finally, the drawback of the MRTD and wavelet–Galerkin approaches to yield elongated simulations for the steady state is outbalanced by its implementation in a strictly confined sector. Thus, the overall duration needed for the hybrid scheme is drastically decreased, compared with the individual application of each method.

Additional connection conditions. Recalling Figure 3.20, some alternative, yet fairly important, manipulations can be conducted. Especially in the vicinity of the interface, the update of **E** field components in the MRTD area entails the knowledge of magnetic scaling and wavelet coefficients one cell to the left within the FDTD region. For their computation, the FDTD-acquired **H** field quantities may be plugged as an input to a recursive fast wavelet transform [76]. Therefore, the magnetic field in the particular cells is decomposed and engaged for the update of **E** unknown values at the FDTD/MRTD boundary. When the opposite—i.e., FDTD to MRTD—transition has to be

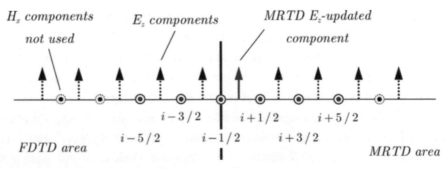

FIGURE 3.20: Graphical depiction of the E_z update procedure in the vicinity of an MRTD/FDTD interface.

contemplated, an inverse fast wavelet transform is applied to provide the unknown FDTD terms via evidence from the MRTD area. It should be stated that both transforms exhibit an optimal complexity of $O(N)$ and are not subject to any involved implementation process.

In the frequent case of an FDTD region enclosing an MRTD one, the exchange of field quantities constitutes an issue of serious concern. So, for the temporal evolution of all FDTD-lattice **E** nodes, whose stencil enters into the MRTD region, it is considered adequate to calculate the corresponding values one cell inside this area, as depicted in Figure 3.21. In this manner, all tangential **E** components along the interface of the FDTD domain may be retrieved through this efficient connection approach. Furthermore, because the MRTD-obtained tangential fields in the above area are treated as boundary conditions for the FDTD method, the overall problem can be reduced to the independent derivation of nodal terms along the MRTD interface, which is literally the objective of the inverse fast wavelet transform. On the contrary, for the time advancing of **H** MRTD components, their **E** FDTD analogues extending over one cell in the FDTD region are handled by means of the fast wavelet transform. Should this procedure surpasses the domain's outer limits, zero field values are imposed. Such a preference is completely justified for both closed and unbounded EMC arrangements because the MRTD (like the FDTD) technique employs a cross-oriented stencil for all mesh points of its tessellation. Addressing similar transforms for the electric MRTD terms as well and using (3.89) and (3.90) for the **H** fields lead to the extraction of every tangential electromagnetic quantity along the boundary of the MRTD area. Observe that all stages of the connection technique are performed at the same time step, during the calculation of **E** unknowns, without the creation of any oscillatory instability, vector parasite, or loss of convergence.

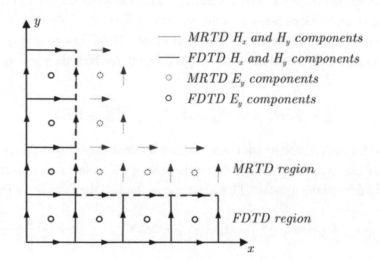

FIGURE 3.21: Location of electric and magnetic field components at an MRTD/FDTD interface.

Finally, the aforesaid concepts are valid for boundaries between wavelet schemes of an arbitrary basis and those derived by the respective scaling functions only, given that grid resolutions in the two regions are decided to be identical.

3.10.4 Combined Second-/Higher-Order FDTD Schemes

This section discusses the possibility of hybridizing a subgridding second-order FDTD procedure with its ordinary (2,4) counterpart [77]. In fact, the subgridding concept is recognized as a competent means for economical models of large-scale EMC applications. The specific method establishes second-order FDTD schemes on a relatively fine lattice to treat structural oddities, whereas the (2,4) formulation is used on coarser tessellations wherever required. Thus, all FDTD implementation aspects, like the excitation patterns, the absorbing boundary conditions, or the near-to-far-field transformation may be promptly realized on both meshes. Moreover, the higher-order approach accomplishes consistent simulations of travelling wave fronts in homogeneous domains, thus leading to precise results devoid of nonphysical artifacts.

To ensure the stability of the algorithm, the ratio between the fine and coarse grid is prefixed to 3:1, whereas their interface contains only electric instead of magnetic field components (Figure 3.22a). Also, two kinds of weighting procedures for the smooth transition and update of the unknown quantities are used near the coarse/fine-lattice boundary. According to the former, electric components are described by

$$\bar{E}_2 = 0.95\bar{E}_2^{\mathrm{F}} + 0.025\left(E_1 + \bar{E}_3\right), \qquad (3.222)$$

with $\bar{E}_2$ as the fine-mesh-weighted electric intensity adjacent to the boundary and $\bar{E}_2^{\mathrm{F}}, \bar{E}_3$ the fields computed by the second-order schemes on the fine grid. Likewise, E_1 signifies the field acquired via the (2,4) approach on the interface, as shown in Figure 3.22a. The second type of weighting relates the coarse-mesh magnetic quantities—neighboring to the boundary and located inside the fine grid—as

$$H_2 = 0.7H_2^{\mathrm{C}} + 0.3\bar{H}_2^{\mathrm{F}} \quad \text{and} \quad \bar{H}_2 = 0.3H_2^{\mathrm{C}} + 0.7\bar{H}_2^{\mathrm{F}}, \qquad (3.223)$$

where C,F denote the components on the coarse and fine lattice evaluated by the (2,4) and the second-order FDTD method, respectively. Of key importance also is the temporal interpolation of the fine-grid electric field values on the interface. This is performed through the third-order formulae of

$$E|^{n+1+1/3} = \frac{2}{9}E|^n - \frac{7}{9}E|^{n+1} + \frac{14}{9}E|^{n+2} \quad \text{and} \quad E|^{n+2+2/3} = -\frac{1}{9}E|^{n+1} + \frac{5}{9}E|^{n+2} + \frac{5}{9}E|^{n+2}$$

$$(3.224)$$

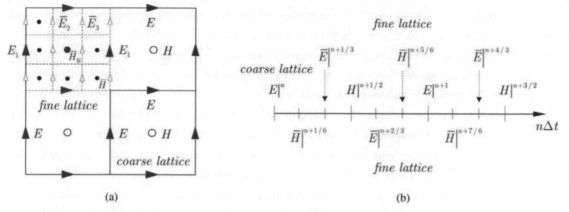

(a)   (b)

FIGURE 3.22: The hybrid second-/higher-order FDTD algorithm. (a) Lattice geometry amd (b) time evolution of electric and magnetic field quantities (the bar indicates fine-mesh field values).

for the time-step depicted in Figure 3.22b. Note that the stability of the hybrid algorithm is controlled in terms of the Courant condition for the (2,4) technique, (2.44), constrained by the safety factor of 20%.

TABLE 3.6: Main stages of the hybrid second-/higher-order FDTD algorithm	
STAGE	**PROCESS TO BE CONDUCTED**
A	Computation of H at $n + 1/2$ for every main mesh node and those of the fine grid via higher-order schemes.
B	Second-order forms are used on the fine mesh to obtain $\bar{H}$ at $n + 1/6$.
C	Second-order schemes on the fine grid and certain interpolations on the interface to derive $\bar{E}$ at $n + 1/3$. Application of (3.222) to weight $\bar{E}$ one cell in the fine mesh.
D	Use of second-order forms on the fine grid to calculate $\bar{H}$ at $n + 3/6$. Weighting of H at $n + 1/2$ via (3.223), placed one coarse-mesh cell in the fine lattice.
E	Update of E at $n + 1$ on the interface in terms of the known H values at $n + 1/2$.
F	Reapplication of stage: A to get $\bar{E}$ at $n + 2/3$, B to obtain $\bar{H}$ at $n + 5/6$, and C to compute $\bar{E}$ at $n + 1$ (the latter are interpolated on the interface by E at $n + 1$).
G	Replace all fine-grid $\bar{E}$ components to their collocated coarse-mesh E constituents.

In brief, the principal stages of the combined procedure, which presumes that all subgrid areas are internal to the coarse-mesh domain, are summarized in Table 3.6. As practical as this concept is for scattering or radiation problems, there are several EMC applications where the opposite arrangement is necessitated. Such cases are the shielding enclosures of large electrical size. Herein, the greatest percentage of the computational space, i.e., the enclosure's interior, is discretized by the (2,4) FDTD schemes, whereas their second-order analogues are implemented near the walls of the structure. The technique, so devised, is formulated in a fashion similar to the presented one, thus attaining an extensive reduction in the total CPU and memory overhead as compared with existing configurations. For its stability, the criterion of equation, now restrained by a safety factor of 50%, is selected. It is emphasized that the spatial interpolation of coarse-grid values on the interface of the two methods is performed by means of customary second-order finite-difference schemes.

3.10.5 Hybridizations via the Boundary Element Method

An area of electromagnetic research where time-domain techniques—especially the FDTD one—cannot be applied in their traditional form, is the class of transient nonlinear eddy current problems. This is basically attributed to the lack of symmetry in Maxwell's equations for quasistatic fields due to the absence of the displacement current term. To overcome such modeling intricacies, a hybrid algorithm that combines an explicit nonstandard finite-differencing concept for the diffusion equation with a boundary element method (BEM) for unbounded regions is discussed [78]. In fact, the analysis yields an entire family of techniques, depending on a set of two optimality parameters that guarantee stability, convergence, and consistency.

For the sake of simplicity and in an attempt to reveal the real weaknesses of existing approaches, let us consider the 1-D diffusion equation for potential ϕ,

$$\beta \frac{\partial \phi}{\partial t} = \frac{\partial^2 \phi}{\partial x^2} , \tag{3.225}$$

with β a prefixed diffusion coefficient. Equation (3.225) may be promptly discretized through the conventional Euler strategy employing central differences for the spatial derivative and forward differences for the temporal derivative. Nonetheless, this approximation is just first-order accurate in time and stable only if $\Delta t \leq 0.5\beta(\Delta x)^2$ for Δt, Δx, the time-step and cell size, respectively. Although the latter condition is not so preventive for several cases of the EMC practice, a second-order discretization, namely, an additional time level, should be pursued to decrease the overall number of time steps. Because the well-known leapfrog integrator is proven unstable, the subsequent category of symmetrical nonstandard finite-difference forms

$$\phi|_i^{n+1} = \alpha_{0,0}\, \phi|_i^n + \alpha_{1,0}\left(\phi|_{i-1}^n + \phi|_{i+1}^n\right) + \alpha_{0,1}\, \phi|_i^{n-1} + \alpha_{1,1}\left(\phi|_{i-1}^{n-1} + \phi|_{i+1}^{n-1}\right), \quad (3.226)$$

is introduced. Note that the differencing schemes in space and time are elegantly implied, rather than a priori decided. Moreover, the latter term of (3.226) involves possible values at nodes not taken into account in customary implementations. To determine the unknown coefficients, $\phi|_{i+k}^{n+p}$ is expressed by means of Taylor expansion series around (i, n), and after the enforcement of (3.225), one arrives at constraints

$$\alpha_{0,0} + 2\alpha_{1,0} + \alpha_{0,1} + 2\alpha_{1,1} = 1 \quad \text{and} \quad \frac{\alpha_{1,0} + \alpha_{1,1}}{1 + \alpha_{0,1} + 2\alpha_{1,1}} = \frac{\Delta t}{\beta(\Delta x)^2} = \lambda \qquad (3.227)$$

Obviously, (3.229) leads to a wide class of approximations owing to parameters $\alpha_{0,1}, \alpha_{1,1}$. For instance, the DuFort–Frankel algorithm is acquired via $\alpha_{0,1} = (1 - 2\lambda)/(1 + 2\lambda)$ and $\alpha_{1,1} = 0$, whereas the symmetrical semi-implicit scheme via $\alpha_{0,1} = -\lambda/(1 + 2\lambda)$ and $\alpha_{0,1} = -0.5\lambda/(1 + 2\lambda)$. However and under the previous circumstances, it is important to stress that none of (3.226) and (3.227) can precisely establish a reliable representation of (3.225), as they actually approximate the wave equation of

$$\frac{\partial \phi}{\partial t} + \varepsilon \frac{\partial^2 \phi}{\partial t^2} = \frac{1}{\beta}\frac{\partial^2 \phi}{\partial x^2}, \quad \text{with} \quad \varepsilon = \frac{(1 - \alpha_{0,1} - 2\alpha_{1,1})\Delta t}{2(1 + \alpha_{0,1} + 2\alpha_{1,1})}, \qquad (3.228)$$

a parasitic term related to the concept of inconsistency and controlled by Δt, Δx. Hence, instead of attempting to eliminate ε by simply opting for $\alpha_{0,1} + 2\alpha_{1,1} = 1$, it is preferable to construct more general types of approximations and evade unusual differencing schemes. Toward this direction, the first step is to retrieve the stability criteria for (3.226) by the von Neumann methodology. So, after the necessary algebra, one gets

$$|\alpha_{1,0}| + 2|\alpha_{1,1}| \leq 1 \quad \text{and} \quad (2\lambda - 1) + (2\lambda + 1)\alpha_{0,1} + 2(2\lambda - 1)\alpha_{1,1} \leq 0, \qquad (3.229)$$

which specify the regions of parameter variation. In particular, if the second relation of (3.229) is valid, i.e., complete independence of time step and cell size occurs, the resulting method is unconditionally stable.

Having decided on the inadequacies of the Euler approach, an alternative framework that exhibits a second-order precision in time is developed. The optimized algorithm, defined by

$$\alpha_{0,1} = 0.5\lambda^{-1} - 1, \quad \alpha_{1,1} = 0.25\lambda^{-1}, \quad \text{if} \quad \lambda > 1/2, \qquad (3.230a)$$

$$\alpha_{0,1} = 1 - 2\lambda, \quad \alpha_{1,1} = \lambda, \quad \text{if} \quad \lambda \le 1/2, \tag{3.230b}$$

remains stable even when the $\lambda \le 1/2$ is violated and possesses an inconsistency with the lowest possible level of $\varepsilon = (\lambda - 1/2)\Delta t$. For the majority of materials in eddy current problems, it is viable to preserve λ within the consistency range, without prolonged simulations. In any case, the slightly augmented number of iterations compared with implicit configurations is overwhelmingly compensated by the ease of (3.232).

As described, the application of an FDTD scheme to low-frequency EMC setups calls for a hybridization of the update mechanism with a mesh truncation strategy in the nonconducting areas where the problem is static. For this aim, based on a 2-D vector potential $\mathbf{A}$ formulation, the generalized time-stepping expression of

$$A|_{i,j}^{n+1} = \alpha_{0,0}\, A|_{i,j}^{n} + \alpha_{1,0}^{x}\left(A|_{i-1,j}^{n} + A|_{i+1,j}^{n}\right) + \alpha_{1,0}^{y}\left(A|_{i,j-1}^{n} + A|_{i,j+1}^{n}\right)$$
$$+ \alpha_{0,1}\, A|_{i,j}^{n-1} + \alpha_{1,1}^{x}\left(A|_{i-1,j}^{n-1} + A|_{i+1,j}^{n-1}\right) + \alpha_{1,1}^{y}\left(A|_{i,j-1}^{n-1} + A|_{i,j+1}^{n-1}\right), \tag{3.231}$$

can be developed. Its coefficients are evaluated through relations similar to (3.230a) or (3.229) and (3.230b), whereas λ has been substituted by $\lambda_x = \Delta t/\beta(\Delta x)^2$ or $\lambda_y = \Delta t/\beta(\Delta y)^2$ in the corresponding x- or y-coefficients and by $\lambda_x + \lambda_y$ in $\alpha_{0,0}, \alpha_{0,1}$. Conversely, the exterior problem is treated via the integral equation

$$A(\mathbf{r}) = \oint_{\partial\Omega}\left[A\frac{\partial G}{\partial n} - G\frac{\partial A}{\partial n}\right]\mathrm{d}l + \iint_{\Omega_s}\mu_0 J_s G \mathrm{d}S, \tag{3.232}$$

for $\partial\Omega$, the outer boundary of conductive region Ω, J_s, the source current density, and G, the 2-D Green's function. To be combined with the interior FDTD-oriented method of (3.231), (3.232) is discretized as

$$-\sum_j q_{i,j} A_j + \sum_j s_{i,j}(A_j - A'_j) + \sum_k r_{i,k} J_k = 0, \tag{3.233}$$

with A_j as the values of A at the boundary nodes, A'_j as the values at the internal nodes (shifted one cell away from the boundary), and J_k as the source current density at each element in the source region Ω_s. Furthermore, the supplementary coefficients $q_{i,j}$ and $r_{i,k}$ in (3.233) are given by

$$q_{i,j} = \frac{L_j \ell_{i,j}}{8\pi}\int_{-1}^{1}\frac{1-\zeta}{[R_{i,j}(\zeta)]^2}\mathrm{d}\zeta + \frac{L_{j-1}\ell_{i,j-1}}{8\pi}\int_{-1}^{1}\frac{1+\zeta}{[R_{i,j-1}(\zeta)]^2}\mathrm{d}\zeta, \tag{3.234}$$

where L_j is the length of edge $\{j, j+1\}$, $\ell_{i,j}$ is the vertical distance of node i from edge $\{j, j+1\}$, ζ is a locally defined coordinate, and $R_{i,j}(\zeta)$ is the distance of node i from any point of edge $\{j, j+1\}$. The calculation of $r_{i,k}$ is fully equivalent to (3.234). Observe that singular entries are depicted by $q_{i,i} = \varphi/2\pi$. Analogously,

$$s_{i,j} = \frac{L_j}{8\pi\Delta} \int_{-1}^{1} (1 - \zeta) \ln[R_{i,j}(\zeta)] \, d\zeta + \frac{L_{i,j-1}}{8\pi\Delta} \int_{-1}^{1} (1 + \zeta) \ln[R_{i,j-1}(\zeta)] d\zeta , \quad (3.235a)$$

for Δ, the cell size to the normal direction, and

$$s_{i,i} = \frac{1}{8\pi\Delta} \left(2L_i \ln L_i - 3L_i + 2L_{i-1} \ln L_{i-1} - L_{i-1} \right) \quad (3.235b)$$

Finally, the discretization and system solution of (3.233)–(3.235) provide the time-advancing equation of

$$\mathbf{A}^n = (\mathbf{Q} - \mathbf{S})^{-1} \left(\mathbf{R}\mathbf{J}^n - \mathbf{Q}\mathbf{A}'^{n-1} \right), \quad (3.236)$$

with the matrices capital lettering associated to the notation of the respective coefficients. Because the matrix inversion in (3.236) is performed once, only simple vector multiplications are involved at each time step. However, in large-scale 3-D EMC problems, the preceding matrices become rather large—because of the inherent discretization policy of BEM—and the required memory storage may become excessive. It is to be stressed that the hybrid FDTD/BEM scheme is applicable to nonlinear situations too. Indeed, the only additional task is the computation of flux density at every cell to trace the exact point on the nonlinear material curve. Such an extension could be proven fairly instructive in the use of conformal or unstructured grids for complex geometries or demanding composite arrangements with many detailed components.

3.10.6 Further Considerations

An apparent conclusion from the inspection of hybrid methodologies, so far discussed, reveals the indubitable tendency of scientific research to take the best possible avail of the merits characterizing each one of the core time-domain algorithms. Indeed, the efforts on this multidisciplinary field—often guided by the need for more reliable discrete models—remain active and ongoing. Therefore, novel combinations continue to emerge and couched in the language of EMC engineering interests. Essentially, there are two issues to consider: (a) given a realistic geometry, how does the proposed algorithm optimize system resources and (b) what is the tactic for the diminution of any inherent numerical drawback or error?

In terms of the FIT, its cooperation with the FDTD method, although not so broadly addressed in the literature, seems to offer some really appealing options, especially for complicated structures and discontinuities. As already demonstrated in Chapter 2, the remarkable capabilities of FIT in the development of curve-fitting lattices and material analysis is likely to upgrade the efficiency of the hybrid scheme in challenging situations, whereas the traditional FDTD configuration may concentrate on the solution of smoothly varying fields. Concerning the interface conditions between the two approaches, no major difficulties are anticipated to arise because the ordinary boundary continuity mechanisms suffice to give accurate results.

To a similar notion, the TLM algorithm can provide several profitable properties when hybridized with FDTD or MRTD formulations. Here, the elegant shunt-node transformation of the former ingredient ensures the correct interpretation of the underlying physical features of electrically large problems with moderate curvatures. Contrarily, the MRTD solver is used for rapid wave interactions near acute components, which cannot be tackled by other second-order time-domain setups. Although the interface connectivity of the resulting scheme is more involved, the final outcomes are not vulnerable to possible oscillatory modes.

The potential of the PSTD technique in its combination with the FETD method or the wavelet–Galerkin procedure must be also stressed [79]. The truly intelligent generation of straight-sided triangular or tetrahedral elements established by FEM may become a very proficient tool for the dispersion-optimized PSTD algorithm, even in the case of nonperiodic domains. Additionally, the sophisticated field representations of wavelets guarantee the drastic alleviation of abrupt singularities in the vicinity of troublesome interfaces and simultaneously reinforce the stability profile of the hybrid method. This benefit is predominantly useful when one wants to discretize fine geometrical characteristics, which are smaller than the basic FDTD cell or iteratively repeated to a considerable extent throughout the whole computational space. Finally, recent works have presented some interesting ways for the cooperation of PSTD and FDTD [81, 81] or ADI-FDTD [82] techniques in their conformal variant as well as several alternative realizations for more specialized problems [83–88]. The schemes, so created, can be applied in mixed-scale problems, where electrically fine structures coexist with electrically large and comparatively homogeneous regions.

REFERENCES

1. J.-M. Jin, *The Finite Element Method in Electromagnetics*, 2nd ed. New York: John Wiley and Sons, 2002.

2. A. Bossavit, *Computational Electromagnetism: Variational Formulations, Complementarity, Edge Elements*. San Diego, CA: Academic Press, 1998.

3. J. Volakis, K. Sertel, and B. C. Usner, *Frequency Domain Hybrid Finite Element Methods for Electromagnetics*, San Rafael, CA: Morgan & Claypool Publishers, 2006, doi:10.2200/S00038 ED1V01Y200606CEM010.

4. J.-F. Lee, R. Lee, A. C. Cangellaris, "Time-domain finite-element methods," *IEEE Trans. Antennas Propagat.*, vol. 45, no. 3, pp. 430–442, Mar. 1997.

5. A. C. Cangellaris, C. C. Lin, and K. K. Mei, "Point-matched time-domain finite element methods for electromagnetic radiation and scattering," *IEEE Trans. Antennas Propagat.*, vol. AP-35, no. 10, pp. 1160–1173, Oct. 1987, doi:10.1109/TAP.1987.1143981.

6. M. Feliziani and F. Maradei, "Point matched finite element-time domain method using vector elements," *IEEE Trans. Magn.*, vol. 30, no. 5, pp. 3184–3187, Sept. 1994, doi:10.1109/20.312614.

7. Z. S. Sacks and J.-F. Lee, "A finite-element time-domain method using prism elements for microwave cavities," *IEEE Trans. Electromagn. Compat.*, vol. 37, no. 4, pp. 519–527, Nov. 1995.

8. E. Abenius and F. Edelvik, "Thin sheet modeling using shell elements in the finite-element time-domain method," *IEEE Trans. Antennas Propagat.*, vol. 54, no. 1, pp. 28–34, Jan. 2006, doi:10.1109/TAP.2005.861554.

9. N. Venkatarayala, R. Lee, G. Yeow, L. Le-Wei, "Hanging variables in finite element time domain method with hexahedral edge elements," *Proc. 17th Int. Zurich Symp. Electromagn. Compat.*, pp. 184–187, 2006.

10. J.-F. Lee and Z. S. Sacks, "Whitney elements time domain (WETD) methods," *IEEE Trans. Magn.*, vol. 31, no. 3, pp. 1325–1329, Mar. 1995.

11. B. He and F. L. Teixeira, "A sparse and explicit FETD via approximate inverse Hodge (mass) matrix," *IEEE Microwave Wireless Compon. Lett.*, vol. 16, no. 6, pp. 348–350, June 2006.

12. Z. Lou and J.-M. Jin, "A new explicit time-domain finite-element method based on element-level decomposition," *IEEE Trans. Antennas Propagat.*, vol. 54, no. 10, p. 2990–2998, Oct. 2006, doi:10.1109/TAP.2006.882178.

13. D. Jiao and J.-M. Jin, "A general approach for the stability analysis of the time-domain finite-element method for electromagnetic simulations," *IEEE Trans. Antennas Propagat.*, vol. 50, no. 11, p. 1624–1632, Nov. 2003.

14. R. N. Rieben, G. H. Rodrigue, and D. A. White, "A high order mixed vector finite element method for solving the time dependent Maxwell equations on unstructured grids," *J. Comp. Phys.*, vol. 204, pp. 490–519, 2005.

15. T. V. Yioultsis, N. V. Kantartzis, C. S. Antonopoulos, and T. D. Tsiboukis, "A fully explicit Whitney element-time domain scheme with higher-order vector finite elements for three dimensional high frequency problems," *IEEE Trans. Magn.*, vol. 34, no. 5, pp. 3288–3291, Sept. 1998, doi:10.1109/20.717772.

16. N. K. Madse, and R. W. Ziolkowski, "A three-dimensional modified finite volume technique for Maxwell's equations," *Electromagnetics*, vol. 10, pp. 147–161, 1990.

17. R. Holland, V. P. Cable, and L. C. Wilson, "Finite-volume time-domain (FVTD) techniques for EM scattering," *IEEE Trans. Electromagn. Compat.*, vol. 33, no. 4, pp. 281–294, Nov. 1991, doi:10.1109/15.99109.

18. A. Mohammadian, V. Shankar, and W. F. Hall, "Application of time-domain finite-volume method to some radiation problems in two and three dimensions," *IEEE Trans. Magn.*, vol. 27, pp. 3841–3844, 1991, doi:10.1109/20.104939.

19. D. Schiavoni, M. Borlino Cappio, M. Giunta, R. Pompani, R. De Leo, and G. Pierucci, "Finite volume time domain technique for evaluation of scattering in GHz–THz cell: Model, comparison with measurements and literature references," *Proc. IEEE Int. Symp. Electromagn. Compat.*, pp. 602–607, 1997, doi:10.1109/ISEMC.1997.667750.

20. P. Bonnet, X. Ferrieres, B. L. Michielsen, P. Klotz, J. L. Roumiguières, "Finite-volume time domain method," in *Time Domain Electromagnetics*, S. M. Rao, Ed. San Diego, CA: Academic Press, 1999, doi:10.1016/B978-012580190-4/50011-2.

21. C. Fumeaux, D. Baumann, and R. Vahldieck, "A generalized local time-step scheme for efficient FVTD simulations in strongly inhomogeneous meshes," *IEEE Trans. Microwave Theory Tech.*, vol. 52, no. 3, pp. 1067–1076, Mar. 2004, doi:10.1109/TMTT.2004.823595.

22. D. Firsov, J. LoVetri, O. Jeffrey, V. Okhmatovski, C. Gilmore, and W. Chamma, "High-order FVTD schemes for Maxwell's equations," *ACES J.*, vol. 22, pp. 71–82, 2007.

23. M. Fujii and W. R. J. Hoefer, "A three-dimensional Haar wavelet-based multiresolution analysis similar to the FDTD method—Derivation and application," *IEEE Trans. Microwave Theory Tech.*, vol. 46, no. 12, pp. 2463–2475, Dec. 1998.

24. M. Krumpholz and L. P. B. Katehi, "MRTD: New time-domain schemes based on multiresolution analysis," *IEEE Trans. Microwave Theory Tech.*, vol. 44, no. 4, pp. 555–571, Apr. 1999.

25. N. Bushyager and M. Tentzeris, *MRTD (Multi Resolution Time Domain) Method in Electromagnetics*. San Rafael, CA: Morgan & Claypool Publishers, 2005, doi:10.2200/S00009ED1V 01Y200508CEM002.

26. C. D. Sarris and L. P. B. Katehi, "Fundamental gridding related dispersion effects in MRTD schemes," *IEEE Trans. Microwave Theory Tech.*, vol. 49, no. 12, pp. 2248–2257, Dec. 2001.

27. T. Dogaru and L. Carin, "Application of Haar-wavelet-based multiresolution time-domain," *IEEE Trans. Microwave Theory Tech.*, vol. 50, no. 6, pp. 774-784, June 2001.

28. E. M. Tentzeris, A. C. Cangellaris, L. P. B. Katehi, and J. F. Harvey, "Multiresolution time-domain (MRTD) adaptive schemes using arbitrary resolutions of wavelets," *IEEE Trans. Microwave Theory Tech.*, vol. 50, no. 2, pp. 501–516, Feb. 2002, doi:10.1109/22.982230.

29. S. Braun and P. Russer, "A low-noise multiresolution high-dynamic ultra-broad-band time-domain EMI measurement system," *IEEE Trans. Microwave Theory Tech.*, vol. 53, no. 11, pp. 3354–3363, Nov. 2005.

30. S. Barmada, A. Musolino, M. Raugi, "Wavelet-based time-domain solution of multiconductor transmission lines with skin and proximity effect," *IEEE Trans. Electromagn. Compat.*, vol. 47, no. 4, pp. 774–780, Nov. 2005, doi:10.1109/TEMC.2005.857868.

31. Q. H. Liu, "The PSTD algorithm: A time-domain method requiring only two cells per wavelength," *Microwave Opt. Technol. Lett.*, vol. 15, no. 3, pp. 158–165, 1997, doi:10.1002/(SICI)1098-2760(19970620)15:3<158::AID-MOP11>3.0.CO;2-3.

32. Q. H. Liu and G. Zhao, "Review of the PSTD methods for transient electromagnetics," *Int. J. Numer. Model.: Electron. Networks Devices Fields*, vol. 22, pp. 299–321, 2004.

33. B. Yang and J. S. Hesthaven, "A pseudospectral method for time-domain computation of electromagnetic scattering by bodies of revolution," *IEEE Trans. Antennas Propagat.*, vol. 47, no. 1, pp. 132–141, Jan. 1999.

34. Y. J. Fan, B. L. Ooi, and M. S. Leong, "Fast multipole accelerated pseudospectral time-domain algorithm," *IET Microwave Antennas, Propagat.*, vol. 1, no. 3, pp. 763–769, June 2007, doi:10.1049/iet-map:20060257.

35. T. Namiki, "A new FDTD algorithm based on ADI method," *IEEE Trans. Microwave Theory Tech.*, vol. 47, no. 10, pp. 2003–2007, Oct. 1999, doi:10.1109/22.795075.

36. F. Zheng, Z. Chen, and J. Zhang, "A finite-difference time-domain method without the Courant stability conditions," *IEEE Microwave Guided Wave Lett.*, vol. 9, no. 11, pp. 441–443, Nov. 1999, doi:10.1109/75.808026.

37. J. S. Kole, M. T. Figge, and H. De Raedt, "Unconditionally stable algorithms to solve the time-dependent Maxwell equation," *Phys. Rev. E*, vol. 64, no. 6, pp. 066705(1)–066705(14), Dec. 2001, doi:10.1103/PhysRevE.64.066705.

38. S. G. García, T.-W. Lee, and S. C. Hagness, "On the accuracy of the ADI-FDTD method," *IEEE Antennas Wireless Propagat. Lett.*, vol. 1, no. 1, pp. 31–34, 2002, doi:10.1109/LAWP.2002.802583.

39. A. P. Zhao, "Analysis of the numerical dispersion of the 2-D alternating-direction implicit FDTD method," *IEEE Trans. Microwave Theory Tech.*, vol. 50, no. 4, pp. 1156–1164, Apr. 2002.

40. M. Darms, R. Schuhmann, H. Spachmann, and T. Weiland, "Dispersion and asymmetry effects of ADI-FDTD," *IEEE Microwave Wireless Compon. Lett.*, vol. 12, no. 12, pp. 491–493, Dec. 2002.

41. G. Sun and C. W. Trueman, "Analysis and numerical experiments on the numerical dispersion of two-dimensional ADI-FDTD," *IEEE Antennas Wireless Propagat. Lett.*, vol. 2, pp. 78–81, 2003.

42. S. Staker, C. Holloway, A. Bhobe, and M. Piket-May, "ADI formulation of the FDTD method: Algorithm and material dispersion implementation," *IEEE Trans. Electromagn. Compat.*, vol. 45, pp. 156–166, May 2003, doi:10.1109/TEMC.2003.810815.

43. S. Wang, F. L. Teixeira, and J. Chen, "An iterative ADI-FDTD with reduced splitting error," *IEEE Microwave Wireless Compon. Lett.*, vol. 15, no. 2, pp. 92–94, Feb. 2005.

44. W. Fu and E. L. Tan, "Unconditionally stable ADI-FDTD method including passive lumped elements," *IEEE Trans. Electromagn. Compat.*, vol. 48, no. 4, pp. 661–668, Nov. 2006.

45. N. V. Kantartzis, D. L. Sounas, C. S. Antonopoulos, and T. D. Tsiboukis, "A wideband ADI-FDTD algorithm for the design of double negative metamaterial-based waveguides and antenna structures," *IEEE Trans. Magn.*, vol. 43, no. 4, pp. 1329–1333, 2007.

46. Q. L. Liu and G. Zhao, "The unconditionally stable pseudospectral time-domain (PSTD) method," *IEEE Microwave Wireless Compon. Lett.*, vol. 13, no. 11, pp. 475–477, Nov. 2003.

47. S. D. Gedney and U. Navsariwala, "A unconditionally stable finite element time-domain solution of the vector wave equation," *IEEE Microwave Guided Wave Lett.*, vol. 5, no. 10, pp. 332–334, Oct. 1995.

48. J. Chen and J. Wang, "A novel WCS-FDTD method with weakly conditional stability," *IEEE Trans. Electromagn. Compat.*, vol. 49, no. 2, pp. 419–426, May 2007.

49. M. Movahhedi, A. Abdipour, A. Nentchev, M. Dehghan, S. Selberherr, "Alternating-direction implicit formulation of the finite-element time-domain method," *IEEE Trans. Microwave Theory Tech.*, vol. 55, no. 6, pp. 1322–1331, June 2007.

50. R. E. Mickens, *Nonstandard Finite Difference Models of Differential Equations*. Singapore: World Scientific, 1994.

51. J. B. Cole, "A high accuracy FDTD algorithm to solve microwave propagation and scattering problems on a coarse grid," *IEEE Trans. Microwave Theory Tech.*, vol. 43, no. 9, pp. 2053–2058, Sept. 1995, doi:10.1109/22.414540.

52. N. V. Kantartzis and T. D. Tsiboukis, "A higher order nonstandard FDTD-PML method for the advanced modeling of complex EMC problems in generalized 3-D curvilinear coordinates," *IEEE Trans. Electromagn. Compat.*, vol. 46, no. 1, pp. 2–11, Feb. 2004

53. T. Ohtani, K. Taguchi, T. Kashiwa, and Y. Kanai, "Overlap algorithm for the nonstandard FDTD method using nonuniform mesh," *IEEE Trans. Magn.*, vol. 43, no. 4, pp. 1317–1320, Apr. 2007.

54. J. Fang, *Time Domain Finite Difference Computation for Maxwell's Equations*. PhD thesis, Univ. California, Berkeley, 1989.

55. J. L. Young, D. Gaitonde, and J. S. Shang, "Toward the construction of a fourth-order difference scheme for transient EM wave simulation: Staggered grid approach," *IEEE Trans. Antennas Propagat.*, vol. 45, no. 11, pp. 1573–1580, Nov. 1997, doi:10.1109/8.650067.

56. E. Turkel, "High-order methods," in *Advances in Computational Electrodynamics: The Finite-Difference Time-Domain Method*, A. Taflove, Ed. Norwood, MA: Artech House, 1998, pp. 63–110.

57. K. Lan, Y. Liu, and W. Lin, "A higher order (2,4) scheme for reducing dispersion in FDTD algorithms," *IEEE Trans. Electromagn. Compat.*, vol. 41, no. 2, pp. 160–165, May 1999.

58. H. M. Jurgens and D. W. Zingg, "Numerical solution of the time-domain Maxwell equations using high-accuracy finite-difference methods," *SIAM J. Sci. Comput.*, vol. 22, pp. 1675–1696, 2000.

59. A. Yefet and P. G. Petropoulos, "A staggered fourth-order accurate explicit finite difference scheme for the time-domain Maxwell's equations," *J. Comp. Phys.*, vol. 168, no. 2, pp. 286–315, Apr. 2001, doi:10.1006/jcph.2001.6691.

60. S. V. Georgakopoulos, C. R. Birtcher, C. A. Balanis, and R. A. Renaut, "Higher-order finite-difference schemes for electromagnetic radiation, scattering, and penetration, part I: Theory," *IEEE Antennas Propagat. Mag.*, vol. 44, 134–142, Feb. 2002, doi:10.1109/74.997945.

61. H. Spachmann, R. Schuhmann, and T. Weiland, "Higher order explicit time integration schemes for Maxwell's equations," *Int. J. Numer. Model.*, vol. 15, no. 5–6, pp. 419–437, Sept.–Dec. 2002.

62. K. P. Prokopidis and T. D. Tsiboukis, "Higher-order FDTD (2,4) scheme for accurate simulations in lossy dielectrics," *IEE Electron. Lett.*, vol. 39, no. 11, pp. 835–836, 2003.

63. T. T. Zygiridis and T. D. Tsiboukis, "Low-dispersion algorithms based on the higher order (2,4) FDTD methods," *IEEE Trans. Microwave Theory Tech.*, vol. 52, no. 4, pp. 1321–1327, Apr. 2004.

64. T. T. Zygiridis and T. D. Tsiboukis, "Optimized three-dimensional FDTD discretizations of Maxwell's equations on Cartesian grids," *J. Comput. Phys.*, vol. 226, pp. 2372–2388, 2007.

65. Z. Xie, C.-H. Chan, and B. Zhang, "An explicit fourth-order orthogonal curvilinear staggered grid FDTD method for Maxwell's equations," *J. Comp. Phys.*, vol. 175, no. 2, pp. 739–763, Dec. 2002.

66. X.-D. Liu, S. Osher, and T. Chan, "Weighted essentially non-oscillatory schemes," *J. Comp. Phys.*, vol. 115, no. 1, pp. 200–212, Nov. 1994.

67. Y.-X. Ren, M. Liu, and H. Zhang, "A characteristic-wise hybrid compact-WENO scheme for solving hyperbolic conservation laws," *J. Comp. Phys.*, vol. 192, no. 2, pp. 365–386, Dec. 2003.

68. Y. Xing and C.-W. Shu, "High order finite difference WENO schemes with the exact conservation property for the shallow water equations," *J. Comp. Phys.*, vol. 208, no. 1, pp. 206–227, Sept. 2005.

69. N. V. Kantartzis, T. D. Tsiboukis, and E. E. Kriezis, "An explicit weighted essentially non-oscillatory time-domain algorithm for the 3-D EMC applications with arbitrary media," *IEEE Trans. Magn.*, vol. 42, no. 4, pp. 803–806, 2006.

70. T. Rylander, F. Edelvik, A. Bondeson, and D. Riley, "Advances in hybrid FDTD-FE techniques," in *Computational Electromagnetics: The Finite Difference Time-Domain Method*, 3rd ed., A. Taflov, and S. C. Hagness, Eds. Norwood, MA: Artech House, 2005, pp. 907–953.

71. R. B. Wu and T. Itoh, "Hybrid finite-difference time-domain modeling of curved surfaces using tetrahedral elements," *IEEE Trans. Antennas Propagat.*, vol. 45, pp. 227–232, 1997.

72. K. S. Yee and J. S. Chen, "Conformal hybrid finite difference time domain and finite volume time domain," *IEEE Trans. Antennas Propagat.*, vol. 42, no. 10, pp. 1450–1455, Oct. 1994.

73. Y. Mingwu, Y. Chen, R. Mittra, "Hybrid finite-difference/finite-volume time-domain analysis for microwave integrated circuits with curved PEC surfaces using a nonuniform rectangular grid," *IEEE Trans. Microwave Theory Tech.*, vol. 48, no. 6, pp. 969–975, June 2000, doi:10.1109/22.846728.

74. X. Ferrieres, J.-P. Parmantier, S. Bertuol, A. R. Ruddle, "Application of a hybrid finite difference/finite volume method to solve an automotive EMC problem," *IEEE Trans. Electromagn. Compat.*, vol. 46, no. 4, pp. 624–634, Nov. 2004.

75. N. V. Kantartzis, T. I. Kosmanis, T. V. Yioultsis, and T. D. Tsiboukis, "A nonorthogonal higher-order wavelet-oriented FDTD technique for 3-D waveguide structures on generalised curvilinear grids," *IEEE Trans. Magn.*, vol. 37, no 5, pp. 3264–3268, Sep. 2001, doi:10.1109/20.952591.

76. C. D. Sarris, *Adaptive Mesh Refinement for Time-Domain Numerical Electromagnetics*. San Rafael, CA: Morgan & Claypool Publishers, 2006.

77. S. V. Georgakopoulos, R. A. Renaut, C. A. Balanis, and C. R. Birtcher, "A hybrid fourth-order FDTD utilizing a second-order FDTD subgrid," *IEEE Microwave Wireless Compon. Lett.*, vol. 11, no. 11, pp. 462–464, Nov. 2001.

78. T. V. Yioultsis, K. S. Charitou, C. S. Antonopoulos, and T. D. Tsiboukis, "A finite difference time domain scheme for transient eddy current problems," *IEEE Trans. Magn.*, vol. 37, no. 5, pp. 3145–3149, Sept. 2001.

79. Y. J. Fan, B. L. Ooi, and M. S. Leong, "A novel hybrid TDFEM-PSTD method with stability consideration," *Microwave Opt. Technol. Lett.*, vol. 47, no. 2, pp. 195–197, 2005, doi:10.1002/mop.21121.

80. W. K. Leung and C. H. Chan, "Combining the FDTD and PSTD methods," *Microwave Opt. Technol. Lett.*, vol. 23, pp. 249–254, 1999, doi:10.1002/(SICI)1098-2760(19991120)23:4<249::AID-MOP18>3.0.CO;2-O.

81. Q. Li, Y. Chen, and C. K. Li "Hybrid PSTD–FDTD technique for scattering analysis," *Microwave Opt. Technol. Lett.*, vol. 34, no. 1, pp. 19–24, 2002, doi:10.1002/mop.10361.

82. M. Chai, T. Xiao, G. Zhao, Q. H. Liu, "A hybrid PSTD/ADI-CFDTD method for mixed-scale electromagnetic problems," *IEEE Trans. Antennas Propagat.*, vol. 55, no. 5, pp. 1398–1406, May 2007.

83. C. Feng and Z. Shen, "A hybrid FD-MoM technique for predicting shielding effectiveness of metallic enclosures with apertures," *IEEE Trans. Electromagn. Compat.*, vol. 47, no. 3, pp. 456–462, Aug. 2005, doi:10.1109/TEMC.2005.851726.

84. W. Shao, B.-Z. Wang, and Z.-J. Yu, "Space-domain finite-difference and time-domain moment method for electromagnetic simulation," *IEEE Trans. Electromagn. Compat.*, vol. 48, no. 1, pp. 10–18, Feb. 2006, doi:10.1109/TEMC.2005.861376.

85. B. K. Li, F. Liu, and S. Cozier, "High-field magnetic resonance imaging with reduced field/tissue RF artifacts—A modeling study using hybrid MoM/FEM and FDTD technique," *IEEE Trans. Electromagn. Compat.*, vol. 48, no. 4, pp. 628–633, Nov. 2006.

86. J. Hu and Z. Nie, "Improved electric field integral equation (IEFIE) for analysis of scattering from 3-D conducting structures," *IEEE Trans. Electromagn. Compat.*, vol. 49, no. 3, pp. 644–648, Aug. 2007.

87. C. J. Railton, D. L. Paul, and S. Dumanli, "The treatment of thin wire and coaxial structures in lossless and lossy media in FDTD by the modification of assigned material parameters," *IEEE Trans. Electromagn. Compat.*, vol. 48, no. 4, pp. 654–660, Nov. 2006.

88. M. de Magistris, L. De Tommasi, A. Maffucci, and G. Miano, "Low-order identification of interconnects with the generalized method of characteristics," *IEEE Trans. Electromagn. Compat.*, vol. 49, no. 3, pp. 670–676, Aug. 2007.

CHAPTER 4

Principal Implementation Issues of Time-Domain EMC Simulations

4.1 INTRODUCTION

The success of a numerical simulation, as broadly acknowledged, is the blend of several significant factors and correct parameter selections. To this deduction, time-domain techniques could not constitute an exception; on the contrary, they are actually one of the most ideal fields where the above remarks come into practice. Notwithstanding the indisputable maturity of these methods, there is still a range of things to be addressed for the improvement of their modeling competence and efficient implementation. After all, the construction of robust mathematical equivalents for realistic EMC applications is about the art of accurate spatial/temporal approximation, and such a path is strewn with pitfalls. For instance, the issues of finite discretization, limited computational domain, and restricted machine precision retain the character of "open problems", despite the development of various inspired advances. Therefore, it becomes apparent that to comply with the demanding aspects of contemporary structures and avoid deceptive outcomes, every realization detail must be scrupulously chosen in accordance with the algorithm executing the analysis.

Having presented the basic theoretical principles of diverse time-domain methodologies in the previous chapters, our interest will now concentrate on the configuration characteristics as well as on some important hints regarding their enhancement. First of all, the topic of *excitation* is carefully investigated, with the most popular schemes presented in a general way to facilitate satisfactory arrangements for every technique. Realizing the influence of *dispersion-error mechanisms* in discrete models, the next section discusses two powerful concepts for their suppression, based on artificial anisotropy and angle optimization. Subsequently, the main categories of *absorbing boundary conditions (ABCs)* along with a detailed description on the revolutionary perfectly matched layer (PML) for several approaches is performed, and the perspective of using *curvilinear* or *nonorthogonal lattices* is also examined. To encompass all possible configuration challenges, *frequency-dependent* and *anisotropic media* are explored in terms of the most preferable methods, whereas the study of *surface impedance boundary* and *interface conditions* provides viable means for notable system savings.

4.2 ELECTROMAGNETIC WAVE EXCITATION SCHEMES

The well-posed advancement of electromagnetic fields in the interior of a computational domain is tightly related to the wave source settings enforced at the beginning of (or during) every simulation [1–5]. In fact, such settings act like a connecting ring between the continuous and discrete state because they transfer the past profile of a field and introduce the necessary premises for the realistic future interaction of its components. Thus, a cautiously configured excitation is the basis for a successful numerical study. After all, it can be easily understood that any irregularity incorporated via an incident wave is likely to propagate (and not eliminated) in the lattice, despite the efficiency of the selected method. Not to mention the growth of nonphysical and below-cutoff reactive modes that may affect the stability of any discretizing strategy. Nowadays, EMC simulations tend to deal with devices of constantly increasing complexity, which subsequently demand the elaborate representation of nonlinear circuit-type sources. However, the punctual modeling of these sources should be executed with regard to computer means. This implies that the researcher must use only a limited amount of electric/magnetic quantities to describe the incident conditions and hence minimize the additional burden for their performance.

This section investigates the aspects of free-space and guiding-structure excitation. Focusing on the FDTD-based methods, due to their prevalent role in time-domain EMC models, the analysis—wherever required—discusses the appropriate modifications or extensions for the other algorithms as well. In this context, the early plane-wave scheme is presented and then the simple hard/soft source models are examined. The next category postulates a more sophisticated discipline, which is deemed ideal for open-region illuminations, i.e., the scattered- and total/scattered-field formulations, that enable the analysis of laborious problems without demanding an excessive number of cells for wave impingement. Finally, the section closes with the excitation of particular structures and some contemporary ideas for the enhancement of existing schemes.

4.2.1 Plane-Wave Illumination

The structural simplicity of the early problems enabled the adoption of a simple excitation scheme: a plane wave propagating from infinity. This was the first source condition of the FDTD method, and as deduced from the scientific literature, it became a handy tool especially in far-field calculations, where plane waves are the dominant electromagnetic forms. During its extended use, this type of excitation was outlined by several important rules. At the outset, a source of this type should support a variety of field polarizations, time-domain envelopes, temporal intervals, and directions of traveling. Second, the incident condition must permit the unperturbed entrance of the illuminating wave into the domain and avoid any obstacles for the period of reciprocal field interactions. Its general idea is implemented through the insertion of the incident wave as an initial condition at every electric and magnetic quantity in the grid, retaining the one-half time-step leapfrog offset between **E** and **H** vectors. Apparently, the sort of material occupying the mesh has a decisive influence on the determination of the corresponding spatial shift. So, in the occasion of a dielectric medium whose constitutive properties do not depend upon frequency variations, the only change is the reduced phase velocity that controls wave motion. By selecting a modified temporal increment (according to the new speed), the procedure returns to its original free-space arrangement. However, when dispersive media are considered, spatial shifts can no longer be productively computed because the continuous frequency change prevents the use of a fixed choice. Actually, the essential attitude of a plane-wave source is to create a numerical counterpart that allows the propagation of reflected modes in the grid. Unfortunately, when applications become more complicated, some inevitable difficulties arise, such as the unduly amount of CPU time for the simulation of elongated pulses and the distortion of incident waves when launched at oblique angles. These shortcomings consume large amounts of memory resources without producing reliable outcomes, thus limiting the usage of this source condition to coarse lattices. Adequate mitigation, nevertheless, comes from the development of other more mature excitation techniques, presented in the rest of this section.

4.2.2   Hard and Soft Source Excitation

The aforementioned weaknesses of plane-wave illumination led to two local schemes able to propagate a multitude of different pulses or wave fronts, namely, the hard and soft source processes. The former is established by merely associating the preferred time function to a certain electric or magnetic field component in the mesh. So, the resulting numerical wave is radiated directly into the structure, following a fully symmetrical route relative to source position. For example, let us consider a 3-D application excited by a continuous sinusoid frequency f_0 on H_x component. Then, it can be written that

$$H_x\big|_{pt}^{n+1/2} = H_0 \, \sin\left[2\pi f_0(n + 1/2)\Delta t\right], \tag{4.1}$$

with $pt=(i_s,j_s,k_s)$ as the presumed source point. Another very popular (and often regarded as a reference) source is the Gaussian pulse with a finite dc content and adjustable spectrum. It is centered at time step n_c as

$$E_y\big|_{pt}^{n} = E_0 \, e^{-[(n-n_c)/n_d]^2}, \tag{4.2}$$

herein for E_y, and has an inverse exponential decay of n_d time steps. To alleviate abrupt transitions during the ignition of the pulse, it is advisable to pick $n_c \geq 3n_d$. Evidently, various combinations such as bandpass Gaussian pulses or cosine waves can be devised. Moreover, time stepping may continue until all transients fade away and a steady state is achieved. Nonetheless, this is not always the case because time advancing for the late-time impulse response will eventually obligate some reflected waves to return to the source grid point *pt* where the electric or magnetic field is defined. As no extra precautions are taken to evade this situation, the hard source generates spurious vectors at *pt*. This anomaly is transmitted everywhere in the lattice via the update expressions, and the simulation is severely contaminated. In fact, such a difficulty should have been anticipated, as the hard source abstraction is not physical. Indeed, the artificial imposition of a time function at a specific location violates the rules of consistency in the FDTD region, and if ignored, it is likely to inhibit the traveling of reflected energy toward infinity and the proper modeling of the incident wave.

These remarks resulted in the development of the soft source approach, whose name is indicative for its purpose [1, 5]. Attempting to lessen the discrepancies around the source point, this scheme adds the value of the incident pulse at any time step to the update equation of the individual field component. Hence,

$$H_x|_{pt}^{n+1/2} = \text{usual FDTD update} + \text{pulse}, \quad \text{with} \quad \text{pulse} = H_0 \sin\left[2\pi f_0(n + 1/2)\Delta t\right] \quad (4.3)$$

As observed, (4.3) hardly affects the computational overhead, concurrently offering a critical improvement to initial conditions. The reason is that by means of a hard source, a propagating pulse will encounter its new value and be reflected because it looks like a metal wall to the FDTD method. On the other hand, with the soft source, the pulse will just pass through *pt* as an active part. However, one should be aware of possible instabilities because (4.3) deviates from the symmetrical relations (2.23) and (2.24) at the source point.

A second idea for suppressing the reflective nature of a hard source is to exclude it from the computer code after the pulse has decayed potentially to zero and substitute it with a regular expression. The only concern is to know in advance the spectrum of the incident wave to compute the exact time step of its removal. Such an approach though is not applicable to continuous waveforms, which create a lot of reflections—due to energy exchange with scattering objects—as it causes the abrupt pause of the excitation.

4.2.3 Scattered-Field Formulation

The underlying principle for this process stems from the separation of total electromagnetic fields into a known incident field and an unknown scattered one throughout the domain. Therefore, it is obtained that

$$\mathbf{E}^{\text{tot}} = \mathbf{E}^{\text{inc}} + \mathbf{E}^{\text{sc}} \quad \text{and} \quad \mathbf{H}^{\text{tot}} = \mathbf{H}^{\text{inc}} + \mathbf{H}^{\text{sc}} \quad (4.4)$$

Only the $\mathbf{E}^{sc}$ quantities, generated on the surface of the interacting structure, need to be computed, and no special manipulations are required for the initial wave. This attribute is deemed crucial because the outgoing scattered quantities may be more adequately attenuated by an ABC, which truncates the infinite region of calculations [2]. To derive the time-advancing equations, consider the electromagnetic scattering induced by an object embedded in free space with arbitrarily varying parameters ε, μ, σ, and ρ'. Inside the media of the scatterer, it holds

$$\nabla \times \mathbf{H}^{tot} - \sigma \mathbf{E}^{tot} = \varepsilon \frac{\partial \mathbf{E}^{tot}}{\partial t} \quad \text{and} \quad \nabla \times \mathbf{E}^{tot} + \rho' \mathbf{H}^{tot} = -\mu \frac{\partial \mathbf{H}^{tot}}{\partial t} \qquad (4.5)$$

for the total fields, whereas the incident ones satisfy free-space relations

$$\nabla \times \mathbf{H}^{inc} = \varepsilon_0 \frac{\partial \mathbf{E}^{inc}}{\partial t} \quad \text{and} \quad \nabla \times \mathbf{E}^{inc} = -\mu_0 \frac{\partial \mathbf{H}^{inc}}{\partial t} \qquad (4.6)$$

From (4.4)–(4.6) and once the suitable subtractions are executed, one acquires the final formulae that can be easily discretized and programmed in the general computer codes. These are

$$\varepsilon \frac{\partial \mathbf{E}^{sc}}{\partial t} = \nabla \times \mathbf{H}^{sc} - \sigma \mathbf{E}^{sc} - \sigma \mathbf{E}^{inc} - (\varepsilon - \varepsilon_0) \frac{\partial \mathbf{E}^{inc}}{\partial t}, \qquad (4.7a)$$

$$-\mu \frac{\partial \mathbf{H}^{sc}}{\partial t} = \nabla \times \mathbf{E}^{sc} + \rho' \mathbf{H}^{sc} + \rho' \mathbf{H}^{inc} + (\mu - \mu_0) \frac{\partial \mathbf{H}^{inc}}{\partial t} \qquad (4.7b)$$

The above expressions offer an efficient means of distinguishing the scattered field and avoid its possible cancellation by the much larger total one. After all, it is the $\mathbf{E}^{sc}$ vector that undergoes the major effects of electromagnetic interactions and the most important: $\mathbf{E}^{sc}$ components constitute the only outgoing waveforms that must be appropriately annihilated. Although the mathematical function that drives the excitation is, in most lossy dielectric EMC applications, relatively simple, the amount of supplementary arithmetic operations is likely to affect the entire algorithmic expenses. Yet, this drawback may be overcome to some extent, if incident fields are precomputed and stored in confined matrices before time marching.

In the occurrence of perfect conductors, (4.7) for the tangential electric field yield $\mathbf{E}^{sc}_{tan} = -\mathbf{E}^{inc}_{tan}$, which clearly states that at the structure's surface the total tangential field is equal to zero for all time steps. This treatment enforces the scattered wave to be locally defined, whereas the incident field follows the geometry of the scatterer in the sense of a hard source positioned at each grid point of its surface. Because $\mathbf{E}^{inc}$ is specified to propagate in the free space, it can be precisely represented by a viable space-time form. Consequently, the incident fields at all points of the perfect conductor's surface are exactly evaluated instead of deliberately imposed to travel through the mesh and probably induce large dispersion errors related to the electrical size. Even so, this obvious merit of the scattered-field approach can turn into a disadvantage when arbitrarily curved scatterers are to be investigated. Expressly, a surface shape not aligned to one of the lattice axes stipulates a careful determination of every cell point in its vicinity with an increased overhead.

The closing stage of the procedure is the choice of the incident source. Note that its argument should include the relative time delay for the coverage of the distance between the origin and the grid point at which the field is desired. A general class of schemes that satisfies these prerequisites has the form of $g[t + (\mathbf{r}_d \cdot \hat{\mathbf{r}}) / v + r_c/v]$, where v is the phase velocity and $\mathbf{r}_d$ the location vector of the calculation point. Compensation factor r_c is a reference distance that confirms the gradual impingement of the excitation on the scattering object without any sudden oscillations. To this extent, a reasonable selection for g is a Gaussian pulse with a constrained bandwidth to exclude unwanted high frequencies and save CPU time. Remember that all these adjustments must be conducted by means of the Courant criterion, which suppresses any risk of noise or instability in the volumetric region containing the dielectric material.

Overall, the scattered-field technique is a conceptually simple and practical tool that supplies fairly punctual results and trustworthy predictions of the field status for a given time interval. However, its deficiencies do restrict its algorithmic universality, notwithstanding several attempts of improvement. Actually, these disadvantages have been the basic motive for the evolution of the next intuitive method.

4.2.4   Total/Scattered-Field Formulation

A key portion in the recognition of the FDTD-oriented approaches is attributed to the total/scattered-field formulation, which can launch demanding waveforms and represent fields in the shadow regions of heavily shielded EMC structures or penetrable materials. According to its central idea, the problem space is divided into an *interior total-field region* and an *exterior scattered-field layer of cells* [1]. The former area, designated as region T, contains the propagating incident and scattered waves as total fields along with the objects of interest embedded at certain locations. Conversely, the latter area, designated as region S, enfolds region T and exclusively accommodates outgoing scattered fields free from any sources. Hence, the conflicts that could arise from the total fields in the infinite space are eliminated, and an ABC can be successfully imposed at the grid's outmost absorbing borders. Again, field decomposition is based on (4.4), with the incident vectors $\mathbf{E}^{inc}$, $\mathbf{H}^{inc}$ known everywhere in the domain and their scattered analogues $\mathbf{E}^{sc}$, $\mathbf{H}^{sc}$ representing the unknowns. The computational interconnection of the two areas is achieved by an artificial interface that plays the role of the linking surface/plane wave source, often known as the Huygens surface. This condition manages to spatially isolate the incident waves in region T while acting like a passing filter for the outgoing scattered modes, accepted to travel unobstructed in region S.

Apart from the suppression of the numerical noise, the preceding lattice division exhibits several benefits, such as its straightforward programming realization, the support of various wave envelopes or angles of incidence, and the evaluation of incident fields solely along the region T/region S surface, which does not have to follow the curvature of any object. Concentrating on the last trait, one may stress that computations will not be as precise as compared with those of the scattered-field approach. This remark is true and occasionally constitutes a point of controversy among the two techniques. For the conclusion to be complete, however, two additional factors should be taken into account: the shape of the interface and the complexity of the objects. Indeed, when an arbitrary structure is to be modeled, any attempt to determine the exact boundaries of its surface will be cumbersome if not impractical. To this scope, the total/scattered-field scheme is fully independent of these details, as it enforces the incident wave only on a simple rectangular surface several cells away to mimic its impingement from infinity. Finally, the presence of region S, as a surrounding volume of scattered waves, comprises a significant aid for near-to-far-field transformations, which weigh electric or magnetic data by free-space Green's functions in an effort to supply different types of radiation patterns.

The development of the algorithm, without loss of generality, will be based on two dimensions, because the 3-D case is a simple extension with straightforward conventions. The interface of regions T and S involves electric and magnetic field vectors that must be time-advanced via the leapfrog rationale and thus necessitates the use of different components on both sides of its border. But when spatial stencils have to cross the above interface, the fields of region T to be included in the update expressions are regarded as total ones, whereas those of region S belong to the scattered kind. To circumvent this difficulty, the already computed incident-field quantities are used in a corrective role.

Let us consider the case of a *TM* mode in the domain of Figure 4.1. Total-field quantities E_z^{tot} and E_y^{tot} are supposed to lie on the interface of regions T and S placed at $y=j_A\Delta y$. As observed to time-march $E_z|_{i,j_A}^{n,\text{tot}}$, one should obtain the $H_x|_{i,j_A\pm 1/2}^{n+1/2,\text{tot}}$ values. Because $(i,j_A+1/2)$ is in region T, the first of H_x^{tot} has already been calculated at time-step $n+1/2$. Nonetheless, no numerical evidence is available for the second magnetic total-field component located at $(i,j_A-1/2)$ because this point lies on region S and only the $H_x|_{i,j_A-1/2}^{n+1/2,\text{sc}}$ term has been updated. Based on the decomposition of (4.4), it is assumed that H_x^{tot} is analyzed as $H_x|_{i,j_A-1/2}^{n+1/2,\text{tot}} = H_x|_{i,j_A-1/2}^{n+1/2,\text{sc}} + H_x|_{i,j_A-1/2}^{n+1/2,\text{inc}}$, hence, leading, for the $(i=i_A,\ldots i_B; j=j_A)$ border side, to

$$E_z|_{i,j_A}^{n+1,\text{tot}} = ep_{i,j_A}^{E_z}\, E_z|_{i,j_A}^{n,\text{tot}} + eq_{i,j_A}^{E_z,\Delta x}\left[H_y|_{i+1/2,j_A}^{n+1/2,\text{tot}} - H_y|_{i-1/2,j_A}^{n+1/2,\text{tot}} \right]$$

$$- eq_{i,j_A}^{E_z,\Delta y}\left[H_x|_{i,j_A+1/2}^{n+1/2,\text{tot}} - H_x|_{i,j_A-1/2}^{n+1/2,\text{sc}} \right] + eq_{i,j_A}^{E_z,\Delta y}\, H_x|_{i,j_A-1/2}^{n+1/2,\text{inc}} \qquad (4.8)$$

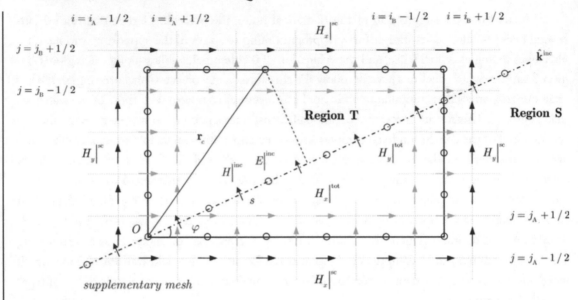

FIGURE 4.1: Component configuration and excitation launch at the interface of the total/scattered-field regions in a 2-D FDTD lattice.

The addition of the last term on the right-hand side of (4.8) transforms the value of scattered-field $H_x\big|_{i,j_A-1/2}^{n+1/2,\,sc}$ to a total-field quantity, attaining the required uniform profile that must be present on both sides of the relation. It is emphasized that this algorithmic correction does not at all affect the leapfrog process. In a similar way, the rest of the E_z terms at the $(i=i_B; j=j_A,...,j_B)$, $(i=i_A,..., i_B; j=j_B)$, and $(i=i_A; j=j_A,...,j_B)$ sides of the interface are accordingly acquired. When numerical simulation reaches the four corner points (i_A,j_A), (i_A,j_B), (i_B,j_A), and (i_B,j_B) of the mesh, E_z quantities are extracted through the overlapping of (4.8) relations, as now two (instead of one) magnetic quantities are found in region S.

Apart from the electric field though, the total/scattered excitation affects the computation of H_x^{sc} and H_y^{sc}, located half a cell outside the interface. For these quantities, E_z^{sc} values positioned half a cell to each side of the magnetic component, are required. Unfortunately, this demands the knowledge of E_z^{sc} on the border where just total fields are available. Again, solution to this trouble comes from (4.4), which yields $E_z\big|_{i,j}^{n,tot} = E_z\big|_{i,j}^{n,sc} + E_z\big|_{i,j}^{n,inc}$. For instance, $H_x\big|_{i,j_A-1/2}^{n+1/2,\,sc}$ at the external side of $(i=i_A,..., i_B; j=j_A)$ edge becomes

$$H_x\big|_{i,j_A-1/2}^{n+1/2,\,sc} = b p_{i,j_A-1/2}^{H_x} H_x\big|_{i,j_A-1/2}^{n-1/2,\,sc} - b q_{i,j_A-1/2}^{H_x,\Delta y}\left[E_z\big|_{i,j_A}^{n,tot} - E_z\big|_{i,j_A-1}^{n,sc}\right] + b q_{i,j_A-1/2}^{H_x,\Delta y} E_z\big|_{i,j_A}^{n,inc}$$

$$(4.9)$$

In this manner, the last term on the right-hand side of (4.9) transforms the value of $E_z\big|_{i,j_{\rm A}}^{\rm n,tot}$ to a scattered field component and fulfills the consistency condition on both sides of the relation. The rest of the magnetic components are similarly obtained at the external sides of the other three edges, as in the electric-field case.

Observing (4.8) and (4.9), it is deduced that for the calculation of the respective fields, one must also find the incident H_x, H_y, and E_z values. For this goal, an arbitrary plane wave whose wavevector $\mathbf{k}_{\rm inc}$ creates an angle φ with the x-axis varying from 0 to $\pi/2$ is briefly used. This excitation develops a numerical wave stemming from the lower left-hand corner of the border, $O(i_{\rm A},j_{\rm A})$, and illuminates the target as in Figure 4.1. Evidently, the resulting wave front encounters electric and magnetic field components after a time delay, which depends on the numerical phase velocity of the incident field $\tilde{\upsilon}(\varphi)$. If s is the amount of mesh cells in the direction of $\mathbf{k}_{\rm inc}$ that exist among O and the perpendicular brought to the wavevector from the location of the field quantity, under examination, the delay $t_{\rm d}$ may be expressed as

$$t_d = \frac{s\Delta}{\tilde{\upsilon}(\varphi)\Delta t}, \quad \text{with} \quad s = \mathbf{r}_c \cdot \hat{\mathbf{k}}_{\rm inc} \quad \text{and} \quad \hat{\mathbf{k}}_{\rm inc} = \cos\varphi\,\hat{\mathbf{x}} + \sin\varphi\,\hat{\mathbf{y}}, \qquad (4.10)$$

where, for simplicity, the domain has been divided into square cells of size $\Delta x = \Delta y = \Delta$ and $\hat{\mathbf{k}}_{\rm inc}$ is the unit incident wavevector. The position vector $\mathbf{r}_c$, which denotes the distance from O to the corresponding point on the interface, is given by $\mathbf{r}_c = (i_c - i_{\rm A})\hat{\mathbf{x}} + (j_c - j_{\rm B})\hat{\mathbf{y}}$, whereas indices i_c, i_c accept the usual half-cell interleaving. Nevertheless, when $\varphi > \pi/2$, the incident wave can no longer be described by the prior $\mathbf{r}_c$ because it does not originate from O anymore. In this situation, alternative coordinate origins may be determined, each one associated with the three remaining corners of the interface border.

The specific technique is proven much more economical than the scattered-field one because as soon as t_d and s are computed, incident $\mathbf{E}$ and $\mathbf{B}$ fields may be readily derived. However, as dimensionality increases, these calculations become considerable and certainly not negligible. The problem, of course, focuses on the connecting condition at the region T/region S interface, especially in the case of large unbounded applications. For this shortcoming to be mitigated, the FDTD approach is realized in two separate grids: the basic lattice (following the problem's geometry) for all computations and a supplementary 1-D excitation mesh placed along the wavevector, so that the origin of the interface coincides with one of this mesh's components. To get a more concrete idea, recall Figure 4.1 where the additional lattice is sketched with a diagonal line toward which the numerically generated wave is supposed to travel. The origin is O and the coinciding component is $E_\ell^{\rm inc}$, with ℓ being the index of the supplementary grid stencils. To combine the two lattices, the quantities of the 1-D mesh are updated in terms of the same spatial and temporal increments that govern the basic tessellation. If the main mesh comprises nonuniform cells, i.e., $\Delta x \neq \Delta y$, the stencil of the auxiliary one is set equal to the smaller value of the two. The next step is to select the point on the interface of the basic grid and calculate its distance s from O. This is geometrically accomplished by projecting the location of interest straight to the excitation mesh and then obtaining s from $\mathbf{r}_c$ and angle of incidence φ through simple trigonometric manipulations. Therefore, the incident field that corresponds at this interface position is acquired via a linear interpolation among the neighboring to point $\ell\Delta h + s$ field quantities on the supplementary lattice. The spot of interpolation is the foot of the previous projection. So, instead of the incident field components directly on the interface, one seeks their projected values on the new grid, thus saving a lot of valuable resources. Moreover, it is observed that incident wave computations entail the presence of a hard source at point $\ell-2$, such as $E|_{l-2}^{n,\text{inc}} = E_0 f(n\Delta t)$, with f being a prefixed time function. Finally, the FDTD equations for the 1-D grid are received from their 2-D analogues, as

$$E|_\ell^{n+1,\text{inc}} = E|_\ell^{n,\text{inc}} + \frac{\Delta t}{S_c \varepsilon_0 \Delta}\left(H|_{\ell-1/2}^{n+1/2,\text{inc}} - H|_{\ell+1/2}^{n+1/2,\text{inc}} \right), \qquad (4.11a)$$

$$H|_{\ell+1/2}^{n+1/2,\text{inc}} = H|_{\ell+1/2}^{n-1/2,\text{inc}} + \frac{\Delta t}{S_c \mu_0 \Delta}\left(E|_\ell^{n,\text{inc}} - E|_{\ell+1}^{n,\text{inc}} \right) \qquad (4.11b)$$

The $S_c = [\tilde{\upsilon}(\varphi = 0°) \,/\, \tilde{\upsilon}(\varphi)] \leq 1$ is a speed-up factor—originally proposed [1]—obtained by means of any iterative process. Although (4.11a) and (4.11b) remain consistent with the whole simulation in the absence of this factor, there are some cases where its use outbalances the numerical phase velocities of the incident wave in the basic and the 1-D grid. Consequently, its primary objective is to correct the slower travel of incident modes in the excitation mesh due to the identical space and time steps.

It is noteworthy to point out that the total/scattered-field formulation can be extended to any waveform described in analytical or semianalytical form. Such excitation requirements arouse at near-field scattering or interaction problems, where both the source and the object reside in the same domain but at a substantial distance from each other. Nonetheless, for real-world structures with a large electrical size, the accuracy of the models should be enhanced. In this case, aside from the appropriate amendments regarding the spatial and temporal increments of the FDTD lattice, the total/scattered-field approach itself must, also, receive a set of simple but effective adjustments, such as the use of a finer discretization for the source grid.

4.2.5   Guiding-Structure Excitation

Modeling of microwave EMC systems has experienced a rapid development in the recent decades. Typical structures of engineering interest have 3-D discontinuities, apertures, cavities, and embedded passive or active circuits that produce complex fields that cannot be easily resolved. Extensive research has proven that the self-consistent simulation of these fields demands a punctual volumetric sampling at subwavelength resolution and a reasonable choice of the excitation procedure. In particular, during the application of the FDTD-based methods to such systems, one of the paramount issues is how to incorporate the incident field and concurrently guarantee a reliable separation between excitation and reflection modes in the time domain. Without an ideal source, ABCs cannot be directly imposed on the near-end terminal (or source) plane because the reflected modes usually reach this plane before excitation is completely launched. Despite the variety of possible means, guiding-structure excitation is less straightforward than free-space sourcing. It is well-acknowledged that the confined dimensions of microwave devices allow the propagation of multimode fields with diverse spatial distributions. Because of the arbitrary curvatures and cross sections of modern components, these distributions in the transverse (relative to the propagation direction) plane may not be numerically implemented when the launch of a single mode is needed. Actually, it is the energy content of the modes that controls our selection. A mainstream (although not always practical) solution to this defect entails a sufficiently long feeding port section between the source plane and the discontinuity. Unfortunately, this treatment creates excessive computational domains that are prohibitive for 3-D problems.

As can be understood, the excitation of a microwave device opts for a rigorous sourcing process, devoid of any artificial cutoff phenomena, undesired wave superposition, and nonphysical retro-reflections from the launching plane. Existing techniques are divided into three categories, as summarized in Table 4.1.

Realization details. Let us refer to the second category and suppose that a Gaussian pulse has been established. As described, a waveform of this type retains a smooth temporal profile, whereas its spectrum is also a Gaussian pulse centered at a certain (commonly the zero) frequency. Moreover, it has to be "large" enough to contain an ample number of stencils for acceptable resolutions. This requirement is, of course, fulfilled by means of the suitable spatial/temporal increments that certify that the initial launch time of the excitation is small but not abrupt. For illustration, a voltage source in a microstrip line stipulates that the vertical electric field, herein E_z, is imposed in a rectangular region beside the port of the structure. Attempting to apply the previous ideas, one might come up with a natural question: what is the treatment of the other field quantities on the source wall? Relative scientific work unveils that these components are usually set to *zero*. However, a frustrating side effect of this convention is the appearance of a sharp magnetic field tangential to the source wall. This, in turn, yields a prominent distortion of the incident wave, which is translated to a magnitude reduction and a negative wave remnant in the pulse. To suppress distortions, a magnetic wall at the excitation plane involving only E_x and E_z components, with the tangential magnetic ones placed at an offset of $\pm \Delta y / 2$, is constructed. If this substitute wall is enforced directly on the source plane, the remaining quantities can be immediately evaluated by the leapfrog scheme.

	TABLE 4.1: Main categories of guiding-structure excitation
A	At the initial time instant, the electromagnetic field is defined inside the entire computational space of the discretized model. This approach is basically used for the analysis of microwave resonators, whereas field distribution at the early moments depends on the experience of the programmer who, for instance, can select a special resonant mode with a set of eigenvalues that will be promptly calculated.
B	The second type of excitation is the incorporation of an electromagnetic field pulse of finite length in space and time. Through this process, arbitrary n-port devices and their individual features can be thoroughly investigated, if the incident, reflected, and transmitted pulses are available for separation. Because of the wideband nature of such pulses—related to their envelope and waveform—transmission properties are easily determined for a preselected frequency window spanning in one analysis cycle.
C	The most popular technique, however, is the one that defines a harmonic field oscillation on a boundary or transverse plane of the structure. It is stressed though that the specific approach is only applicable to classes of problems where a fairly small amount of reflection occurs in the whole space and no disturbance from the inherent discontinuities affects the neighborhood of the source plane. This necessity is usually attained by a significant buffer length of waveguide located between the excitation plane and the first obstacle under study.

Alike in aim but different in realization is the third category. Now, the incident mode is introduced via an electric wall behind the source location, so that the mode propagates only in one direction. The spatial distribution of the pulse must be determined on the launching plane, whereas all of its tangential components should depict real solutions of Maxwell's equations that most closely match the theoretical equivalent of the mode. Ideally, the use of the dominant distribution is preferred, although this is generally not known with adequate accuracy. Furthermore, because of the zero impedance of the setup, reflection from possible discontinuities results in waves that return to the launching plane, interact with the incident fields, and create serious inconsistencies. Therefore, the excitation must be, somehow, "switched off" before any reflected waves reach its location, i.e., the replacement of the corresponding plane by a set of regular cells. Apart from the duration of the pulse, its bandwidth should be also meticulously selected, so as not to contain notable energy content below the cutoff frequency. These deductions point out the high sensitivity of the third category to pulse features and opt for a sufficiently prolonged section to intervene between the source plane and the closest discontinuity. So, a suitable space is provided for the complete propagation and decay of the incident wave earlier than the arrival of any reflection. Despite its efficiency, this abstraction is not always feasible from a computer resources point of view. An alternative scheme that seems to overcome this restriction is discussed below.

Additional improvements. The separation of incident and reflected waves during the numerical analysis of guiding systems is regarded as a vital aspect. For microstrip discontinuities, ordinary excitations induce dc distortions near the incident plane via the boundary conditions. These parasites not only hinder the correct function of the overall method but also influence the traveling waves as well. A practical way to solve this problem is to move the source plane several cells inside the terminal plane of the system and add the excitation as an extra term in the proper update equation. Thus, the lattice is divided into two distinct areas by a fictitious surface. Observing Figure 4.2, it is concluded that the area on the right-hand side of the source plane—containing the discontinuity —involves total fields only. Because the evolution of the original wave is conducted toward one direction, the region on the left-hand side of the interface will exclusively accept the reflected part of the simulation. Hence, the two fields are fully separated, whereas sufficient annihilation of the latter can be attained by the appropriate ABC, now simultaneously applied with the excitation [3]. Apparently, the pulse plane should act as a connecting condition with a triple role: (a) the precise spatial derivative approximation through neighboring values at the interface, (b) the profitable generation of the desired incident wave, and (c) the transparent action for all outgoing reflected modes.

Returning to Figure 4.2, the whole domain is separated in two subdomains by means of the source plane at j_{sc}. For a given excitation, $E_z\big|_{i,\,j_{sc},\,k}^{n,\,\mathrm{inc}}$, the update expression on the interface becomes

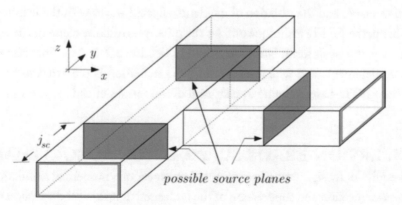

possible source planes

FIGURE 4.2: Division of a general microwave system containing a discontinuity into two parts by the transverse source plane.

$$E_z|_{i,j_{sc},k}^{n+1} = ep_{i,j_{sc},k}^{E_z} E_z|_{i,j_{sc},k}^{n} + eq_{i,j_{sc},k}^{E_z,\Delta x} \left[H_y|_{i+1/2,j_{sc},k}^{n+1/2} - H_y|_{i-1/2,j_{sc},k}^{n+1/2} \right]$$

$$- eq_{i,j_{sc},k}^{E_z,\Delta y} \left[H_x|_{i,j_{sc}+1/2,k}^{n+1/2} - H_x|_{i,j_{sc}-1/2,k}^{n+1/2} \right] + E_z|_{i,j_{sc},k}^{n,inc} \qquad (4.12)$$

Note that the equations for the other quantities remain unchanged on the source plane. Because of its universal nature, (4.12) can handle all types of EMC guiding structures. Most importantly, there is no need to zero our original source or remove it from the mesh nor must we lengthen the distance between the pulse plane and the first obstacle. In contrast, the excitation may now operate on a constant basis and close to any structural oddity. These merits lead to remarkable savings without disturbing the validity of the results. After all, the omitted supplementary section is just a simple artifact that does not carry any meaningful data.

4.2.6   Alternative Configurations

Because of the multitude of time-domain methodologies, there are some cases where the excitation scheme could receive different forms particularly when the incident field has to be launched in a region discretized by a non-FDTD-oriented technique. In situations of this kind, the analysis should follow the distinct rules that hold for the respective method, which are normally easy to comprehend and implement. Therefore, initial field conditions for the FIT are inserted in the domain directly through the basic (2.99), (2.100), or (2.123) as additional terms in the corresponding matrix vectors. No other assumption has to be conducted because the general character of these expressions can handle almost any type of wave front. Similarly, the TLM algorithm inserts the excitation function in the computational space via (2.127), either for shunt (2.74) and (2.75) or series (2.76) and (2.77) nodes. It is to be mentioned that both methods share the same structural concepts for the excitation planes and the isolation of incident/reflected waves with the leapfrog approaches. On the other hand, the FETD technique utilizes the principles of finite elements to excite the modeled setup, namely, the weak formulation described in Section 3.2.1. As a final remark, for hybrid time-domain arrangements and whenever feasible, the excitation is preferred to be configured in the region of the FDTD-based scheme to take avail of the various effective means presented in this section.

4.3   DISPERSION-ERROR SUPPRESSION ALGORITHMS

The next and equivalently significant implementation aspect of a numerical simulation in the time domain is the quantification and suppression of the undesired nonphysical dispersion errors that can lead to totally misleading results, despite the general efficacy of the computational scheme. Based on the theoretical framework of Chapters 2 and 3, it is easily deduced that, among the diverse techniques, the FDTD-oriented ones are the most vulnerable to this structural deficiency, with the FIT, the WENO-TD, the PSTD as well as the MRTD schemes being sufficiently more resilient. This is, actually, reflected in the impressive research regarding the discretization policy and tessellation topology of the former approaches in an attempt to derive viable conceptions for the drastic anni-hilation of the shortcoming [6–20]. As a consequence, this section refers basically to the construc-tion of dispersion-reduction formulations for the FDTD-like methods in conjunction with their application to complicated EMC problems. Toward this direction, the investigation concentrates on two robust and popular methodologies: the direct embedding of artificial anisotropy and the development of numerical filtering along with angle-optimized schemes.

4.3.1   Incorporation of Artificial Anisotropy

The present method is founded on the systematic acceleration of wave propagation by introducing a set of optimal anisotropy parameters into the FDTD algorithm [10]. For their evaluation, a detailed Fourier-mode analysis is implemented, which covers several EMC cases and achieves a decrease of dispersion errors by a factor of 2 to 7, depending on the shape of the elementary cell. Although the whole process refers to a simple frequency, the efficiency of this scheme may also successfully cover various wideband inhomogeneous problems.

Dispersion reduction in two dimensions. Assume an electrically anisotropic medium with its relative permittivity diagonal tensor defined as $\bar{\bar{\varepsilon}}_r = \text{diag}\{\varepsilon_x, \varepsilon_y\}$ and the *TE* mode being respon-sible for the electromagnetic radiation in the domain. Following the guidelines of Section 2.2.4, the stability condition and the dispersion relation of this configuration become

$$\upsilon \Delta t \leq \left[\frac{1}{\varepsilon_y (\Delta x)^2} + \frac{1}{\varepsilon_x (\Delta y)^2} \right]^{-1/2}, \tag{4.13a}$$

$$\frac{1}{(\upsilon \Delta t)^2} \sin^2 \left(\frac{\omega \Delta t}{2} \right) = \frac{1}{\varepsilon_y (\Delta x)^2} \sin^2 \left(\frac{\tilde{k}_x \Delta x}{2} \right) + \frac{1}{\varepsilon_x (\Delta y)^2} \sin^2 \left(\frac{\tilde{k}_y \Delta y}{2} \right), \tag{4.13b}$$

for $\tilde{k}_x, \tilde{k}_y$ the components of the numerical wavevector $\tilde{k}$ and $v = 1/\sqrt{\mu \varepsilon}$. The action of (4.13b) does not affect the *TM* mode, whose corresponding expressions are dually derived through the sub-stitution, in (4.13), of the relative permittivities with the relative permeabilities. Let us, now, for-mulate the dispersion relation in a nondimensionalized (independent of physical dimensions) way. Therefore, the shape parameter Z of the FDTD cell and the spatial resolution coefficient R in terms of wavelength λ and cell size are defined as

$$Z = \Delta x / \Delta y \text{ and } R = \lambda / \sqrt{(\Delta x)^2 + (\Delta y)^2} \qquad (4.14)$$

Moreover, through the downgrading weight $q \le 1$, the Courant criterion of (4.13a) is transformed to

$$v \Delta t = q \left[\frac{1}{\varepsilon_y (\Delta x)^2} + \frac{1}{\varepsilon_x (\Delta y)^2} \right]^{-1/2} \Rightarrow v \Delta t = q \frac{\Delta y Z \sqrt{\varepsilon_x \varepsilon_y}}{\sqrt{\varepsilon_x + \varepsilon_y Z^2}} \qquad (4.15)$$

Further use of (4.14) gives $\lambda = R \Delta y \sqrt{1 + Z^2}$ and because $\omega = 2\pi v / \lambda$, it can be obtained that

$$\frac{\omega \Delta t}{2} = \pi \frac{v \Delta t}{\lambda} = \frac{q \pi Z \sqrt{\varepsilon_x \varepsilon_y}}{R \sqrt{1 + Z^2} \sqrt{\varepsilon_x + \varepsilon_y Z^2}} \qquad (4.16)$$

If $\tilde{\lambda}$ is the numerical wavelength and $\tilde{k} = 2\pi / \tilde{\lambda}$, then $\tilde{k} \Delta y = (\lambda / \tilde{\lambda})(2\pi / R \Delta y \sqrt{1 + Z^2})$. The critical quantity, here, is the ratio of velocities $A \equiv \tilde{v} / v = \tilde{\lambda} / \lambda$, which finally yields $\tilde{k} \Delta y = (2\pi / A R \sqrt{1 + Z^2})$. Replacement of A and (4.14)–(4.16) modifies the dispersion relation of (4.13b) to the subsequent equation

$$\frac{\varepsilon_x + \varepsilon_y Z^2}{\varepsilon_x \varepsilon_y q^2 Z^2} \sin^2 \left(\frac{q \pi Z \sqrt{\varepsilon_x \varepsilon_y}}{R \sqrt{1 + Z^2} \sqrt{\varepsilon_x + \varepsilon_y Z^2}} \right) = \frac{1}{\varepsilon_x} \sin^2 \left(\frac{\pi \sin \theta}{A R \sqrt{1 + Z^2}} \right)$$

$$+ \frac{1}{\varepsilon_y Z^2} \sin^2 \left(\frac{\pi Z \cos \theta}{A R \sqrt{1 + Z^2}} \right), \qquad (4.17)$$

with $\tilde{k}_x = \tilde{k} \cos \theta$ and $\tilde{k}_y = \tilde{k} \sin \theta$, which is easily solved for A, given a set of coefficients Z, q, R, ε_x, ε_y, and θ. To calculate the optimal correction parameters, one should initially evaluate ε_x and ε_y, which guarantee $A = 1$ along the coordinate directions of $\theta = 0°$ and $\theta = 90°$. Thus, setting $A = 1$, $\theta = 90°$, and isolating ε_x in the right-hand side of (4.17) yield

$$\varepsilon_x = \frac{\varepsilon_x \varepsilon_y q^2 Z^2}{\varepsilon_x + \varepsilon_y Z^2} \left[\sin^2 \left(\frac{\pi}{R \sqrt{1 + Z^2}} \right) \middle/ \sin^2 \left(\frac{q \pi Z \sqrt{\varepsilon_x \varepsilon_y}}{R \sqrt{1 + Z^2} \sqrt{\varepsilon_x + \varepsilon_y Z^2}} \right) \right] \qquad (4.18a)$$

Similarly, if $A = 1$ and $\theta = 0°$, ε_y is acquired from

$$\varepsilon_y = \frac{\varepsilon_x \varepsilon_y q^2}{\varepsilon_x + \varepsilon_y Z^2} \left[\sin^2 \left(\frac{\pi Z}{R \sqrt{1 + Z^2}} \right) \middle/ \sin^2 \left(\frac{q \pi Z \sqrt{\varepsilon_x \varepsilon_y}}{R \sqrt{1 + Z^2} \sqrt{\varepsilon_x + \varepsilon_y Z^2}} \right) \right] \qquad (4.18b)$$

Division of (4.18a) by (4.18b) results in

$$\frac{\varepsilon_x}{\varepsilon_y} = Z^2 \left[\sin^2 \left(\frac{\pi}{R\sqrt{1 + Z^2}} \right) \middle/ \sin^2 \left(\frac{\pi Z}{R\sqrt{1 + Z^2}} \right) \right] \qquad (4.19)$$

Furthermore, if we solve (4.19) for $\boldsymbol{\varepsilon}_x$ and insert the outcome into (4.18a),

$$\varepsilon_y = \frac{R^2(1 + Z^2)}{\pi^2 q^2 Z^2} \left(1 + \frac{b_2^2}{b_1^2} \right) \arcsin \left(\frac{q b_1 b_2}{\sqrt{b_1^2 + b_2^2}} \right)^2, \qquad (4.20)$$

with $b_1 = \sin\left[\pi/ \left(R\sqrt{1 + Z^2} \right) \right]$ and $b_2 = \sin\left[\pi Z/ \left(R\sqrt{1 + Z^2} \right) \right]$. So, $\boldsymbol{\varepsilon}_x$ can then be directly found from (4.19). In essence, the basic goal in this technique is to specify the maximum deviation of A from unity and then redefine ε_x and ε_y to slow down the wave velocity by half of this maximum. A crucial remark to this step is that the maximum A does not depend much on Z and not at all on q [10]. This seems to be the ideal selection of R in units of diagonal length of the cell. Hence, at first place, an acceptable approximation of the maximum A is calculated via $q = Z = 1$ in (4.17), (4.19), (4.20), and next, the value of $\theta = 45°$ is set back in (4.17) to derive

$$A_{\max} = \pi \left\{ 2R \arcsin \left[\frac{\sin(\pi/ R\sqrt{2})}{\sqrt{2}} \right] \right\}^{-1} \qquad (4.21)$$

In this sense, the maximum deviation of A from unity is now $b_3 = A_{\max} - 1$. Setting $A = 1 - 0.5b_3$ along axial directions provides the optimal ε_x and ε_y parameters. To substantiate the worth of (4.21), a simple EMC problem with an H_z excitation is taken into account to measure the numerical phase velocity. This is attained by looking for the time instants during which a certain part of the incident waveform reaches two preset monitoring points being situated at distances $\mathbf{r}$ and $2\mathbf{r}$ from the source, respectively. Knowing the spatial difference between the two points and the required time, velocity is promptly computed. Figure 4.3 shows the dispersion-error diminution for the relatively coarse resolution of $Z = R = 5$, $q = 0.99$.

Extension to three dimensions. Suppose that the aforesaid material is both electrically and magnetically anisotropic, i.e., its relative constitutive tensors are $\bar{\bar{\varepsilon}}_r = \mathrm{diag}\{\varepsilon_x, \varepsilon_y, \varepsilon_z\}$, $\bar{\bar{\mu}}_r = \mathrm{diag}\{\mu_x, \mu_y, \mu_z\}$. Using the Fourier mode of (2.32) as an eigenfunction of the discrete FDTD differential operators, it is obtained

$$\frac{\partial F}{\partial t} \rightarrow j \frac{2}{\Delta t} \sin \left(\frac{\omega \Delta t}{2} \right) F \equiv j K_t F \text{ and } \frac{\partial F}{\partial u} \rightarrow -j \frac{2}{\Delta u} \sin \left(\frac{\tilde{k}_u u}{2} \right) F \equiv j K_u F, \qquad (4.22)$$

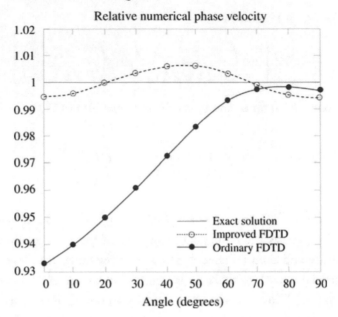

FIGURE 4.3: Relative numerical phase velocity for an ordinary and an improved 2-D FDTD simulation.

where $u = x, y, z$ and $F = E, H$. Through (4.22), Maxwell's curl equations receive the following form

$$\mu_0 K_t \bar{\bar{\mu}}\, \mathbf{H} = \bar{\bar{\mathbf{K}}}\mathbf{E} \\ \varepsilon_0 K_t \bar{\bar{\varepsilon}}\, \mathbf{E} = -\bar{\bar{\mathbf{K}}}\mathbf{H}, \qquad \text{with} \qquad \bar{\bar{\mathbf{K}}} = \begin{bmatrix} 0 & K_z & -K_y \\ -K_z & 0 & K_x \\ K_y & -K_x & 0 \end{bmatrix} \qquad (4.23)$$

Merging Ampère's and Faraday's laws in (4.23), field vector $\mathbf{E}$ must satisfy the eigenvalue problem of

$$\bar{\bar{\varepsilon}}^{-1}\bar{\bar{\mathbf{K}}}\bar{\bar{\mu}}^{-1}\bar{\bar{\mathbf{K}}}\mathbf{E} = \Lambda\mathbf{E} = -(K_t^2/v^2)\mathbf{E} \qquad (4.24)$$

Generally, matrix Λ has two different nonzero eigenvalues. If, additionally, $\bar{\bar{\varepsilon}} = \bar{\bar{\mu}}$, (4.24) has a double eigenvalue given by $-(\varepsilon_y^{-1}\varepsilon_z^{-1}K_x^2 + \varepsilon_x^{-1}\varepsilon_z^{-1}K_y^2 + \varepsilon_x^{-1}\varepsilon_y^{-1}K_z^2)$, and the dispersion relation becomes

$$\frac{1}{(v\Delta t)^2}\sin^2\left(\frac{\omega\Delta t}{2}\right) = \frac{1}{\varepsilon_y\varepsilon_z(\Delta x)^2}\sin^2\left(\frac{\tilde{k}_x\Delta x}{2}\right)$$

$$+ \frac{1}{\varepsilon_x\varepsilon_z(\Delta y)^2}\sin^2\left(\frac{\tilde{k}_y\Delta y}{2}\right) + \frac{1}{\varepsilon_x\varepsilon_y(\Delta z)^2}\sin^2\left(\frac{\tilde{k}_z\Delta z}{2}\right), \qquad (4.25a)$$

with a stability criterion of

$$v\Delta t \leq \left[\frac{1}{\varepsilon_y \varepsilon_z (\Delta x)^2} + \frac{1}{\varepsilon_x \varepsilon_z (\Delta y)^2} + \frac{1}{\varepsilon_x \varepsilon_y (\Delta z)^2} \right]^{-1/2} \qquad (4.25b)$$

Let us, again, define $Z_y = \Delta x / \Delta y$, $Z_z = \Delta x / \Delta z$ and $R = \lambda / \sqrt{(\Delta x)^2 + (\Delta y)^2 + (\Delta z)^2}$ as the dimensionless parameters of the approach and $q \leq 1$ as the Courant downgrading coefficient [10]. Like in the 2-D formulation, the appropriate algebra leads to

$$\frac{\varepsilon_x + \varepsilon_y Z_y^2 + \varepsilon_z Z_z^2}{\varepsilon_x \varepsilon_y \varepsilon_z q^2} \sin^2 \left(\frac{q\pi \sqrt{\varepsilon_x \varepsilon_y \varepsilon_z}}{R\sqrt{1 + Z_y^{-2} + Z_z^{-2}} \sqrt{\varepsilon_x + \varepsilon_y Z_y^2 + \varepsilon_z Z_z^2}} \right)$$

$$= \frac{1}{\varepsilon_y \varepsilon_z} \sin^2 \left(\frac{\pi \sin\theta \cos\varphi}{AR\sqrt{1 + Z_y^{-2} + Z_z^{-2}}} \right)$$

$$+ \frac{Z_y^2}{\varepsilon_x \varepsilon_z} \sin^2 \left(\frac{\pi \sin\theta \sin\varphi}{ARZ_y \sqrt{1 + Z_y^{-2} + Z_z^{-2}}} \right)$$

$$+ \frac{Z_z^2}{\varepsilon_x \varepsilon_y} \sin^2 \left(\frac{\pi \cos\theta}{ARZ_z \sqrt{1 + Z_y^{-2} + Z_z^{-2}}} \right), \qquad (4.26)$$

where θ and φ are the polar and azimuth propagation angles, respectively, and A is the ratio of numerical to physical velocity in a certain direction. The improvement of (4.26) is verified via the next example in which the H_z component acts as the excitation standing for a point magnetic dipole source. Figure 4.4 compares the phase velocities with and without the anisotropy correction on the $x = 0$ plane. A coarse lattice of $R = 5$ is used, and cells have $Z_y = 1.5$ and $Z_z = 3$. Clearly, the dispersion error is significantly suppressed and fluctuates within a $\pm 1\%$ range from the ideal value instead of the much larger deviations of the simple FDTD method.

Because a major advantage of time-domain algorithms is their ability to conduct wideband calculations, one might be skeptical about the efficiency of artificial anisotropy as a general purpose tool. Its optimal performance at a single frequency is, indeed, a restrictive factor. Nevertheless, inspecting the dispersion relation (4.26) with constant anisotropy parameters and a varying resolution, it is deduced that error suppression remains optimal in an adequately broad range of frequencies around the central value of R needed for the whole arrangement. Thus, for frequencies higher than the central one, the maximum dispersion error is always smaller than its ordinary FDTD counterpart. A rule of thumb recommends the incorporation of anisotropy parameters associated with the highest frequency of a given band and not the central value. Finally, if the initial frequency starts

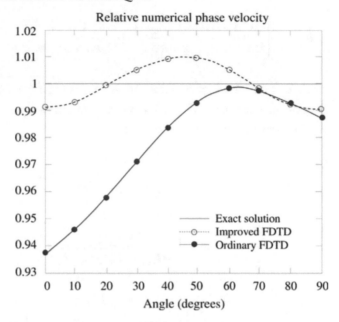

FIGURE 4.4: Relative numerical phase velocity for an ordinary and an improved 3-D FDTD simulation.

in the area of dc, the usual FDTD algorithm could be safely used for the lowest subband and the corrected one for the higher subbands.

4.3.2   Numerical Filtering and Angle-Optimized Schemes

The second category of techniques for the reduction of numerical dispersion lies on simple filtering notions that modify the traditional FDTD coefficients. To accomplish a zero average phase error at a specified frequency, the wave velocity is changed, so that the average numerical wavenumber (overall propagation angles) is equal to its natural value. Through a careful wavelength selection, sizeable enhancements for a fairly extensive bandwidth can be derived without any complexity to the entire simulation. This treatment, also, entails coarser resolutions for constant accuracy thresholds, thus leading to considerable system savings.

   Optimal FDTD coefficients. The extraction of the advanced approximating coefficients starts from a 2-D *TM* polarized steady-state plane wave traveling in an open region occupied by a homogeneous and isotropic medium. The analytical solution of the problem is $\mathrm{E} = (-\sin\theta\,\hat{x} + \cos\theta\,\hat{y})$ $\exp[\,j(\omega t - xk\cos\theta - yk\sin\theta)$, whereas for a stable modeling in terms of a cubic mesh ($\Delta x = \Delta y = \Delta$), the Courant limit requires that $0 < L_{\mathrm{C}} \leq 1$ with $L_{\mathrm{C}} = v\Delta t\sqrt{2}/\Delta$. In this context, the numerical dispersion relation reads

$$\frac{2}{L_C^2}\sin^2\left(\frac{2\Delta}{\omega\Delta t}\right) = \sin^2\left(\frac{\tilde{k}\Delta}{k\lambda}\pi\cos\theta\right) + \sin^2\left(\frac{\tilde{k}\Delta}{k\lambda}\pi\sin\theta\right) \tag{4.27}$$

To solve an equation such as (4.27) for $\tilde{k}$ at a given angle $\theta \in [0, \pi/2]$, it is realized that (a) the phase is $\pi/2$ periodic and (b) the best results are always derived at $\theta = \pi/4$, with the worst ones produced at $\theta = 0$ and $\theta = \pi/2$. Exact solution is achieved when $\theta = \pi/4$ and $L_C = 1$. Over all other angles and Courant values, $\tilde{k} > k$ and any effort to lower L_C enforce the phase error curve to deviate even farther from zero. Because of this shortcoming, it is desirable to decrease the numerical wavenumber, so that its average at all θ equals to the physical value k. A prompt way to fulfill this aim is to replace the numerical phase velocity $\tilde{v}$ with a somewhat larger compensated propagation speed $\tilde{v}_s$ as $\tilde{v}_s = \tilde{v}_r$ $\tilde{v}$. Then, the optimal $\tilde{v}_s$ is selected to confirm that the modified dispersion curve $\tilde{k}_s(\theta)$ coincides with k in the average sense, namely,

$$k = \frac{1}{2\pi}\int_0^{2\pi}\tilde{k}_s(\theta)\,\mathrm{d}\theta \tag{4.28}$$

By increasing $\tilde{v}$ to $\tilde{v}_s$, the dispersion curve is additionally shifted and seems to become more isotropic. It can be proven that for the demanding situation of $\Delta/\lambda < 0.2$, $\tilde{v}_r$ is given by

$$\tilde{v}_r = \frac{\sqrt{2}\sin\left[\pi L_C\Delta/(\sqrt{2}\lambda)\right]}{L_C\sqrt{\sin^2[\pi(\Delta/\lambda)\cos\theta] + \sin^2[\pi(\Delta/\lambda)\sin\theta]}}, \tag{4.29a}$$

at $\theta = \pi/8$. The analogous relation for the 3-D case takes the form of

$$\tilde{v}_r = \frac{\sqrt{3}\sin\left[\pi L_C\Delta/(\sqrt{3}\lambda)\right]}{L_C\sqrt{\sin^2[\pi(\Delta/\lambda)\cos\theta] + \sin^2[\pi(\Delta/\lambda)\sin\theta\cos\varphi] + \sin^2[\pi(\Delta/\lambda)\sin\theta\sin\varphi]}}, \tag{4.29b}$$

at $\theta = \pi/8$ and $\varphi = 4\pi/27$. Despite their single-frequency formulation, expressions (4.29a) and (4.29b) manage to drop off the phase error over a large bandwidth for several EMC (shielding or immunity) applications.

Angle-optimized FDTD method. Extending the ideas of numerical filtering, this paragraph deals with an alternative that can offer a sufficient alleviation of dispersion errors through the control of its angular sector [14]. The main update equations that will be implemented, herein, include two extra degrees of freedom χ_1 and χ_2 that need to be found. In particular, for a 2-D *TE* wave in the frequency domain

$$-j\omega\varepsilon\, E_x|_{I,J} = \frac{1}{\Delta y}\left[\chi_1 + \chi_2\left(\frac{\omega\Delta y}{c}\right)^2\right]\left(H_z|_{I,J+1/2} - H_z|_{I,J-1/2}\right), \qquad (4.30a)$$

$$-j\omega\varepsilon\, E_y|_{I,J} = \frac{1}{\Delta x}\left[\chi_1 + \chi_2\left(\frac{\omega\Delta y}{c}\right)^2\right]\left[H_z|_{I-1/2,J} - H_z|_{I+1/2,J}\right], \qquad (4.30b)$$

$$-j\omega\mu\, H_z|_{I,J} = \frac{1}{\Delta y}\left[\chi_1 + \chi_2\left(\frac{\omega\Delta y}{c}\right)^2\right]\left(E_x|_{I,J+1/2} - E_x|_{I,J-1/2}\right)$$

$$+ \frac{1}{\Delta x}\left[\chi_1 + \chi_2\left(\frac{\omega\Delta x}{c}\right)^2\right]\left(E_y|_{I-1/2,J} - E_y|_{I-1/2,J}\right), \qquad (4.30c)$$

where, just for (4.30a)–(4.30c), the capital indices have been used to avoid confusions with the complex root j. Factors χ_1 and χ_2 regulate phase—correcting time derivatives of higher order—and therefore are constructed to impose artificial dispersion effects that balance lattice deficiencies. Provided that the medium is homogeneous, (4.30a)–(4.30c) are transformed in the time domain to obtain

$$E_x|_{i,j}^{n+1} = E_x|_{i,j}^{n} + \frac{\Delta t}{\varepsilon\Delta y}\left[G_{yx}\left(H_z|_{i,j+1/2}^{n+1/2} - H_z|_{i,j-1/2}^{n+1/2}\right) + \chi_2\left(H_z|_{i,j-3/2}^{n+1/2} - H_z|_{i,j+3/2}^{n+1/2}\right)\right.$$

$$\left. + \chi_2\Delta_{yx}^2\left(H_z|_{i+1,j-1/2}^{n+1/2} - H_z|_{i+1,j+1/2}^{n+1/2} + H_z|_{i-1,j-1/2}^{n+1/2} - H_z|_{i-1,j+1/2}^{n+1/2}\right)\right], \qquad (4.31a)$$

$$E_y|_{i,j}^{n+1} = E_y|_{i,j}^{n} + \frac{\Delta t}{\varepsilon\Delta x}\left[G_{xy}\left(H_z|_{i-1/2,j}^{n+1/2} - H_z|_{i+1/2,j}^{n+1/2}\right) + \chi_2\left(H_z|_{i+3/2,j}^{n+1/2} - H_z|_{i-3/2,j}^{n+1/2}\right)\right.$$

$$\left. + \chi_2\Delta_{yx}^2\left(H_z|_{i+1/2,j+1}^{n+1/2} - H_z|_{i-1/2,j+1}^{n+1/2} + H_z|_{i+1/2,j-1}^{n+1/2} - H_z|_{i-1/2,j-1}^{n+1/2}\right)\right], \qquad (4.31b)$$

$$H_z|_{i,j}^{n+1/2} = H_z|_{i,j}^{n-1/2} + \frac{\Delta t}{\mu\Delta y}\left[G_{yx}\left(E_x|_{i,j+1/2}^{n} - E_x|_{i,j-1/2}^{n}\right) + \chi_2\left(E_x|_{i,j-3/2}^{n} - E_x|_{i,j+3/2}^{n}\right)\right.$$

$$\left. + \chi_2\Delta_{yx}^2\left(E_x|_{i-1,j-1/2}^{n} - E_x|_{i-1,j+1/2}^{n} + E_x|_{i+1,j-1/2}^{n} - E_x|_{i+1,j+1/2}^{n}\right)\right]$$

$$+ \frac{\Delta t}{\mu\Delta x}\left[G_{xy}\left(E_y|_{i-1/2,j}^{n} - E_y|_{i+1/2,j}^{n}\right) + \chi_2\left(E_y|_{i+3/2,j}^{n} - E_y|_{i-3/2,j}^{n}\right)\right.$$

$$\left. + \chi_2\Delta_{xy}^2\left(E_y|_{i+1/2,j+1}^{n} - E_y|_{i-1/2,j+1}^{n} + E_y|_{i+1/2,j-1}^{n} - E_y|_{i-1/2,j-1}^{n}\right)\right], \qquad (4.31c)$$

with $G_{uv} = \chi_1 + \chi_2(3 + 2\Delta_{uv}^2)$ and $\Delta_{uv} = \Delta u / \Delta v$ for $u, v = x, y, z$. It is stressed that the 3-D case shares a similar establishment except for introducing more degrees of freedom χ to consider all mesh directions.

Next, analysis focuses on the selection of optimal angles and frequency bandwidths that enable dispersion error suppression together with (4.31). Presuming a monochromatic wave so that $E_x|^n = \mathcal{E}_x \exp(j\omega n\Delta t)$, $E_y|^n = \mathcal{E}_y \exp(j\omega n\Delta t)$, and $H_z|^{z+1/2} = \mathcal{H}_z \exp[(j\omega(n+1/2)\Delta t]$, (4.31a)–(4.31c) turn into

$$\sin\left(\frac{\omega\Delta t}{2}\right)\mathcal{E}_x = \frac{\Delta t}{\varepsilon\Delta y}R_x\mathcal{H}_z, \quad \sin\left(\frac{\omega\Delta t}{2}\right)\mathcal{E}_y = -\frac{\Delta t}{\varepsilon\Delta x}R_y\mathcal{H}_z, \quad (4.32a)$$

$$\sin\left(\frac{\omega\Delta t}{2}\right)\mathcal{H}_z = \frac{\Delta t}{\mu\Delta y}R_x\mathcal{E}_x - \frac{\Delta t}{\mu\Delta x}R_y\mathcal{E}_y \quad (4.32b)$$

$$R_x = \chi_2 \sin\left(3\tilde{k}_y\Delta y/2\right) - \left[\chi_1 + 3\chi_2 + 4\chi_2\Delta_{yx}^2 \sin^2\left(\tilde{k}_x\Delta x/2\right)\right]\sin\left(\tilde{k}_y\Delta y/2\right),$$

with

$$R_y = \chi_2 \sin\left(3\tilde{k}_x\Delta x/2\right) - \left[\chi_1 + 3\chi_2 + 4\chi_2\Delta_{xy}^2 \sin^2\left(\tilde{k}_y\Delta y/2\right)\right]\sin\left(\tilde{k}_x\Delta x/2\right)$$

Substitution of (4.32a) into (4.32b), for a square FDTD grid, concludes to the dispersion relation

$$\left[\frac{\sqrt{2}}{L_C}\sin\left(\frac{L_C\tilde{k}\Delta}{2\sqrt{2}}\right)\right]^2 = R_x^2 + R_y^2 \Rightarrow err = \left[\frac{\sqrt{2}}{L_C}\sin\left(\frac{L_C\tilde{k}\Delta}{2\sqrt{2}}\right)\right]^2 - \left(R_x^2 + R_y^2\right) \quad (4.33)$$

For a predetermined pair of incidence angle θ and L_C, the phase velocity error may be zeroed if, moreover, χ_1 and χ_2 are appropriately picked. Hence, one is able to tackle the equivalent problem of $err = 0$. Using $\tilde{k}\Delta = 2\pi\Delta/\lambda$, (4.33) is expanded in series as a function of $Q = \Delta/\lambda$,

$$err = \left(1 - \chi_1^2\right)\pi^2Q^2 + \frac{\pi^4Q^4}{12}\left\{\chi_1[3(\chi_1 - 32\chi_2) + \chi_1\cos(4\theta)] - 2L_C^2\right\}$$

$$+ \frac{\pi^6Q^6}{180}\left[720\chi_1\chi_2 - 5\chi_1^2 - 2880\chi_2^2 + 2L_C^4 - 3\chi_1(\chi_1 - 80\chi_2)\cos(4\theta)\right] \quad (4.34)$$

To acquire χ_1 and χ_2, the first two terms in e (4.34) are zeroed, i.e., $\chi_1 = 1$ and $\chi_2 = \left[3 - 2L_C^2 + \cos(4\theta)\right]$. For $0 \leq \chi_2 \leq 1/24$, the minimum L_C for a stable simulation is 0.75. This deduction along with the role of θ in χ_2 is utilized for the annihilation of dispersion errors at any angle in the domain [14].

Despite the simplicity of monochromatic problems, practical EMC applications involve broadband excitations, an issue that obligates the optimization of the dispersion error around a central frequency for diverse incidence angles. Actually, monomials Q^m in (4.34) can be viewed as a basis of an infinite-dimensional linear space, and (4.34) can be considered as an expansion of the dispersion error in this space. So, according to algebraic analysis, it is possible to opt for another basis to expand err, such as the $\{(Q-Q_0)^m\}$, where Q_0 is the central frequency. In this manner, numerical dispersion can be adequately controlled and the most substantial: its frequency response tends to modification. Let us, first, rewrite err of (4.34) as

$$err = Q^2[d_1, 0, d_3, 0, d_5] \cdot \mathbf{V}_Q^T, \tag{4.35}$$

with

$$d_1 = \left(1 - \chi_1^2\right)\pi^2, \quad d_3 = \pi^4\left\{\chi_1[3(\chi_1 - 32\chi_2) + \chi_1\cos(4\theta)] - 2L_C^2\right\}/12,$$

$$d_5 = \pi^6\left[720\chi_1\chi_2 - 5\chi_1^2 - 2880\chi_2^2 + 2L_C^4 - 3\chi_1(\chi_1 - 80\chi_2)\cos(4\theta)\right]/180,$$

$$\mathbf{V}_Q = \left[1, Q, Q^2, Q^3, Q^4\right]$$

Proceeding to the $\{(Q-Q_0)^m\}$ basis, it holds that

$$\mathbf{V}_{(Q-Q_0)}^T = \mathbf{A}_V \cdot \mathbf{V}_Q^T$$

$$\mathbf{V}_{(Q-Q_0)} = \left[1, (Q - Q_0), (Q - Q_0)^2, (Q - Q_0)^3, (Q - Q_0)^4\right]$$

for
$$\mathbf{A}_V = \begin{bmatrix} 1 & 0 & 0 & 0 & 0 \\ -Q_0 & 1 & 0 & 0 & 0 \\ Q_0^2 & -2Q_0 & 1 & 0 & 0 \\ -Q_0^3 & 3Q_0^2 & -3Q_0 & 1 & 0 \\ Q_0^4 & -4Q_0^3 & 6Q_0^2 & -4Q_0 & 1 \end{bmatrix},$$

and as a consequence, (4.35) is transformed to

$$err = Q^2[d_1, 0, d_3, 0, d_5] \cdot \mathbf{A}_V^{-1}\mathbf{V}_{(Q-Q_0)}^T = Q^2[d_1', d_2', d_3', d_4', d_5'] \cdot \mathbf{V}_{(Q-Q_0)}^T \tag{4.36}$$

Because, for the computation of χ_1, χ_2, the zeroing of d'_1, d'_2 is sufficient, only the first two elements presented, i.e.,

$$d'_1 = \pi^2 \left[1 - \chi_1^2 \left(36 - 9\pi^2 Q_0^2 + \pi^4 Q_0^4 \right) /36 \right]$$

$$+ \pi^4 Q_0^2 \left\{ 720(\pi^2 Q_0^2 - 2)\chi_1\chi_2 - 30L_C^2 + 2\pi^2 Q_0^2 \left(L_C^4 - 1440\chi_2^2 \right) \right.$$

$$\left. - 3\chi_1 \left[\left(\pi^2 Q_0^2 - 5 \right) \chi_1 - 80\pi^2 Q_0^2 \chi_2 \right] \cos(4\theta) \right\} /180 \tag{4.37a}$$

$$d'_2 = \pi^4 Q_0 \left\{ \left(45 - 10\pi^2 Q_0^2 \right) \chi_1^2 + 1440 \left(\pi^2 Q_0^2 - 1 \right) \chi_1\chi_2 - 30L_C^2 + 4\pi^2 Q_0^2 \left(L_C^4 - 1440\chi_2^2 \right) \right.$$

$$\left. + 3\chi_1 \left[\left(5 - 2\pi^2 Q_0^2 \right) \chi_1 + 160\pi^2 Q_0^2 \chi_2 \right] \cos(4\theta) \right\} \tag{4.37b}$$

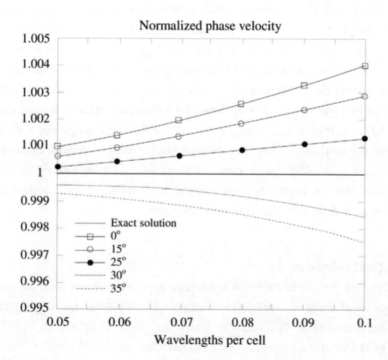

FIGURE 4.5: Normalized phase velocity of an angle-optimized FDTD scheme adjusted at 27.5°.

Now, the error is given by $err = Q^2[d'_3 (Q - Q_0)^2 + d'_4 (Q - Q_0)^3 + d'_5 (Q - Q_0)^4$. Note that because since the filter is set to be of second order and around the central frequency, err is dominated by the $Q^2 d'_3 (Q - Q_0)^2$ term. Suppose, also, that the optimized angle is θ_{opt} and $\theta = \theta_{opt} + \Delta\theta$. Then, the variation of d'_1, d'_2 will be analogous to $\sin(4\theta)\Delta\theta$ and the dispersion error very smooth around θ_{opt}. In fact, err is almost flat at $\theta_{opt} = 0°$ and periodic because of $\cos(4\theta)$. Accordingly, whenever it is optimized for some angle θ_{opt}, other seven angles $(\pi/2 \pm \theta_{opt}, \pi \pm \theta_{opt}, 3\pi/2 \pm \theta_{opt}, 2\pi - \theta_{opt})$ are simultaneously optimized. To validate the merits of the technique, the normalized phase velocity for an FDTD simulation is examined. For this configuration, $\chi_1 = 0.999843$ and $\chi_2 = 0.026875$, whereas the central frequency is such that $Q_0 = 0.08$. Figure 4.5 exhibits the promising improvements because almost no dispersion is generated for a wide range of frequencies.

4.4   ABSORBING BOUNDARY CONDITIONS

One of the major motives for the selection of a time-domain method to solve an EMC problem is its proficiency in the treatment of infinite regions. As a matter of fact, when numerically handling a model originally established on an unbounded domain, it is compulsory to impose the pertinent ABC to terminate the endless mesh while retaining a well-posed and stable simulation. At the outer boundary, so created, the unknown scalar or vector field function should satisfy an operator equation for the coupling of the inner computational area to the free-space region that extends to infinity. This is accomplished by designing the particular operator in a way that guarantees the absorption (with trivial reflections) of outward traveling waves that arrive at the boundary from the interior. Because most of the techniques render a linear mapping to unbounded domains in respect to memory and CPU requirements, it is obvious that any endeavor for space reduction is very important. Thus, ABCs comprise the most popular means for the manipulation of boundary-value situations without the need of auxiliary (and perhaps burdensome) practices. Bearing in mind these considerations, the analysis of the present section focuses on the brief description of the two basic ABC categories and their applicability to the truncation of infinite EMC models, discussing their formulation in terms of different time-domain algorithms.

4.4.1   Analytical Schemes

The early attempts on the confinement of open-region problems have been the analytical ABCs that utilize simple mathematical concepts to absorb the larger portion of outgoing electromagnetic waves. Essentially, they are classified into two kinds: the *global* and the *local*. Being the eldest, the representatives of the former type are nonlocal in character, namely, they relate the value of a field component at each boundary point to all other points on the same boundary. Moreover, they are chiefly defined either as an infinite modal expansion with unknown coefficients or as a boundary integral equation, hence providing an exact formulation of the required open-space coupling. Although these ABCs are proven to be trustworthy in their fundamental role, they do have a serious disadvantage: their integral transforms and the linear system of equations resulting from the discretization of the nonlocal operator lead to dense matrices, which are rather inefficient for arbitrarily structured spaces with involved objects and fine material features.

To ameliorate the previous difficulties, most time-domain simulations started to adopt the second type of ABCs, based on the proper modification of a partial differential equation applied to the artificial boundary. This concept enabled the imposition of the ABCs closer to the object under study, thus saving a significant amount of cells for the free space. Their expressions are not only affected by the geometry of the boundary but also by the governing laws of the problem. Among the diverse disciplines, the notion of one-way wave equation seems to be the most efficient one. Generally, local ABCs are mathematically robust as well as numerically affordable, and therefore, they can successfully cooperate with several time-domain techniques. Some of the most indicative members of the class are the Engquist–Majda [21], Bayliss–Turkel [22], Lindman [23], Mur [24], and Higdon [25] ABCs along with the Liao [26] multitransmitting operators and the Grote–Keller approach [27]. Their reflection coefficients may fluctuate between −25 and −40 dB, depending on the configuration, but they deteriorate rather quickly in the case of near-grazing incidence or evanescent waves.

To point out the performance of analytical ABCs, a set of examples regarding the reflection coefficient as a function of incident angle and grid resolution is given in Figure 4.6 for Mur, Bayliss–Turkel, Higdon, and Liao schemes. From the outcomes, one can deduce that the specific techniques are fairly competent for a relatively large range of angles, even when the mesh is chosen to be rather coarse. Such a conclusion is deemed important because analytical ABCs are very economical, which is due to their one-way wave equation implementation, therefore, they can provide rapid and affordable solutions at satisfactory levels.

Apart from these techniques, another category of ABC enhancement processes has been launched. Two characteristic algorithms are the superabsorption [28] and the complementary operators method (COM) [29]. The former incorporates a mutually error-canceling idea that uses the same ABC to both electric and magnetic quantities on and near the boundaries. On the other hand, the COM suppresses first-order reflections regarding propagating and evanescent waves through two boundary operators that are complementary in their action. Apparently, two distinct runs of the computer code are demanded at the expense of an additional, yet not at all high-priced, system cost. In fact, because of its absorptive competences, the COM allows the imposition of outer boundaries closer to the objects, reimbursing for the extra simulation time. Consequently, a notably confined domain is created without the need of excessive free-space discretizations.

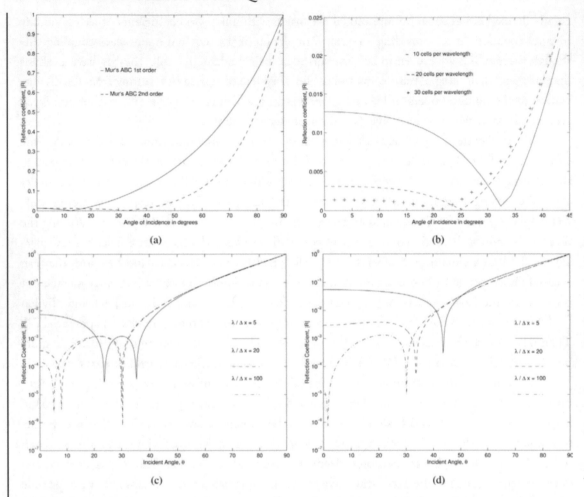

FIGURE 4.6: Reflection coefficient versus angle of incidence and different lattice resolutions for (a) Mur, (b) Bayliss–Turkel, (c) Higdon, and (d) Liao ABCs.

4.4.2   PML Realizations

Indisputably, a foremost breakthrough in the area of computational electromagnetics that revitalized the scientific interest in the use of absorbing materials as ABCs has been the introduction of the PML [30–32]. Its development stemmed from the deduction that the most prominent problem with existing artificial absorbers has been their incapacity to function properly over a broad range of incidence angles. Thus, augmenting the degrees of freedom available to offer the desired match, the PML wraps the core domain with a layer of a fictitious absorbing medium, backed by PEC walls, which forms a "numerical" anechoic chamber. Such a conception is achieved through the splitting of certain (depending on the dimensionality) field components and the assignment of different electric and magnetic losses to each of them. Then, by aptly regulating this set of constitutive parameters in the layer, the vacuum–PML interface is rendered completely transparent for any frequency, polarization, and angle of incidence, at least from a theoretical point of view. In addition, the transmitted waves inside the PML share the same phase velocity and characteristic impedance with their incident counterparts, but they accept very fast attenuations in the normal direction. Almost immediately after its first presentation, PML became (and continues to be) a topic of meticulous research [1] that targets PML's overall advancement and correction of potential defects. In this sense, an assortment of important contributions may be recognized, such as complex coordinate stretching [33, 34] and unsplit field [35, 36] techniques, curvilinear [37–39] or nonorthogonal [40, 41] arrangements, generalized material-based methods [42, 43], and complex-shifted frequency formulations [44, 45]. In the following paragraphs, various split- and unsplit-field PML absorbers in their most typical, for EMC setups, realization are derived for the family of FDTD-oriented approaches, the FETD scheme, and the PSTD algorithm.

FDTD-based implementations. The split-field form of Maxwell's curl laws along the x-direction, is

$$\gamma_x \mu \frac{\partial H_x}{\partial t} + \Psi_x \mu H_x = \frac{\partial E_y}{\partial z} - \frac{\partial E_z}{\partial y} \quad \text{and} \quad \gamma_x \varepsilon \frac{\partial E_x}{\partial t} + \Psi_x \varepsilon E_x = \frac{\partial H_z}{\partial y} - \frac{\partial H_y}{\partial z}, \quad (4.38)$$

with analogous expressions toward y, z and γ_x, Ψ_x quantities that define the inherent traits of the coordinate system. Discretizing (4.38), one obtains

$$\gamma_x^- \mu^- \left.\frac{\partial^+ H_x^-}{\partial t}\right|_B^{n-1/2} + \Psi_x^- \mu^- H_x^- \big|_B^{n+1/2} = \left(\frac{\partial^+ E_y^+}{\partial z} - \frac{\partial^+ E_z^+}{\partial y}\right)\Bigg|_A^n \qquad (4.39a)$$

$$\gamma_x^+ \varepsilon^+ \left.\frac{\partial^- E_x^+}{\partial t}\right|_A^{n-1} + \Psi_x^+ \varepsilon^+ E_x^+ \big|_A^n = \left(\frac{\partial^- H_z^-}{\partial y} - \frac{\partial^- H_y^-}{\partial z}\right)\Bigg|_B^{n-1/2} \qquad (4.39b)$$

where the $+$, $-$ superscripts denote the use of forward/backward finite-difference approximations for the discrete state, whereas $A = (i, j, k)$ and $B = (i + 1/2,\ j + 1/2,\ k + 1/2)$. So, when a forward field vector rests on an (i, j, k) node, its components are linked to the spatial vectors placed in front of the particular node. For instance, the x component of $\mathbf{E}^+$ at A is connected with node $(i + 1/2, j, k)$ and that of $\mathbf{H}^-$ at B with $(i,\ j + 1/2, k + 1/2)$. These aspects are, subsequently, applied to a plane wave

$$\mathbf{E}^+ \big|_A^n = \mathbf{E}_0^+ \, e^{-j\omega n \Delta t} \quad \text{and} \quad \mathbf{H}^- \big|_B^{n-1/2} = \mathbf{H}_0^- \, e^{-j\omega(n-1/2)\Delta t} \quad \text{for } j^2 = -1, \qquad (4.40)$$

and $\partial^{\pm}/\partial t \Rightarrow \pm j\Psi_t^{\pm}$, with $\Psi_t^{\pm} = \omega \sin(\omega \Delta t/2)\, e^{\pm j\omega\Delta t/2}$. So,

$$
j\Psi_t^+ P_x^- \mu^- H_x^-\big|_B = \left(\frac{\partial^+ E_y^+}{\partial z} - \frac{\partial^+ E_z^+}{\partial y}\right)\Bigg|_A^n
$$

$$
j\Psi_t^- P_x^+ \varepsilon^+ E_x^+\big|_A = \left(\frac{\partial^- H_z^-}{\partial y} - \frac{\partial^- H_y^-}{\partial z}\right)\Bigg|_B^{n-1/2}
$$

(4.41)

for $P_x = \gamma_x + j\Psi_x/\Psi_t$. Repetition of the analysis for the other components and combination of the results give the vector forms of

$$
j\Psi_t^+ \mu^- \mathbf{H}^-\big|_B = \nabla_s^+ \times \mathbf{E}^+\big|_A \quad\text{and}\quad j\Psi_t^- \varepsilon^+ \mathbf{E}^+\big|_A = -\nabla_s^- \times \mathbf{H}^-\big|_B,
$$

(4.42)

with

$$
\nabla_s^{\pm} = \hat{\mathbf{x}}\frac{1}{P_x^{\mp}}\frac{\partial^{\pm}}{\partial x} + \hat{\mathbf{y}}\frac{1}{P_y^{\mp}}\frac{\partial^{\pm}}{\partial y} + \hat{\mathbf{z}}\frac{1}{P_z^{\mp}}\frac{\partial^{\pm}}{\partial z}.
$$

If (4.40) are expressed by means of $\mathbf{E}^+\big|_A = \mathbf{E}_0^+ e^{j\mathbf{k}\cdot\mathbf{r}}$ and $\mathbf{H}^+\big|_B = \mathbf{H}_0^+ e^{j\mathbf{k}\cdot\mathbf{r}}$, their plugging in (4.42) leads to

$$
\Psi_t^+ \mu^- \mathbf{H}_0^- = \mathbf{L}_s^+ \times \mathbf{E}_0^+ \quad\text{and}\quad \Psi_y^- \varepsilon^+ \mathbf{E}_0^+ = -\mathbf{L}_s^- \times \mathbf{H}_0^-
$$

(4.43)

for

$$
\mathbf{L}_s^{\pm} = \frac{L_x^{\pm}}{P_x^{\mp}}\hat{\mathbf{x}} + \frac{L_y^{\pm}}{P_y^{\mp}}\hat{\mathbf{y}} + \frac{L_z^{\pm}}{P_z^{\mp}}\hat{\mathbf{z}},
$$

and $L_u^{\pm} = k_u \operatorname{sinc}(k_u \Delta u/2)e^{\pm jk_u\Delta u/2}$ for $u \in (x, y, z)$. It should be emphasized that the PML losses are controlled independently along x, y, and z directions via the different selections of stretched coordinates P_u.

As an illustration, consider the 2-D TE case, where field splitting entails that $H_z = H_{zx} + H_{zy}$ with

$$
\frac{\partial(H_{zx} + H_{zy})}{\partial x} = -j\omega\varepsilon_2\left(1 + \frac{\sigma_x}{j\omega\varepsilon_2}\right)E_y, \qquad \frac{\partial(H_{zx} + H_{zy})}{\partial y} = j\omega\varepsilon_2\left(1 + \frac{\sigma_y}{j\omega\varepsilon_2}\right)E_x,
$$

(4.44a)

$$
\frac{\partial E_y}{\partial x} = -j\omega\mu_2\left(1 + \frac{\rho_x'}{j\omega\mu_2}\right)H_{zx}, \qquad \frac{\partial E_x}{\partial y} = j\omega\mu_2\left(1 + \frac{\rho_y'}{j\omega\mu_2}\right)H_{zy},
$$

(4.44b)

the modified Maxwell's equations in the PML medium, designated as 2 (free space is medium 1). Choosing $P_u = 1 + \sigma_u / j\omega\varepsilon_2$ and $P_u^* = 1 + \rho_u' / j\omega\mu_2$ for $u = x, y$, it is acquired that

$$E_x = -H_0 T \frac{\beta_{2,y}}{\omega\varepsilon_2} \sqrt{\frac{P_y^*}{P_y}} \Phi, \qquad E_y = H_0 T \frac{\beta_{2,x}}{\omega\varepsilon_2} \sqrt{\frac{P_x^*}{P_x}} \Phi, \; H_z = H_0 T\Phi, \qquad (4.45)$$

for $\Phi = \exp\left\{-j\left[(P_x P_x^*)^{1/2}\beta_{2,x}x + (P_y P_y^*)^{1/2}\beta_{2,y}y\right]\right\}$ the harmonic propagation term, T the transmission coefficient, $\beta_{x,2}^2 + \beta_{y,2}^2 = k_2^2$ and $\beta_u = 2\pi/\lambda_u$. To apply the phase-matching condition at the vacuum–PML interface, the propagation constants along the y-direction should be equal, i.e., $P_y P_y^* = 1 \Rightarrow \sigma_y = \rho_y' = 0$ [30]. Hence, it holds that $\beta_{2,y} = \beta_{1,y} = k_1 \sin \theta_i$, with θ_i the angle of incidence. Moreover, the reflection coefficient of the total wave impingement is found to be, after some algebra,

$$R = \left(\frac{\beta_{1,x}}{\omega\varepsilon_1} - \frac{\beta_{2,x}}{\omega\varepsilon_2}\sqrt{\frac{P_x^*}{P_x}}\right)\left(\frac{\beta_{1,x}}{\omega\varepsilon_1} + \frac{\beta_{2,x}}{\omega\varepsilon_2}\sqrt{\frac{P_x^*}{P_x}}\right)^{-1} = T - 1 \qquad (4.46)$$

Enforcing $\varepsilon_1 = \varepsilon_2$, $\mu_1 = \mu_2$, and $P_x = P_x^*$, one concludes to $k_1 = k_2$, $\eta_1 = (\mu_2/\varepsilon_2)^{1/2} = (\mu_1/\varepsilon_1)^{1/2} = \eta_2$, and $\sigma_x/\varepsilon_1 = \rho_x'/\mu_1$, which, consecutively, yield $\beta_{1,x} = \beta_{2,x}$ and the desired result of $R = 0$. In this manner, the final form of H_z quantity in the interior of the PML is

$$H_z = H_0 e^{-j(\beta_{1,x}x + \beta_{1,y}y)} e^{-\sigma_x \eta_1 \cos \theta_i x}, \qquad (4.47)$$

with similar formulae for E_x and E_y, as well. These expressions can then be easily discretized via any FDTD-like technique to establish the leapfrog procedure. For example, Figure 4.7a compares the local error of a 100×50 FDTD PML implementation (along a line parallel to the vacuum–layer interface) with that of a Mur ABC, whereas in Figure 4.7b, the same error for various absorber depths is depicted. Notice the order (around 10^{-7}) in the latter figure, a fact that verifies the remarkable PML absorption rate. Also, the symmetry in the distribution of the local error is illustrated in Figure 4.8 with two snapshots of its variation.

Proceeding to the generalized unsplit-field development of the absorber, presume an infinite 3-D domain, V, in spherical coordinates (r, θ, φ). For the objectives of the analysis, V is separated into two regions such that $V = V_D \cup V_{PML}$, with V_D as the computational space and V_{PML} as the area of the curvilinear PML. Commencing from an isotropic, inhomogeneous, and dielectric medium without losses, Maxwell's equations become

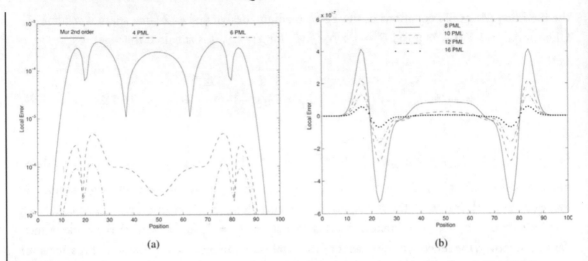

(a)   (b)

FIGURE 4.7: Local error along a line parallel to the vacuum–absorber interface for a 2-D FDTD open-region problem truncated by (a) Mur and two PMLs and (b) four PMLs with different depths.

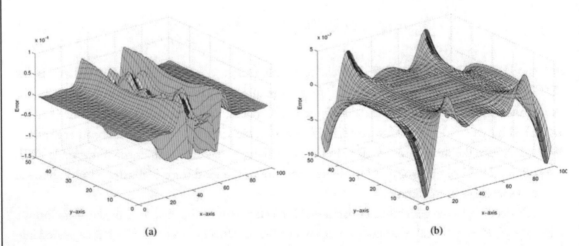

(a)   (b)

FIGURE 4.8: Two transverse-cut snapshots of the PML local error for a depth of (a) six and (b) eight cells.

$$j\omega\varepsilon\bar{\mathbf{E}} = \bar{\nabla} \times \bar{\mathbf{H}}, \bar{\nabla} \cdot (\varepsilon\bar{\mathbf{E}}) = 0, \qquad -j\omega\mu\bar{\mathbf{H}} = \bar{\nabla} \times \bar{\mathbf{E}}, \bar{\nabla} \cdot (\mu\bar{\mathbf{H}}) = 0 \tag{4.48}$$

Herein, V_D is a sphere of radius a, and V_{PML}—extending from a to infinity—is the region to be terminated till radius b by a perfect electric conducting wall. The advanced PML is constructed in terms of the correct scaling between the independent (barred) and dependent variables in (4.48), which conserve the initial field propagation in V_D and launch the suitable coordinate stretching in V_{PML}. This approach manages to control the appearance of highly dispersive reflective modes because of the curvilinear outer boundaries and suppresses their exponential increase in the layer [39]. Thus, for the above medium,

$$\bar{\mathbf{G}} = \begin{cases} \mathbf{G}, & \text{for } \bar{\mathbf{r}} \in V_D \\ \mathrm{diag}\{\zeta_r, \xi_r^{-1}, \xi_r^{-1}\}\mathbf{G}, & \text{for } \bar{\mathbf{r}} \in V_{PML} \end{cases} \tag{4.49}$$

where $\bar{\mathbf{r}} = \mathrm{diag}\{\xi_r, 1, 1\}\mathbf{r}$ and $\mathbf{r} = (r, \theta, \varphi)^T$. Plugging (4.49) in (4.48) yields

$$j\omega\varepsilon\Xi \cdot \mathbf{E} = \nabla \times \mathbf{H}, \qquad -j\omega\mu\Xi \cdot \mathbf{H} = \nabla \times \mathbf{E}, \tag{4.50}$$

for Ξ the material tensor; a diagonal matrix $\Xi = \mathrm{diag}\{\xi_r\zeta_r, \zeta_r^{-1}, \zeta_r^{-1}\}$. Moreover, Gauss laws are written as

$$\nabla \cdot (\varepsilon\Xi \cdot \mathbf{E}) = 0, \qquad \nabla \cdot (\mu\Xi \cdot \mathbf{H}) = 0 \tag{4.51}$$

The system of (4.50) and (4.51) is causal hyperbolic and thus not easily prone to oscillatory wave fronts that create severe late-time instabilities. In the relations previously reported,

$$\zeta_r(r, \omega) = \left[1 + \frac{\sigma_r(r)}{j\omega}\right]^{-1}, \qquad \xi_r(r, \omega) = 1 + \frac{\rho_r'(r)}{j\omega}, \tag{4.52}$$

with $\sigma_r(r) = \sigma_r^{max}[(r - a)/b]^T$; $\tau \geq 0$, $\bar{\sigma}_r(r) = r^{-1}\int_a^r \sigma_r(s)\,ds$, V_{PML} limited to a sphere of radius $a + b$, σ_r^{max} a real parameter that tunes the layer's absorption rate, and $b = s\Delta r$ as its depth extending to s cells. Substitution of (4.52) in Faraday's law provides

$$-j\omega\mu\xi_r^2\zeta_r H_r = (\nabla \times \mathbf{E})_r, \qquad -j\omega\mu\zeta_r^{-1}H_\theta = (\nabla \times \mathbf{E})_\theta, \qquad -j\omega\mu\zeta_r^{-1}H_\varphi = (\nabla \times \mathbf{E})_\varphi \tag{4.53}$$

The temporal integration of H_r quantities in (4.53) is performed by the magnetic flux density $B_r = \mu\xi_r\zeta_r H_r$, as

$$-j\omega B_r - \bar{\sigma}_r(r)B_r = (\nabla \times \mathbf{E})_r, \qquad j\omega B_r + \sigma_r(r)B_r = j\omega H_r + \mu\bar{\sigma}_r(r)H_r \tag{4.54a}$$

In a similar way,

$$-j\omega B_\theta = (\nabla \times \mathbf{E})_\theta, \qquad j\omega B_\theta = j\omega H_\theta + \mu\bar{\sigma}_r(r)H_\theta, \qquad (4.54b)$$

$$-j\omega B_\varphi = (\nabla \times \mathbf{E})_\varphi, \qquad j\omega B_\varphi = j\omega H_\varphi + \mu\bar{\sigma}_r(r)H_\varphi \qquad (4.54c)$$

It should be mentioned that (4.54) are fully consistent, and the update relations—obtained via the discretization strategy of any FDTD-like time-domain method—are very accurate, stable, and sufficiently convergent. To verify these claims, consider the global error of three spherical PMLs terminating a 75×240 infinite domain with a PEC spherical scatterer centered at its origin. As observed, the selection of the depth does affect the absorber's performance, which, however, is always superior to that of the reference Sommerfeld-type ABC [1].

Development of PMLs for the FETD/WETD method. Assume the termination of an open-region 3-D domain by a PML, inside which electric field vector $\mathbf{E}$ is expressed, according to stretched-coordinate theory, as

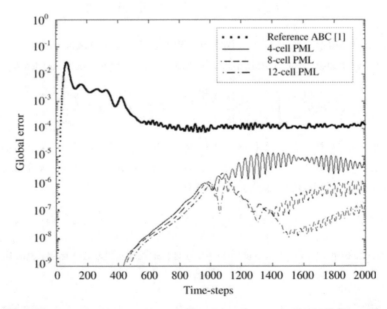

FIGURE 4.9: Global error versus time steps produced by three spherical PMLs and a reference ABC in a $75 \cdot 240$ lattice containing a PEC scatterer centered at the origin.

$$\mathbf{E} = \left[\left(1 + \tfrac{\sigma_x}{\varepsilon} \mathrm{int_t}\right) E_x, \left(1 + \tfrac{\sigma_y}{\varepsilon} \mathrm{int_t}\right) E_y, \left(1 + \tfrac{\sigma_z}{\varepsilon} \mathrm{int_t}\right) E_z \right]^{\mathrm{T}}, \qquad (4.55)$$

where $\mathrm{int_t}$ denotes temporal integration and $\sigma_x, \sigma_y, \sigma_z$ are the conductivities of the absorber [46, 47]. In this context, the unknown $\mathbf{E}$ vector in the PML is given by

$$\varepsilon \bar{\bar{\mathrm{K}}} * \frac{\partial^2 \mathbf{E}}{\partial t^2} + \nabla \times \left(\frac{1}{\mu} \bar{\bar{\mathrm{K}}}^{-1} \right) * \nabla \times \mathbf{E} = 0, \qquad (4.56)$$

with the asterisk signifying the operation of convolution in the time domain and diagonal tensor

$$\bar{\bar{\mathrm{K}}} = \mathrm{diag} \left\{ P_y P_z P_x^{-1}, P_x P_z P_y^{-1}, P_x P_y P_z^{-1} \right\}, \qquad (4.57)$$

for $P_u = a + \sigma_u / (j\omega \varepsilon_0)$ the stretching variables and $u = x, y, z$. To solve (4.56) by means of the FETD/WETD method, the methodology of section 3.2.1 is applied, while, for simplicity, parameter a is set equal to unity. Thus, after the use of the appropriate impedance boundary condition on the backing wall of the PML, the desired weak-form solution of (4.56) fulfills

$$\int_\Omega \left\{ \varepsilon \mathbf{w}_i \bar{\bar{\mathrm{K}}} * \frac{\partial^2 \mathbf{E}}{\partial t^2} + \nabla \times \mathbf{w}_i \left(\frac{1}{\mu} \bar{\bar{\mathrm{K}}}^{-1} \right) * \nabla \times \mathbf{E} + \mathbf{w}_i \frac{\partial \mathbf{J}}{\partial t} \right\} \mathrm{d}\Omega$$

$$+ \iint_S \Gamma \hat{\mathbf{n}} \times \mathbf{w}_i \cdot \left(\hat{\mathbf{n}} \times \frac{\partial \mathbf{E}}{\partial t} \right) \mathrm{d}S \qquad (4.58)$$

with Ω the truncated computational domain, S the surface of the outer boundary, $\mathbf{w}_i$ the required set of vector basis functions, $\hat{\mathbf{n}}$ the outward normal unit vector, $\mathbf{J}$ the electric current density, and Γ the coefficient that determines the total impedance. Expanding the stretched electric field by means of (3.9) and presuming a nonconstant variation of $\sigma_x, \sigma_y, \sigma_z$ losses within each element e of the mesh, the subsequent ordinary differential equation for the entire space, comprising M elements, is generated

$$\sum_{e=1}^{M} \left[\Lambda^e \frac{\mathrm{d}^2 \mathbf{u}_c}{\mathrm{d} t^2} + \left(\Lambda^e_p + \mathbf{M}^e \right) \frac{\mathrm{d} \mathbf{u}_c}{\mathrm{d} t} + \left(\Lambda^e_q + \mathbf{N}^e \right) \mathbf{u}_c + \sum_{v=x,y,z} \left(\kappa^e_v + \delta^e_v \right) + \mathbf{f}^e \right] = 0, \quad (4.59)$$

where $\mathbf{u}_c$ represents the optimal solution vector. Also, the elements of square matrices $\Lambda^e, \Lambda^e_p, \mathbf{M}^e, \Lambda^e_q, \mathbf{N}^e$, are

$$\Lambda^e_{ij} = \varepsilon \left\langle \mathbf{w}_i, \mathbf{w}_j \right\rangle_{\Omega^e}, \qquad \Lambda^e_{u,ij} = \varepsilon \left\langle \mathbf{w}_i, \bar{\bar{\xi}}_u \cdot \mathbf{w}_j \right\rangle_{\Omega^e}, \qquad \text{for } u = x,y,z,p,q \qquad (4.60a)$$

$$M_{ij}^e = \Gamma \langle \hat{\mathbf{n}} \times \mathbf{w}_i, \hat{\mathbf{n}} \times \mathbf{w}_j \rangle_S, \qquad N_{ij}^e = \mu^{-1} \langle \nabla \times \mathbf{w}_i, \nabla \times \mathbf{w}_j \rangle_{\Omega^e}, \qquad (4.60b)$$

with $\langle \, . \, , \, . \, \rangle_{\Omega^e, S}$ featuring integration over the element volume Ω^e or the outer surface S, respectively. Additionally, vectors $\boldsymbol{\kappa}_v^e, \Delta_v^e$, which must be evaluated at every time step, are obtained via

$$\boldsymbol{\kappa}_v^e = \varepsilon \left\langle \mathbf{w}_i, \sum_{j=1}^{N} \zeta_v \bar{\bar{\xi}}_v \cdot \mathbf{w}_j \right\rangle_{\Omega^e}, \quad \delta_v^e = \frac{1}{\mu} \left\langle \nabla \times \mathbf{w}_i, -\sum_{j=1}^{N} \zeta_v \bar{\bar{\xi}}_{s,v} \cdot \nabla \times \mathbf{w}_j \right\rangle_{\Omega^e}, \qquad (4.61)$$

for $v = x, y, z$ and N the number of expansion functions in relation to (3.9). In (4.60a), (4.60b), and (4.61)

$$\bar{\bar{\xi}}_p = \varepsilon^{-1} \mathrm{diag} \left\{ (\sigma_y + \sigma_z - \sigma_x), (\sigma_x + \sigma_z - \sigma_y), (\sigma_x + \sigma_y - \sigma_z) \right\},$$

$$\bar{\bar{\xi}}_q = \varepsilon^{-2} \mathrm{diag} \left\{ \sigma_x^2 - \sigma_x (\sigma_y + \sigma_z) + \sigma_y \sigma_z, \, \sigma_y^2 - \sigma_y (\sigma_x + \sigma_z) \right.$$
$$\left. + \sigma_x \sigma_z, \, \sigma_z^2 - \sigma_z (\sigma_x + \sigma_y) + \sigma_x \sigma_y \right\},$$

$$\bar{\bar{\xi}}_x = -\varepsilon^{-2} \mathrm{diag} \left\{ \sigma_x^2 - \sigma_x (\sigma_y + \sigma_z) + \sigma_y \sigma_z, \, 0, \, 0 \right\},$$

$$\bar{\bar{\xi}}_y = -\varepsilon^{-2} \mathrm{diag} \left\{ 0, \, \sigma_y^2 - \sigma_y (\sigma_x + \sigma_z) + \sigma_x \sigma_z, \, 0 \right\},$$

$$\bar{\bar{\xi}}_z = -\varepsilon^{-2} \mathrm{diag} \left\{ 0, 0, \, \sigma_z^2 - \sigma_z (\sigma_x + \sigma_y) + \sigma_x \sigma_y \right\},$$

$$\bar{\bar{\xi}}_{s,x} = \mathrm{diag} \left\{ 0, \, \frac{\sigma_y - \sigma_x}{\sigma_z - \sigma_x}, \, \frac{\sigma_z - \sigma_x}{\sigma_y - \sigma_x} \right\}, \quad \bar{\bar{\xi}}_{s,y} = \mathrm{diag} \left\{ \frac{\sigma_x - \sigma_y}{\sigma_z - \sigma_y}, \, 0, \, \frac{\sigma_z - \sigma_y}{\sigma_x - \sigma_y} \right\}$$

$$\bar{\bar{\xi}}_{s,z} = \mathrm{diag} \left\{ \frac{\sigma_x - \sigma_z}{\sigma_y - \sigma_z}, \, \frac{\sigma_y - \sigma_z}{\sigma_x - \sigma_z}, \, 0 \right\}$$

are diagonal tensors, and

$$\mathbf{f}_i^e = \left\langle \mathbf{w}_i, \frac{\partial \mathbf{J}}{\partial t} \right\rangle_{\Omega^e}, \qquad \zeta_v = \frac{\sigma_v}{\varepsilon} e^{-\sigma_v t / \varepsilon} \mathbf{u}_{\mathrm{unit}} * \mathbf{u}_c,$$

is the excitation and absorption vector (computed recursively), with $\mathbf{u}_{unit}$ the unit-step function. In fact, only ζ_v have to be computed at each time step because all spatial evidence can be precalculated and stored. Therefore, the temporal update inside the PML area may be carried out successfully.

An alternative realization of PMLs for the FETD or the WETD technique may be accomplished through the anisotropic modeling of their media composition [48]. Keeping the notation of the preceding analysis the same and denoting as $\bar{\bar{P}} = \mathrm{diag}\{P_x, P_y, P_z\}$ the diagonal matrix of stretching variables, it is derived that $\nabla \times (\bar{\bar{P}} \cdot \mathbf{w}_i) = P_i \nabla \times \mathbf{w}_i$, where P_i is related to the direction of the associated vector basis function $\mathbf{w}_i$. Consequently, the weak form of the governing equations becomes

$$\sum_i E_i \int_{\Omega^e} \left[\left(\mu_r^{-1} \bar{\bar{K}}^{-1} P_i \nabla \times \mathbf{w}_i \right) \cdot \left(P_j \nabla \times \mathbf{w}_i \right) - \omega^2 \varepsilon_r \varepsilon_0 \mu_0 \bar{\bar{K}} P_i \mathbf{w}_i \cdot P_j \mathbf{w}_j \right] d\Omega = 0 , \quad (4.62)$$

with every term containing two additional terms: the first one referring to the stiffness matrix and the second to the mass matrix of the FE configuration. More specifically, the second quantity in (4.62) may be written as

$$-\omega^2 \varepsilon_r \varepsilon_0 \mu_0 \bar{\bar{K}} P_i \mathbf{w}_i \cdot P_j \mathbf{w}_j = -\omega^2 \varepsilon_r \varepsilon_0 \mu_0 P_x P_y P_z \mathbf{w}_i \cdot \mathbf{w}_j, \quad (4.63)$$

because it is nonzero only when $\mathbf{w}_i$ and $\mathbf{w}_j$ are along the same direction. On the other hand, the first quantity in (4.62) is proven to be equivalent, after the suitable manipulations, to a certain matrix form that incorporates the edge lengths of each element and the losses of the PML, under construction, toward the x, y, z direction. Observe that at corner regions, the material tensor must be carefully modified to evade the superposition of outgoing waves. For example, when $\sigma_x \neq 0$ and $\sigma_y = \sigma_z = 1$, (4.57) reduces to $\bar{\bar{K}} = \mathrm{diag}\{P_x^{-1}, P_x, P_x\}$. Finally, temporal update is conducted via an inverse Fourier transform, whose key aim is to convert all frequency-dependent parts into the respective convolutions or integrations in the time domain.

For the analysis to be complete, consider a 3-D computational space containing a scatterer with a cylindrical cross-section. The domain is discretized into 256,378 finite elements, whereas its infinite ends are terminated by the FETD PML, described above (see Figure 4.9). Figure 4.10 indicates the global error as a function of time for different spatial distributions (i.e., $n = 1$, linear; $n = 2$, square; $n = 3$, cubic; and $n = 4$, quadratic) of the losses inside the layer. From the numerical results, one can deduce that after a specific temporal interval (around 150 time steps), the notable performance of the absorber is practically stabilized for $n = 2, 3, 4$.

Unsplit-field PML derivation for the PSTD algorithm. As already explained in Section 3.5, an important obstruction in the application of the PSTD method is the wraparound effect during the treatment of periodic structures. To this problem, drastic mitigation is provided by the PML absorber, especially in its unsplit-field variant, which decreases memory needs and attains large

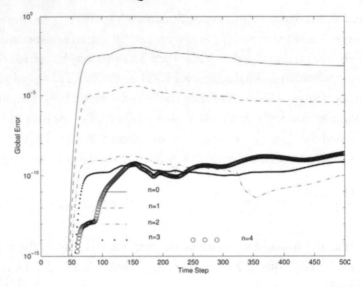

FIGURE 4.10: Global error versus time steps induced by a FETD PML for different spatial distributions of the losses.

absorption rates [49]. The development begins with Maxwell's equation and the assignment of the correct material parameters to every field component. For instance, the evolution of H_x quantity in a 3-D domain is controlled by the choice of the relative magnetic permeability μ_r in the corresponding x-component in Faraday's law. So, a selection of

$$\mu_r = \left(a_y + \frac{\rho'_y}{j\omega\mu_0} \right) \left(a_z + \frac{\rho'_z}{j\omega\mu_0} \right) \left(a_x + \frac{\rho'_x}{j\omega\mu_0} \right)^{-1}, \tag{4.64}$$

for a_u tunable parameters with $u = x, y, z$, leads to

$$j\omega H_x \left(a_y + \frac{\rho'_y}{j\omega\mu_0} \right) \left(a_z + \frac{\rho'_z}{j\omega\mu_0} \right) = -\frac{1}{\sqrt{\mu_0\varepsilon_0}} \left(a_x + \frac{\rho'_x}{j\omega\mu_0} \right) (\nabla \times \mathbf{E})_x \tag{4.65}$$

Expanding (4.65) and transforming the outcome in the time domain, one gets

$$a_y a_z \frac{\partial H_x}{\partial t} + \frac{a_y\rho'_z + a_z\rho'_y}{\mu_0} H_x + \frac{\rho'_y\rho'_z}{\mu_0^2} \int_0^{(n+1/2)\Delta t} H_x \, dt =$$

$$-\frac{1}{\sqrt{\mu_0\varepsilon_0}} \left[a_x (\nabla \times \mathbf{E})_x + \frac{\rho'_x}{\mu_0} \int_0^{n\Delta t} (\nabla \times \mathbf{E})_x \, dt \right] \tag{4.66}$$

Moreover, application of the well-known matching condition $\sigma / \varepsilon_0 = \rho' / \mu_0$ and the definition of coefficient $d_u = u /$ (PML depth) in (4.66) yields the H_x time-update expression in the absorber

$$H_x|_{i,j,k}^{n+1/2} = b_{\mathrm{A}}\,H_x|_{i,j,k}^{n-1/2} - b_{\mathrm{B}}\mathrm{Int}_{H_x}^{n-1/2} - b_{\mathrm{C}}\,(\nabla \times \mathbf{E})_x^n - b_{\mathrm{D}}\mathrm{Int}_{(\nabla \times \mathbf{E})_x}^{n-1} \qquad (4.67)$$

where $b_{\mathrm{A}} = \dfrac{a_y\,a_z - a_y\,d_z - a_z\,d_y}{b_{\mathrm{E}}}$, $b_{\mathrm{B}} = \dfrac{4d_y\,d_z}{b_{\mathrm{E}}}$, $b_{\mathrm{C}} = \dfrac{a_x + d_x}{b_{\mathrm{E}}}$, $b_{\mathrm{D}} = \dfrac{2d_x\Delta t}{b_{\mathrm{E}}\sqrt{\mu_0\varepsilon_0}}$, and $b_{\mathrm{E}} = a_y\,a_z + a_y\,d_z + a_z\,d_y + 2d_y\,d_z$. To this extent, integral quantities in (4.66) have the form of

$$\int_0^{(n+1/2)\Delta t} H_x\,\mathrm{d}t = \Delta t\,\mathrm{Int}_{H_x}^{n-1/2} + 0.5\Delta t\,H_x|_{i,j,k}^{n+1/2} \quad \text{and} \quad \mathrm{Int}_{H_x}^{n-1/2} = \mathrm{Int}_{H_x}^{n-3/2} + H_x|_{i,j,k}^{n-1/2},$$

$$\int_0^{n\Delta t} (\nabla \times \mathbf{E})_x\,\mathrm{d}t = \Delta t\,\mathrm{Int}_{(\nabla \times \mathbf{E})_x}^{n-1} + 0.5\Delta t(\nabla \times \mathbf{E})_x^n \quad \text{and} \quad \mathrm{Int}_{(\nabla \times \mathbf{E})_x}^{n-1} = \mathrm{Int}_{(\nabla \times \mathbf{E})_x}^{n-2} + (\nabla \times \mathbf{E})_x^{n-1},$$

with similar equations holding for the other components. It is stressed that (4.67) does not exhibit any oscillations (at least nontrivial ones) at dissimilar interfaces, whereas its accuracy and stability are very satisfactory.

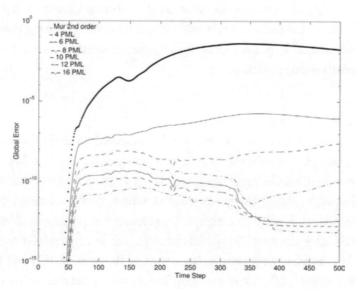

FIGURE 4.11: Global error versus time steps created by several PSTD PMLs with variable depths.

As an indication, let us focus on Figure 4.11, which provides the global error of diverse PSTD PML realizations regarding the layer's depth. For the sake of comparison, the Mur ABC is included as well. Evidently, the PMLs so constructed overwhelm the analytical scheme, attaining absorption rates up to the level of 10^{-13}.

4.5   CONFORMAL MODELING, CURVILINEAR GRIDS, AND NONORTHOGONAL TESSELLATIONS

The establishment of the pertinent coordinate system, according to the EMC problem of interest, is a crucial issue in the development of rigorous discretized models and robust lattices with acceptable convergence rates. The most common preference is, of course, the Cartesian system, whose mathematical consistency allows the extraction of well-posed operators for several physical phenomena. Because of the orthogonality of the resultant meshes and the uniform spacing of their stencils, the computation of electric and magnetic field vectors is significantly simplified. However, when the grid does not conform to the shape of an object or the geometric variation of its surface is rather abrupt, conventional tessellations are unable to provide a systematic handling. This disadvantage is more prominent in the FDTD-based methods, where the staircase approximation does not suffice and thus concludes to erroneous simulations, no matter how fine the resolution of the lattice is. On the contrary, the FIT, the FETD/WETD, the FVTD, and the WENO-TD techniques, owing to their fully generalized topological formulation and adaptive gridding, can support almost any type of coordinate system and structural irregularity. To compete these algorithms, the FDTD approach, together with the other staircase-like schemes described in Chapter 3, has been extended to curvilinear coordinates, as the fairly wide scientific literature reveals. Bearing the above in mind, it is the purpose of this section to present some of the most considerable formulations in the area of conformal and nonorthogonal modeling.

4.5.1   Conformal Meshing

This type of grid construction uses the conventional orthogonal discretization rules everywhere in the computational domain, except for the areas that are immediately neighboring to curved surfaces, boundaries, or discontinuities. For these areas, a special management is, of course, required. Thus, their cells are deformed to follow the geometrical variations of the aforesaid curvatures. In this framework, Maxwell's time-dependent curl equations cannot be approximated via the well-known logic of time-domain schemes. Actually, spatial discretization in these cells is conducted by means of integral forms that describe the governing laws, such as the ones of (2.1) and (2.2), and comply with the content of Section 2.2.3. The resultant update equations contain contour as well as surface integrals and refer only to the deformed cells [1, 50]. In this manner, very precise solutions may be attained at the expense of a slight burden increase, on condition that all curved shapes, cross sections, and fine details have been carefully investigated.

Furthermore, the use of conformal meshes for 3-D EMC shielding/immunity structures or reverberating chambers with perfect electric conducting curvatures grants several benefits, among which the enhanced accuracy at coarser grid resolutions as compared with typical realizations, is deemed the most notable. Toward this direction, conformal time-domain algorithms may be sometimes proven more efficient than their unstructured-lattice counterparts because away from the localized peculiarity, they use the simple update equations without affecting the overall numerical phase velocity and the stability features. However, their limits start to emerge as soon as geometrical irregularities become totally arbitrary and obtain a global character. Then, cell deformation is rather difficult, and the number of distorted cells should be much larger because of the far more meticulous gridding treatment entailed by the hardly predictable attributes of the objects in the domain. Satisfactory alleviation of these weaknesses is accomplished through the methods of the subsequent paragraphs.

4.5.2   Curvilinear Time-Domain Schemes

A chief trait of the algorithms fitting in this class is the use of dual-cell tessellations with a reasonably flexible stencil handling of spatial operators. Based on a covariant/contravariant vector field flux concept, a set of parametric explicit approximators is derived, whereas the consistent and hyperbolic nature of Maxwell's equations is definitely retained. For cumbersome media interfaces, convergence is guaranteed by the well-characterized application of continuity and ABCs. On the other hand, temporal integration involves either the leapfrog or the Runge–Kutta approach that accept several excitation schemes. So, mesh errors are kept at low levels and stability does not demand any particular implementations.

Before describing the main formulation stages of a curvilinear time-domain technique, it would be instructive to provide some important items of vector analysis. Assume a 3-D region that is sufficiently defined by the general curvilinear coordinate system (u, v, w), where any vector $\mathbf{F}$ can be decomposed into three components with regard to the contravariant $\hat{\mathbf{a}}^u$, $\hat{\mathbf{a}}^v$, $\hat{\mathbf{a}}^w$, or the covariant $\hat{\mathbf{a}}_u$, $\hat{\mathbf{a}}_v$, $\hat{\mathbf{a}}_w$, basis, as

$$\mathbf{F} = \sum_{\xi=u,v,w} (\hat{\mathbf{a}}_\xi \cdot \mathbf{F})\hat{\mathbf{a}}^\xi = \sum_{\xi=u,v,w} f_\xi \hat{\mathbf{a}}^\xi = \sum_{\zeta=u,v,w} (\hat{\mathbf{a}}^\zeta \cdot \mathbf{F})\hat{\mathbf{a}}_\zeta = \sum_{\zeta=u,v,w} f^\zeta \hat{\mathbf{a}}_\zeta \qquad (4.68)$$

The f_ξ and f^ζ terms are, respectively, the covariant and contravariant components of $\mathbf{F}$, which fulfill the $\hat{\mathbf{a}}_\xi \cdot \hat{\mathbf{a}}^\zeta = \Delta_{\xi\zeta}$ identity ($\Delta_{\xi\zeta}$ is the delta of Kronecker). Metric coefficients $g_{\zeta\zeta}$, $g^{\zeta\zeta}$ of the system are acquired by $g_{\zeta\zeta} = g_{\xi\xi} = \hat{\mathbf{a}}_\zeta \cdot \hat{\mathbf{a}}_\xi$, $g^{\zeta\zeta} = g^{\xi\xi} = \hat{\mathbf{a}}^\zeta \cdot \hat{\mathbf{a}}^\xi$, and the determinant (Jacobian) of the covariant tensor $g^{1/2} = \hat{\mathbf{a}}_{\xi+1} \cdot (\hat{\mathbf{a}}_{\xi+2} \times \hat{\mathbf{a}}_{\xi+3})$, for $\xi + 1$, $\xi + 2$ and $\xi + 3$ signifying a cyclic u, v, w permutation. In this sense, $\mathbf{F}$ components are linked to each other through $f_\xi = g_{\xi\xi} f^\zeta$ and $f^\zeta = g^{\zeta\zeta} f_\xi$. Exploiting this background, all three fundamental vector operators can be extracted. Thus, the *gradient* of a scalar ϕ is given by

$$\nabla \phi = g^{-1/2} \sum_{\xi=u,v,w} \frac{\partial \phi}{\partial \xi} \hat{\mathbf{a}}^\xi \tag{4.69}$$

Moreover, the *divergence* of $\mathbf{F}$ uses contravariant quantities and reads

$$\nabla \cdot \mathbf{F} = g^{-1/2} \sum_{\zeta=u,v,w} \frac{\partial}{\partial \zeta} \left(g^{1/2} \mathbf{F} \cdot \hat{\mathbf{a}}^\zeta \right), \tag{4.70}$$

whereas its *curl* receives the form of

$$\nabla \times \mathbf{F} = g^{-1/2} \sum_{\xi=u,v,w} \frac{\partial}{\partial \xi} \left(g^{1/2} \hat{\mathbf{a}}^\xi \times \mathbf{F} \right) \tag{4.71}$$

It is stressed that to guarantee solution consistency and algorithmic convergence for the prior coordinates, one must be very careful because irregular selections may generate severe modal parasites. Such cases are frequent in large applications, such as EMC test facilities or biomedical problems that require multifrequency excitations. Hence, the goal focuses on the construction of two dual lattices that assign electric and magnetic components to a family of adaptive topologies capable of supporting physical phase velocities.

To this goal, consider a 3-D domain, defined by the curvilinear system ($u = i\Delta u$, $v = j\Delta v$, $w = k\Delta w$) and separated into two dual grids [40, 51, 52]. The center of every primary cell is placed at (i, j, k) node, with the secondary ones mutually centered at the primary mesh vertices ($i - 1/2$, $j - 1/2$, $k - 1/2$). Evenly, the center of each cell face is determined by shifting one of its indices along half a stencil. This implies that covariant h_ξ and contravariant h^ξ magnetic components are placed at primary face centers and e_ξ, e^ζ electric quantities at the corresponding edge centers. The curvilinear time-domain method engages the flux notion for the vectors through the faces given by $f^{(\zeta)} = g^{1/2} f^\zeta$ for $f = e, h$. Then, f_ξ are expressed in terms of f^ζ and linear operator $\mathcal{FL}^{(\zeta)}[.]$ is introduced as $f^{(\zeta)} = \mathcal{FL}^{(\zeta)}\left[f_\xi\right]$, establishing a local connection of f_ξ with the adjacent $f_{\xi+1}$, $f_{\xi+2}$ values, multiplied by $\mathcal{G}^{\zeta\zeta} = g^{1/2} g^{\zeta\zeta}$. For example, the $\mathcal{FL}^{(u)}[h_u]$ yields

$$\mathcal{FL}^{(u)}\left[h_u\big|_{i-1/2,j,k}^{n+1/2}\right] = \mathcal{G}_{i-1/2,j,k}^{uu}\, h_u\big|_{i-1/2,j,k}^{n+1/2} + \frac{1}{4}\left[\mathcal{G}_{i,j,k}^{uv}\left(h_v\big|_{i,j-1/2,k}^{n+1/2} + h_v\big|_{i,j+1/2,k}^{n+1/2}\right)\right.$$

$$+\, \mathcal{G}_{i-1,j,k}^{uv}\left(h_v\big|_{i-1,j-1/2,k}^{n+1/2} + h_v\big|_{i-1,j+1/2,k}^{n+1/2}\right)$$

$$+\, \mathcal{G}_{i,j,k}^{uw}\left(h_w\big|_{i,j,k-1/2}^{n+1/2} + h_w\big|_{i,j,k+1/2}^{n+1/2}\right)$$

$$\left. +\, \mathcal{G}g_{i-1,j,k}^{uw}\left(h_w\big|_{i-1,j,k-1/2}^{n+1/2} + h_w\big|_{i-1,j,k+1/2}^{n+1/2}\right)\right] \qquad (4.72)$$

A significant feature of (4.72) is its fully conservative profile, which has an important effect on the simulated outcomes. Plugging these abstractions into Maxwell's equations, one concludes to

$$(\mathbf{I} + 0.5\mathbf{Y}_t)\mathbf{E}_{cv}^{n+1} = (\mathbf{I} - 0.5\mathbf{Y}_t)\mathbf{E}_{cv}^{n} - \varepsilon^{-1}\mathbf{J}_{cv}^{n+1/2} + \bar{\bar{G}}^{\mathrm{H}}\mathcal{D}\left[\mathbf{H}_{cv}^{n+1/2}\right] + \mathbf{S}_{\mathrm{E}}^{n+1/2}, \qquad (4.73a)$$

$$(\mathbf{I} + 0.5\mathbf{R}_t)\mathbf{H}_{cv}^{n+1/2} = (\mathbf{I} - 0.5\mathbf{R}_t)\mathbf{H}_{cv}^{n-1/2} - \mu^{-1}\mathbf{M}_{cv}^{n} - \bar{\bar{G}}^{\mathrm{E}}\mathcal{D}\left[\mathbf{E}_{cv}^{n}\right] + \mathbf{S}_{\mathrm{H}}^{n}, \qquad (4.73b)$$

where $\mathcal{D}$ is the curl counterpart of the spatial operator according to a specific FDTD-based algorithm, $\mathbf{E}_{cv} = [e_u\, e_v\, e_w]^T$, $\mathbf{H}_{cv} = [h_u\, h_v\, h_w]^T$ are matrices of the covariant electric/magnetic quantities, $\mathbf{J} = \sigma\mathbf{E}$, $\mathbf{M} = \rho'\mathbf{H}$ the respective conduction current densities, $\mathbf{Y}_t$, $\mathbf{R}_t$ constitutive matrices of every material, and $\mathbf{G}^{\mathrm{H}}$, $\mathbf{G}^{\mathrm{E}}$ suitable metric tensors. Finally, matrices $\mathbf{T}_{\mathrm{E}}$ and $\mathbf{T}_{\mathrm{H}}$ collect all temporal derivatives of higher order, owing to operators $\mathcal{D}[.]$ and facilitate the total time-marching process.

4.5.3   Generalized Nonorthogonal Methodologies

In the case of completely arbitrary surface curvatures or cross sections, the use of nonorthogonal unstructured meshes appears to be a very proficient choice [53–65]. Essentially, the configuration, described in this paragraph, may be viewed as a generalization of the FDTD method, and therefore, its extraction will be instructive for every related time-domain scheme [63]. Again, the computational region is divided into a dual-grid ensemble with both lattices comprising polyhedra (usually hexahedra), whose edges connect the barycenters of adjacent primary/secondary cells. For this arrangement, electric components are located along primary-cell edges, whereas magnetic components along secondary-cell ones. The resulting discretization strategy reexpresses Ampère's and Faraday's laws—via the proper surface and line integrals—as

$$\sum_{j=1}^{L_i^e} \mathbf{q}_j \cdot \mathbf{H}_{i,j}^{n+1/2} = A_i\left[\frac{1}{\Delta t}\left(\mathbf{D}_i^{n+1} - \mathbf{D}_i^{n}\right) + \frac{\sigma_{\mathrm{A}}}{2\varepsilon_{\mathrm{A}}}\left(\mathbf{D}_i^{n+1} - \mathbf{D}_i^{n}\right) + \mathbf{J}_i^{n+1/2}\right]\cdot\hat{\mathbf{n}}_q, \qquad (4.74a)$$

$$\sum_{j=1}^{L_i^e} \mathbf{r}_j \cdot \mathbf{E}_{i,j}^n = -\frac{A_i}{\Delta t}\left(\mathbf{B}_i^{n+1/2} - \mathbf{B}_i^{n-1/2}\right) \cdot \hat{\mathbf{n}}_r, \qquad (4.74b)$$

where L_i^e is the number of edges forming the ith secondary face in (4.74a) or its primary counterpart in (4.74b) of area A_i, vectors $\mathbf{q}_j$ and $\mathbf{r}_j$ denote the length of the jth edges that surround the secondary and primary face, respectively, and σ_A, ε_A represent the average loss and dielectric parameters of the media occupying the domain.

Isolating the most recent values of $\mathbf{D}$ and $\mathbf{B}$ in (4.74), the subsequent update relations are derived

$$\mathbf{D}_i^{n+1} \cdot \hat{\mathbf{n}}_q = \frac{1}{2\varepsilon_A + \sigma_A \Delta t}\left[(2\varepsilon_A - \sigma_A \Delta t)\mathbf{D}_i^{n+1} + 2\varepsilon_A \Delta t\left(\frac{1}{A_i}\sum_{j=1}^{L_i^e}\mathbf{q}_j \cdot \mathbf{H}_{i,j}^{n+1/2} - \mathbf{J}_i^{n+1/2}\right)\right] \cdot \hat{\mathbf{n}}_q,$$

$$(4.75a)$$

$$\mathbf{B}_i^{n+1/2} \cdot \hat{\mathbf{n}}_r = \mathbf{B}_i^{n+1/2} \cdot \hat{\mathbf{n}}_r - \frac{\Delta t}{A_i}\sum_{j=1}^{L_i^e}\mathbf{r}_j \cdot \mathbf{E}_{i,j}^n \qquad (4.75b)$$

Because edge vectors are not necessarily parallel to $\mathbf{q}_j$ and $\mathbf{r}_j$, all perpendicular fields should be projected on the complementary mesh edge, thus mandating the use of an auxiliary relation. For instance, to accomplish the required projection on a secondary edge, an interpolation among all magnetic flux densities perpendicular to neighboring faces must be considered with two distinct prerequisites: (a) the projected flux is divergence-free in a medium without charges and (b) numerical stability is globally conserved. Taking into account that every face is shared by N^f cells and each face is formed by L^e edges connecting P^e nodes, magnetic fluxes assigned to the ith node and the jth cell are evaluated via the system of

$$\mathbf{B}_{i,j} \cdot \mathbf{A}_r = \mathbf{B} \cdot \mathbf{A}_r, \ \mathbf{B}_{i,j} \cdot \mathbf{A}_r|_{i,j} = \mathbf{B} \cdot \mathbf{A}_r|_{i,j}, \qquad \mathbf{B}_{i,j} \cdot \mathbf{A}_r|_{i+1,j} = \mathbf{B} \cdot \mathbf{A}_r|_{i+1,j}, \qquad (4.76)$$

with $\mathbf{A}_r = A_i\hat{\mathbf{n}}_r$ the vector areas such as in the FVTD algorithm (see Section 3.3.1). Observe that $\mathbf{B}_{i,j}$ does not play the role of an interpolation of the total field. On the contrary, it is a local value assigned to the (i,j)th corner shared by the face of interest, as graphically depicted in Figure 4.12. This process is repeated for each of the P^e nodes of the face and for every one of the N^f cells. In this manner, the interpolated magnetic flux density—projected on a secondary-cell edge—is given by

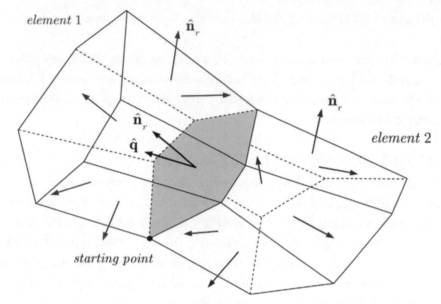

FIGURE 4.12: Two neighboring primary cells where every face is accessed by a secondary edge along the direction of unit vector $\hat{\mathbf{q}}$.

$$\mathbf{B} \cdot \hat{\mathbf{q}} = \left[\left(\sum_{j=1}^{Nf} \sum_{i=1}^{L_j^e} \left| s_{i,j}^r \right| \right)^{-1} \sum_{j=1}^{Nf} \sum_{i=1}^{L_j^e} \left| s_{i,j} \right| \mathbf{B}_{i,j} \right] \cdot \hat{\mathbf{q}} \ , \qquad (4.77a)$$

for $s_{i,j}^r = \mathbf{A}_r \cdot \left(\mathbf{A}_r \big|_{i,j} \times \mathbf{A}_r \big|_{i+1,j} \right)$ certain weighting coefficients and $\hat{\mathbf{q}}$ the unit vector directed toward the aforesaid secondary edge. Evidently, magnetic field intensity $\mathbf{H}$ is computed via the $\mathbf{H} = \mathbf{B} / \mu$ constitutive law. In an absolutely dual way, the interpolated electric flux density—projected on a primary-cell edge—reads

$$\mathbf{D} \cdot \hat{\mathbf{r}} = \left[\left(\sum_{j=1}^{Nf} \sum_{i=1}^{L_j^e} \left| s_{i,j}^q \right| \right)^{-1} \sum_{j=1}^{Nf} \sum_{i=1}^{L_j^e} \left| s_{i,j} \right| \mathbf{D}_{i,j} \right] \cdot \hat{\mathbf{r}} \ , \qquad (4.77b)$$

where, now, $s_{i,j}^q = \mathbf{A}_q \cdot \left(\mathbf{A}_q \big|_{i,j} \times \mathbf{A}_q \big|_{i+1,j} \right)$ and $\mathbf{A}_q = A_i \hat{\mathbf{n}}_q$. Again, electric field intensity $\mathbf{E}$ is calculated by the dual $\mathbf{E} = \mathbf{D} / \varepsilon$ relation. Note that (4.77a) and (4.77b) do satisfy Gauss' law in discrete form, as in the usual FDTD method, whereas their stability limit follows the gist of the Courant criterion.

The nonorthogonal methodology, described above, can adopt almost any polyhedron as a space cell and, therefore, it can handle significantly complex EMC geometries. However, serious attention must be drawn on the construction of the appropriate computer codes, especially in the case of indexing and variable storage.

4.6 FREQUENCY-DEPENDENT MEDIA

It is commonplace in real-world EMC applications that the permittivity and permeability of one or more materials to depend on frequency variations. In this case, simplistic assumptions of constant values within a certain spectrum are likely to yield misleading results. Not to mention that such choices violate the principle of causality, deviating from the actual physical meaning. Moreover, a nonunity constant relative permittivity is analogous to deeming the charge polarization in the material—also known as *dispersive*—instantaneous and in full proportion to the induced electric field. Actually, none of these conventions is correct because, as already presented elsewhere in this book, instead of having the plain constitutive laws for electric/magnetic intensities and fluxes, it is reasonable to use the tensor form of ε, μ to interpret different behaviors in diverse directions. Note that the off-diagonal terms in these tensors indicate the amount of coupling from one direction to another.

To precisely model media dispersion in a time-domain method, it is mandatory to introduce the so-called dispersive formulations, which can be roughly classified into three categories: (a) the *piecewise recursive convolution* method [66], (b) the *Z-transform* technique [67], and (c) the *auxiliary differential equation* algorithm [1, 68, 69]. In all these approaches, the analysis is performed in the frequency domain, and then, through an inverse Fourier transform and the selected discretization strategy, the frequency-dependent properties are launched in the time domain. So, the constitutive relations between the fields and the properties of a medium, $\mathbf{D}(\omega) = \varepsilon(\omega)\mathbf{E}(\omega)$ and $\mathbf{B}(\omega) = \mu(\omega)\mathbf{H}(\omega)$ may be promptly inverse transformed into the time domain by

$$\mathbf{D}(t) = \int_0^t \varepsilon(\tau)\mathbf{E}(t-\tau)\,\mathrm{d}\tau \quad \text{and} \quad \mathbf{B}(t) = \int_0^t \mu(\tau)\mathbf{H}(t-\tau)\,\mathrm{d}\tau \qquad (4.78)$$

Regularly, constitutive relations are expressed in the more detailed form of

$$\mathbf{D}(\omega) = \varepsilon_0\mathbf{E}(\omega) + \mathbf{P}(\omega) \quad \text{and} \quad \mathbf{H}(\omega) = \mu_0^{-1}\mathbf{B}(\omega) - \mathbf{Q}(\omega), \qquad (4.79)$$

where $\mathbf{P}(\omega) = \chi_e(\omega)\varepsilon_0\mathbf{E}(\omega)$ and $\mathbf{Q}(\omega) = \chi_m(\omega)\mathbf{H}(\omega)$ are the electric and magnetic polarization currents, whereas $\chi_e(\omega)$ and $\chi_m(\omega)$ are the corresponding susceptibilities. Hence, media properties can be described by

$$\varepsilon(\omega) = \varepsilon_0 \left[\varepsilon_\infty + \chi_e(\omega) \right] \qquad \text{and} \qquad \mu(\omega) = \mu_0 \left[\mu_\infty + \chi_m(\omega) \right], \qquad (4.80)$$

with ε_∞ and μ_∞ as the relative permittivity and permeability at infinite frequency, respectively. Using the inverse Fourier transform in (4.79), one gets

$$\mathbf{D}(t) = \varepsilon_0 \varepsilon_\infty \mathbf{E}(t) + \varepsilon_0 \int_0^t \chi_e(\tau) \mathbf{E}(t - \tau) \, d\tau, \qquad (4.81a)$$

$$\mathbf{B}(t) = \mu_0 \mu_\infty \mathbf{H}(t) + \mu_0 \int_0^t \chi_m(\tau) \mathbf{H}(t - \tau) \, d\tau, \qquad (4.81b)$$

in which all fields are meant to be zero before $t = 0$, and $\chi_e(t), \chi_m(t)$ are obtained from

$$\chi_e(t) = \frac{1}{2\pi} \int_0^\infty \chi_e(\omega) e^{j\omega t} \, d\omega \qquad \text{and} \qquad \chi_m(t) = \frac{1}{2\pi} \int_0^\infty \chi_m(\omega) e^{j\omega t} \, d\omega \qquad (4.82)$$

Basically, (4.78)–(4.82) are tailored for three fundamental classes of linear and isotropic material dispersions at the macroscopic scale: (a) the *Debye relaxation*, (b) the *Lorentz resonance*, and (c) the *Drude metal* model. Focusing on the relative dielectric permittivity, the necessary expressions that govern the respective and equivalently designated media are summarized in Table 4.2.

In the relations of Table 4.2, the complex-valued electric susceptibilities have one or more poles m (real for Debye and complex-conjugate pairs for Lorentz) at separate frequencies. Furthermore,

TABLE 4.2: Electric susceptibilities and relative dielectric permittivities for different dispersive media

MATERIAL	ELECTRIC SUSCEPTIBILITY	RELATIVE DIELECTRIC PERMITTIVITY
Debye	$\chi_{e,m}(\omega) = \dfrac{\varepsilon_{s,m} - \varepsilon_{\infty,m}}{1 + j\omega\tau_m} \equiv \dfrac{\Delta\varepsilon_m}{1 + j\omega\tau_m}$	$\varepsilon_r(\omega) = \varepsilon_\infty + \displaystyle\sum_{m=1}^{M} \dfrac{\Delta\varepsilon_m}{1 + j\omega\tau_m}$
Lorentz	$\chi_{e,m}(\omega) = \dfrac{\omega_m^2 \Delta\varepsilon_m}{\omega_m^2 + 2j\omega\delta_m - \omega^2}$	$\varepsilon_r(\omega) = \varepsilon_\infty + \displaystyle\sum_{m=1}^{M} \dfrac{\omega_m^2 \Delta\varepsilon_m}{\omega_m^2 + 2j\omega\delta_m - \omega^2}$
Drude	$\chi_{e,m}(\omega) = -\dfrac{\omega_m^2}{\omega^2 - j\omega\gamma_m}$	$\varepsilon_r(\omega) = \varepsilon_\infty - \displaystyle\sum_{m=1}^{M} \dfrac{\omega_m^2}{\omega^2 - j\omega\gamma_m}$

$\Delta\varepsilon_m = \varepsilon_{s,m} - \varepsilon_{\infty,m}$ signifies the fluctuation of relative permittivity owing to the action of a specific pole, $\varepsilon_{s,m}$ is the static relative permittivity, τ_m is the polar relaxation time, ω_m is the Lorentz or Drude pole frequency, δ_m is the Lorentz damping coefficient, and γ_m the inverse of the Drude pole relaxation time.

Although the above theoretical formulation is more frequently encountered in the FDTD-based methods, they are equivalently efficient for the other time-domain schemes as well because they remain independent of spatial discretization. Consequently, in this section, they will be prescribed in a general framework literally applicable to almost every preference. Moreover, a brief reference on some specialized implementation details for the incorporation of material dispersion in the TLM technique is also performed.

4.6.1   Piecewise Linear Recursive Convolution Techniques

The particular techniques stem from (4.81a), whose discrete form is written as

$$\mathbf{D}^n = \varepsilon_0\varepsilon_\infty\mathbf{E}^n + \varepsilon_0 \int_0^{n\Delta t} \chi_e(\tau)\mathbf{E}(n\Delta t - \tau)\,\mathrm{d}\tau \,, \qquad (4.83)$$

where for the continuous variation of the electric field it is supposed that all components vary linearly between sample points [66]. In this way, for such a fluctuation between $\ell\Delta t$ and $(\ell+1)\Delta t$, one receives

$$\mathbf{E}(t) = \mathbf{E}^\ell + \left(\mathbf{E}^{\ell+1} - \mathbf{E}^\ell\right)\frac{t - \ell\Delta t}{\Delta t} \qquad (4.84)$$

For a more general representation, $\mathbf{E}(t)$ can be expressed as

$$\mathbf{E}(t) = \sum_{\ell=0}^{L-1}\left[\mathbf{E}^\ell + \left(\mathbf{E}^{\ell+1} - \mathbf{E}^\ell\right)\frac{t - \ell\Delta t}{\Delta t}\right]p_\ell(t)\,, \qquad (4.85)$$

with $p_\ell(t)$ as the well-known pulse function. Explicitly, (4.85) guarantees a piecewise linear approximation of $\mathbf{E}(t)$ over the L segments because $p_\ell(t)$ permits the involvement of only one segment for any t. Recalling that (4.83) holds for $\ell\Delta t \le \tau \le (\ell+1)\Delta t$, i.e., segment ℓ, the value of $\mathbf{E}(t)$ in the integrand becomes

$$\mathbf{E}(n\Delta t - \tau) = \sum_{\ell=0}^{n-1}\left[\mathbf{E}^{n-\ell} + \left(\mathbf{E}^{n-\ell-1} - \mathbf{E}^{n-\ell}\right)\frac{\tau - \ell\Delta t}{\Delta t}\right]p_\ell(t) \qquad (4.86)$$

Replacement of (4.86) into (4.83) leads to

$$\mathbf{D}^n = \varepsilon_0\varepsilon_\infty\mathbf{E}^n + \varepsilon_0 \int_0^{n\Delta t} \sum_{\ell=0}^{n-1}\left[\mathbf{E}^{n-\ell} + \left(\mathbf{E}^{n-\ell-1} - \mathbf{E}^{n-\ell}\right)\frac{\tau - \ell\Delta t}{\Delta t}\right]p_\ell(t)\chi_e(\tau)\,\mathrm{d}\tau \,, \quad (4.87)$$

which, if $p_\ell(t)$ is expanded over its range and the sum is interchanged with the integration, yields

$$\mathbf{D}^n = \varepsilon_0 \varepsilon_\infty \mathbf{E}^n + \varepsilon_0 \sum_{\ell=0}^{n-1} \left[\mathbf{E}^{n-\ell} \chi^\ell + \left(\mathbf{E}^{n-\ell-1} - \mathbf{E}^{n-\ell} \right) \tau^\ell \right], \qquad (4.88)$$

for $\chi^\ell = \int_{\ell\Delta t}^{(\ell+1)\Delta t} \chi_e(\tau) d\tau$ and $\tau^\ell = \dfrac{1}{\Delta t} \int_{\ell\Delta t}^{(\ell+1)\Delta t} (\tau - \ell\Delta t) \chi_e(\tau) d\tau$.

To derive an **E**-field update equation, the $\ell = 0$ term is extracted from the sum, and (4.88) gives

$$\mathbf{D}^{n+1} = \varepsilon_0 \tau^0 \mathbf{E}^n + \varepsilon_0 \left(\varepsilon_\infty + \chi^0 - \tau^0 \right) \mathbf{E}^{n+1}$$

$$+ \varepsilon_0 \sum_{\ell=0}^{n-1} \left[\mathbf{E}^{n-\ell} \chi^{\ell+1} + \left(\mathbf{E}^{n-\ell-1} - \mathbf{E}^{n-\ell} \right) \tau^{\ell+1} \right] \qquad (4.89)$$

Therefore, the temporal derivative of electric flux **D** can be computed by

$$\frac{\partial \mathbf{D}}{\partial t} \Rightarrow \frac{\mathbf{D}^{n+1} - \mathbf{D}^n}{\Delta t} = \frac{1}{\Delta t} \left\{ \varepsilon_0 \left(\tau^0 - \varepsilon_\infty \right) \mathbf{E}^n + \varepsilon_0 \left(\varepsilon_\infty + \chi^0 - \tau^0 \right) \mathbf{E}^{n+1} \right.$$

$$\left. - \varepsilon_0 \sum_{\ell=0}^{n-1} \left[\mathbf{E}^{n-\ell} \Delta\chi^\ell + \left(\mathbf{E}^{n-\ell-1} - \mathbf{E}^{n-\ell} \right) \Delta\tau^\ell \right] \right\}, \qquad (4.90)$$

with $\Delta\chi^\ell = \chi^\ell - \chi^{\ell+1}$ and $\Delta\tau^\ell = \tau^\ell - \tau^{\ell+1}$. In this framework, Ampère's law becomes

$$\mathbf{E}^{n+1} = \frac{1}{\varepsilon_\infty + \chi^0 - \tau^0} \left[\left(\varepsilon_\infty - \tau^0 \right) \mathbf{E}^n + \frac{\Delta t}{\varepsilon_0} \nabla \times \mathbf{H}^{n+1/2} + \mathbf{\Psi}^n \right], \qquad (4.91)$$

where $\mathbf{\Psi}^n$, designated as the recursive accumulator, has the form of

$$\mathbf{\Psi}^n = \sum_{\ell=0}^{n-1} \left[\mathbf{E}^{n-\ell} \Delta\chi^\ell + \left(\mathbf{E}^{n-\ell-1} - \mathbf{E}^{n-\ell} \right) \Delta\tau^\ell \right] \qquad (4.92)$$

Observing (4.91), a serious issue arises: the sum must be calculated for all previous electric field values. Such an action, however, can be inhibiting for practical cases with large simulation times. Hence, to circumvent this difficulty, it is assumed that the material impulse response may be expressed in terms of exponentials, thus permitting the profitable redefinition of the sum. To this end, $\mathbf{\Psi}^n$ is written as

$$\Psi^n = \mathbf{E}^{n-1}\Delta\tau^0 + \left(\Delta\chi^0 - \Delta\tau^0\right)\mathbf{E}^n + \sum_{\ell=1}^{n-1}\left[\mathbf{E}^{n-\ell}\Delta\chi^\ell + \left(\mathbf{E}^{n-\ell-1} - \mathbf{E}^{n-\ell}\right)\Delta\tau^\ell\right], \qquad (4.93)$$

which, after a slight index change for ℓ, so as to vary from 0 to $n-2$, leads to

$$\begin{aligned}
\Psi^n = \ &\mathbf{E}^{n-1}\Delta\tau^0 + \left(\Delta\chi^0 - \Delta\tau^0\right)\mathbf{E}^n \\
&+ \sum_{\ell=0}^{n-2}\left[\mathbf{E}^{n-\ell-1}\Delta\chi^{\ell+1} + \left(\mathbf{E}^{n-\ell-2} - \mathbf{E}^{n-\ell-1}\right)\Delta\tau^{\ell+1}\right]
\end{aligned} \qquad (4.94)$$

If, additionally in (4.92), n is substituted with $n-1$

$$\Psi^{n-1} = \sum_{\ell=0}^{n-2}\left[\mathbf{E}^{n-\ell-1}\Delta\chi^\ell + \left(\mathbf{E}^{n-\ell-2} - \mathbf{E}^{n-\ell-1}\right)\Delta\tau^\ell\right] \qquad (4.95)$$

Notice the resemblance of the sum in (4.94) to the right-hand side of (4.95), with the only exception of $\Delta\chi$ and $\Delta\tau$. For certain material types, it is feasible to relate the values of the latter quantities as

$$\Delta\chi^{\ell+1} = K\Delta\chi^\ell \quad \text{and} \quad \Delta\tau^{\ell+1} = K\Delta\tau^\ell, \qquad (4.96)$$

with K as a recursion coefficient determined by the constitutive properties of the medium. Plugging (4.96) into (4.92) and replacing n with $n+1$, the final expression for the accumulator is derived

$$\Psi^{n+1} = \Delta\tau^0\mathbf{E}^n + \left(\Delta\chi^0 - \Delta\tau^0\right)\mathbf{E}^{n+1} + K\Psi^n \qquad (4.97)$$

The above methodology is then ready to be applied in a Debye, Lorentz, or Drude medium on condition that its impulse response has an exponential nature to favor the use of the recursive convolution concept.

4.6.2   Z-Transform Methods

Stemming from signal processing, the Z-transform has been proven a fairly powerful tool for the modeling of dispersive materials [67]. Let us consider the constitutive relation of

$$\mathbf{D}(\omega) = \varepsilon^c(\omega)\mathbf{E}(\omega), \quad \text{for} \quad \varepsilon^c(\omega) = \varepsilon_r + \frac{\varepsilon'}{a + j\omega} \qquad (4.98)$$

a complex dielectric, with a, ε' known parameters that characterize the medium, under study. Obviously, if (4.98) is transformed in the time domain, it will lead to a convolution, i.e.,

$$\mathbf{D}(t) = \int_{-\infty}^{+\infty} \varepsilon(\tau)\mathbf{E}(t-\tau)\,d\tau \tag{4.99}$$

Substitution of (4.99) into (4.98), after the pertinent time-domain transformations, gives

$$\mathbf{D}(t) = \varepsilon_r\mathbf{E}(t) + \int_{-\infty}^{+\infty} \varepsilon' e^{-a\tau} u(\tau)\mathbf{E}(t-\tau)\,d\tau \tag{4.100}$$

Realizing the cumbersome character of (4.100), especially in the case of arbitrary EMC excitations, one can use the Z-transform for (4.98) and acquire

$$\mathbf{D}(z) = \varepsilon(z)\mathbf{E}(z), \quad \text{with} \quad \varepsilon(z) = \varepsilon_r + \frac{\varepsilon'}{1 - z^{-1}e^{-a\Delta t}} \tag{4.101}$$

and z^{-1} playing the role of a delay operator with an interval of Δt. Thus, (4.101) becomes

$$\mathbf{D}(z) = \varepsilon_r\mathbf{E}(z) + \left(1 + \frac{z^{-1}e^{-a\Delta t}}{1 - z^{-1}e^{-a\Delta t}}\right)\varepsilon_1\mathbf{E}(z) = (\varepsilon_r + \varepsilon')\,\mathbf{E}(z) + z^{-1}\mathbf{S}(z), \tag{4.102}$$

where $\mathbf{S}(z) = \dfrac{\varepsilon' e^{-a\Delta t}}{1 - z^{-1}e^{-a\Delta t}}\mathbf{E}(z)$.

Expressions (4.102) and (4.103) are the actual time-marching forms for the $\mathbf{E}$ field because

$$\mathbf{E}(z) = \frac{\mathbf{D}(z) - z^{-1}\mathbf{S}(z)}{\varepsilon_r + \varepsilon'} \quad \text{and} \quad \mathbf{S}(z) = \varepsilon' e^{-a\Delta t}\mathbf{E}(z) + z^{-1}e^{-a\Delta t}\mathbf{S}(z) \tag{4.103}$$

The Z-transform technique avoids complicated integrals, restraining itself to algebraic equations in the Z domain, which are much simpler to implement than those of the piecewise recursive convolution method.

4.6.3 Auxiliary Differential Equation Algorithms

This class of methods attempts to connect the polarization and the electric flux density via a time-domain differential equation, supplementary in action and updated concurrently with the main scheme [1, 68–74]. The analysis is based on Ampère's law, which at any lattice node in the dispersive material, reads

$$\nabla \times \mathbf{H} - \sum_{m=1}^{M} \mathbf{J}_m = \varepsilon_0 \varepsilon_\infty \frac{\partial \mathbf{E}}{\partial t} + \sigma \mathbf{E}, \tag{4.104}$$

with $\mathbf{J}_m$ as the polarization current assigned to the mth pole of the medium's susceptibility function. As a matter of fact, the chief purpose is to derive an efficient formula for $\mathbf{J}_m$ for its simultaneous time-marching with (4.104). In the following, the method is developed for the multiple-pole Debye, Lorentz, and Drude case, where the harmonic (exponential) convention for complex-valued quantities, is assumed.

Debye media. According to (4.104) and Table 4.2, the complex polarization current for the mth pole is

$$\mathbf{J}_m = \varepsilon_0 \Delta \varepsilon_m \left(\frac{j\omega}{1 + j\omega\tau_m} \right) \mathbf{E} \xrightarrow[\text{transform}]{\text{Fourier}} \mathbf{J}_m + \tau_m \frac{\partial \mathbf{J}_m}{\partial t} = \varepsilon_0 \Delta \varepsilon_m \frac{\partial \mathbf{E}}{\partial t} \tag{4.105}$$

The latter Fourier-transformed relation is the desired auxiliary differential equation for the update of $\mathbf{J}_m$. Several time-domain schemes can be utilized for its discretization; however, because of the frequent use of the FDTD algorithm in the relevant EMC problems, (4.105) will receive a central-differencing treatment. Hence,

$$\mathbf{J}_m^{n+1} = A_m \mathbf{J}_m^n + \frac{B_m}{\Delta t} \left(\mathbf{E}^{n+1} - \mathbf{E}^n \right), \text{ with } A_m = \frac{1 - 0.5\tau_m \Delta t}{1 + 0.5\tau_m \Delta t}, B_m = \frac{\varepsilon_0 \Delta \varepsilon_m \Delta t}{\tau_m (1 + 0.5\tau_m \Delta t)} \tag{4.106}$$

As deduced from (4.106), the temporal evolution of $\mathbf{E}^{n+1}$ requires the value of $\mathbf{J}_m^{n+1/2}$ which is obtained from the usual averaging practice, presented in Section 2.2, namely,

$$\mathbf{J}_m^{n+1/2} = 0.5 \left(\mathbf{J}_m^{n+1} + \mathbf{J}_m^n \right) = 0.5 \left[(1 + A_m) \mathbf{J}_m^n + B_m \left(\mathbf{E}^{n+1} - \mathbf{E}^n \right) / \Delta t \right] \tag{4.107}$$

Consequently and after some mathematics, (4.104) is discretized as

$$\mathbf{E}^{n+1} = \frac{1}{2\varepsilon_0 \varepsilon_\infty + B + \sigma \Delta t} \left\{ (2\varepsilon_0 \varepsilon_\infty + B - \sigma \Delta t) \mathbf{E}^n + 2\Delta t \left[\nabla \times \mathbf{H}^{n+1/2} - \frac{1}{2} \sum_{m=1}^{M} (1 + A_p) \mathbf{J}_m^n \right] \right\} \tag{4.108}$$

for $B = \sum_{m=1}^{M} B_m$. Overall, the technique proceeds in a threefold fully explicit way during a time step: (1) the components of $\mathbf{E}^{n+1}$ are computed by means of the already known $\mathbf{E}^n, \mathbf{J}_m^n, \mathbf{H}^{n+1/2}$

values; (2) the novel $\mathbf{J}_m^{n+1}$ components are derived through equation (4.106) and the just-updated $\mathbf{E}^{n+1}$; and (3) the next $\mathbf{H}$-field quantities are acquired from $\mathbf{H}^{n+1/2}$ and $\mathbf{E}^{n+1}$ vectors in terms of the conventional FDTD configuration.

Lorentz media. Working in a similar manner as in the Debye case, the induced complex polarization current for the mth pole is given by

$$\mathbf{J}_m = \varepsilon_0 \Delta\varepsilon_m \omega_m^2 \left(\frac{j\omega}{\omega_m^2 + 2j\omega\delta_m - \omega^2} \right) \mathbf{E} \xrightarrow[\text{transform}]{\text{Fourier}} \omega_m^2 \mathbf{J}_m + 2\delta_m \frac{\partial \mathbf{J}_m}{\partial t} + \frac{\partial^2 \mathbf{J}_m}{\partial t^2} = \varepsilon_0 \Delta\varepsilon_m \omega_m^2 \frac{\partial \mathbf{E}}{\partial t},$$

(4.109)

whereas the discretized variant of (4.109) becomes

$$\mathbf{J}_m^{n+1} = A_m \mathbf{J}_m^n + B_m \mathbf{J}_m^{n-1} + \frac{C_m}{2\Delta t} \left(\mathbf{E}^{n+1} - \mathbf{E}^{n-1} \right),$$

(4.110)

with $A_m = \dfrac{2 - \omega_m^2 (\Delta t)^2}{1 + \delta_m \Delta t}$, $B_m = \dfrac{\delta_m \Delta t - 1}{1 + \delta_m \Delta t}$, and $C_m = \dfrac{\varepsilon_0 \omega_m^2 \Delta\varepsilon_m (\Delta t)^2}{1 + \delta_m \Delta t}$.

The necessary calculation of $\mathbf{J}_m^{n+1}$ for the advance of $\mathbf{E}^{n+1}$ is attained by the average of

$$\mathbf{J}_m^{n+1/2} = 0.5 \left(\mathbf{J}_m^{n+1} + \mathbf{J}_m^n \right) = 0.5 \left[(1 + A_m)\mathbf{J}_m^n + B_m \mathbf{J}_m^{n-1} + C_m \left(\mathbf{E}^{n+1} - \mathbf{E}^{n-1} \right) / (2\Delta t) \right]$$

(4.111)

Hence, the discretization of (4.104) now leads to

$$\mathbf{E}^{n+1} = \frac{1}{2\varepsilon_0 \varepsilon_\infty + C + \sigma\Delta t} \left\{ C\mathbf{E}^{n-1} + (2\varepsilon_0 \varepsilon_\infty - \sigma\Delta t) \mathbf{E}^n + 2\Delta t \right.$$
$$\left. \times \left[\nabla \times \mathbf{H}^{n+1/2} - \frac{1}{2} \sum_{m=1}^{M} \left[(1 + A_p)\mathbf{J}_m^n + B_m \mathbf{J}_m^{n-1} \right] \right] \right\},$$

(4.112)

where $C = 0.5 \sum_{m=1}^{M} C_m$. Again, the update procedure follows a three-stage discipline: (1) the components of $\mathbf{E}^{n+1}$ are derived through the already known $\mathbf{E}^{n-1}$, $\mathbf{E}^n$, $\mathbf{J}_m^{n-1}$, $\mathbf{J}_m^n$, $\mathbf{H}^{n+1/2}$ values; (2) the new $\mathbf{J}_m^{n+1}$ vector is found via (4.110) as well as the just-updated $\mathbf{E}^{n+1}$; and (3) the future $\mathbf{H}$-field components are calculated from $\mathbf{H}^{n+1/2}$ and $\mathbf{E}^{n+1}$ quantities by means of the ordinary FDTD approach.

Drude media. For this class of materials, the complex polarization current is

$$\mathbf{J}_m = -j\omega\varepsilon_0 \left(\frac{\omega_m^2}{\omega^2 - j\omega\gamma_m} \right) \mathbf{E} \xrightarrow[\text{transform}]{\text{Fourier}} \frac{\partial^2 \mathbf{J}_m}{\partial t^2} + \gamma_m \frac{\partial \mathbf{J}_m}{\partial t} = \varepsilon_0 \omega_m^2 \frac{\partial \mathbf{E}}{\partial t} \qquad (4.113)$$

Considering an extra integration with regard to time to evade the second-order temporal derivatives and assuming the same central-differencing scheme as in the other two media types, one gets

$$\mathbf{J}_m^{n+1} = A_m \mathbf{J}_m^n + B_m \left(\mathbf{E}^{n+1} + \mathbf{E}^n \right), \text{ with } A_m = \frac{1 - 0.5\gamma_m \Delta t}{1 + 0.5\gamma_m \Delta t}, \ B_m = \frac{0.5\varepsilon_0 \omega_m^2 \Delta t}{1 + 0.5\gamma_m \Delta t} \qquad (4.114)$$

Moreover, the averaging of $\mathbf{J}_m$ leads to

$$\mathbf{J}_m^{n+1/2} = 0.5 \left(\mathbf{J}_m^{n+1} + \mathbf{J}_m^n \right) = 0.5 \left[(1 + A_m)\mathbf{J}_m^n + B_m \left(\mathbf{E}^{n+1} + \mathbf{E}^n \right) \right], \qquad (4.115)$$

through which the discrete version of (4.104) concludes to

$$\mathbf{E}^{n+1} = \frac{1}{2\varepsilon_0\varepsilon_\infty + B + \sigma\Delta t} \left\{ (2\varepsilon_0\varepsilon_\infty - B - \sigma\Delta t)\mathbf{E}^n + 2\Delta t \left[\nabla \times \mathbf{H}^{n+1/2} - \frac{1}{2} \sum_{m=1}^{M} (1 + A_p)\mathbf{J}_m^n \right] \right\}, \qquad (4.116)$$

for $B = \sum_{m=1}^{M} B_m$ and a completely analogous three-step explicit process as in the Debye case.

4.6.4   Dispersive Materials in TLM Simulations

The incorporation of dispersive media in the TLM method requires the construction of specialized models that are able to comply with the usually strong loss or permittivity dependence on frequency variations [75]. Actually, one can view this dependence as the introduction of a transfer function whose discretization in the frequency domain is easy to perform. Thus, different network equations must be developed with any multiplication translated into a convolution in the time domain that enforces the dispersive parts of constitutive parameters to depend on field values at past time steps. In this context, the first stage of the technique requires the excitation at the current time step in conjunction with the computation of the total voltage. Next, the current term of the state variable is derived through its previous values, whereas the reflected voltages are also calculated. In the case of magnetic permeability dispersion, this is accomplished by inserting series components to supplement the shunt ones in the update expressions. The final stage implements the proper connections to the adjacent TLM segments, and the whole procedure is repeated for the next time step.

4.7   ANISOTROPIC MATERIALS

Anisotropic media whose behavior does not depend on frequency variations have a decisive contribution in the design of modern EMC devices. The key premise in the structure of these materials is the tensorial classification of their constitutive parameters and losses. This indicates that the coupling between the components of $\mathbf{D}$ (or $\mathbf{B}$) and those of $\mathbf{E}$ (or $\mathbf{H}$) is no longer one to one [2, 76]. As by now described, in an isotropic medium, the lattice is meant to account for the transverse character of the electromagnetic field and creates a powerful discretization strategy. Unfortunately, the treatment an anisotropic material is more involved.

Let us conduct the numerical simulation of an arbitrary homogeneous, nondispersive, lossy anisotropic material featured by permittivity tensor $\bar{\bar{\varepsilon}}$, a losses tensor $\bar{\bar{\sigma}}$, and a scalar permeability μ. Then, the constitutive relationship $\mathbf{D} = \bar{\bar{\varepsilon}} \times \mathbf{E}$ and Ohm's law $\mathbf{J}_c = \bar{\bar{\sigma}} \times \mathbf{E}$ can be, respectively, written in matrix notation as

$$
\begin{bmatrix} D_x \\ D_y \\ D_z \end{bmatrix} = \begin{bmatrix} \varepsilon_{xx} & \varepsilon_{xy} & \varepsilon_{xz} \\ \varepsilon_{yx} & \varepsilon_{yy} & \varepsilon_{yz} \\ \varepsilon_{zx} & \varepsilon_{zy} & \varepsilon_{zz} \end{bmatrix} \begin{bmatrix} E_x \\ E_y \\ E_z \end{bmatrix} \quad \text{and} \quad \begin{bmatrix} J_{c,x} \\ J_{c,y} \\ J_{c,z} \end{bmatrix} = \begin{bmatrix} \sigma_{xx} & \sigma_{xy} & \sigma_{xz} \\ \sigma_{yx} & \sigma_{yy} & \sigma_{yz} \\ \sigma_{zx} & \sigma_{zy} & \sigma_{zz} \end{bmatrix} \begin{bmatrix} E_x \\ E_y \\ E_z \end{bmatrix} \tag{4.117}
$$

Therefore, the modified vector representation of Ampère's law is given by

$$
\nabla \times \mathbf{H} = \bar{\bar{\sigma}} \cdot \mathbf{E} + \bar{\bar{\varepsilon}} \cdot \frac{\partial \mathbf{E}}{\partial t}, \tag{4.118}
$$

which, after the usual FDTD-oriented temporal discretization and the necessary rearrangements, provides

$$
\mathbf{E}^{n+1} = \left(\frac{\bar{\bar{\varepsilon}}}{\Delta t} + \frac{\bar{\bar{\sigma}}}{2} \right)^{-1} \cdot \left[\nabla \times \mathbf{H}^{n+1/2} + \left(\frac{\bar{\bar{\varepsilon}}}{\Delta t} - \frac{\bar{\bar{\sigma}}}{2} \right) \cdot \mathbf{E}^n \right] \tag{4.119}
$$

To get the three update electric equations, the first matrix of the right-hand side must be inverted. So,

$$
E_x \big|_{i,j,k}^{n+1} = \frac{1}{2\Delta t \det\{\mathbf{T}\}} \Big[\left(\vartheta_{xx}\vartheta_{zz} - \vartheta_{yz}\vartheta_{zy} \right) W_x \big|_{i,j,k} + \left(\vartheta_{xz}\vartheta_{zy} - \vartheta_{xy}\vartheta_{zz} \right) W_y \big|_{i,j,k}
$$
$$
+ \left(\vartheta_{xy}\vartheta_{yz} - \vartheta_{yy}\vartheta_{xz} \right) W_z \big|_{i,j,k} \Big], \tag{4.120a}
$$

$$
E_y\big|_{i,j,k}^{n+1} = \frac{1}{2\Delta t \det\{\mathbf{T}\}} \Big[\left(\vartheta_{yz}\vartheta_{zx} - \vartheta_{yx}\vartheta_{zz}\right) W_x\big|_{i,j,k} + \left(\vartheta_{xx}\vartheta_{zz} - \vartheta_{xz}\vartheta_{zx}\right) W_y\big|_{i,j,k}
$$
$$
+ \left(\vartheta_{xz}\vartheta_{yx} - \vartheta_{xx}\vartheta_{yz}\right) W_z\big|_{i,j,k} \Big]
$$

$$
E_z\big|_{i,j,k}^{n+1} = \frac{1}{2\Delta t \det\{\mathbf{T}\}} \Big[\left(\vartheta_{yx}\vartheta_{zy} - \vartheta_{yy}\vartheta_{zx}\right) W_x\big|_{i,j,k} + \left(\vartheta_{xy}\vartheta_{zx} - \vartheta_{xx}\vartheta_{zy}\right) W_y\big|_{i,j,k}
$$
$$
+ \left(\vartheta_{xx}\vartheta_{yy} - \vartheta_{xy}\vartheta_{yx}\right) W_z\big|_{i,j,k} \Big], \tag{4.120c}
$$

with

$$
W_x\big|_{i,j,k} = \frac{1}{2\Delta t}\left(\psi_{xx} E_x\big|_{i,j,k}^{n} + \psi_{xy} E_y\big|_{i,j,k}^{n} + \psi_{xz} E_z\big|_{i,j,k}^{n}\right) + \left(\frac{\partial H_z}{\partial y} - \frac{\partial H_y}{\partial y}\right)\bigg|_{i,j,k}^{n+1/2}, \tag{4.121a}
$$

$$
W_z\big|_{i,j,k} = \frac{1}{2\Delta t}\left(\psi_{zx} E_x\big|_{i,j,k}^{n} + \psi_{zy} E_y\big|_{i,j,k}^{n} + \psi_{zz} E_z\big|_{i,j,k}^{n}\right) + \left(\frac{\partial H_y}{\partial x} - \frac{\partial H_x}{\partial y}\right)\bigg|_{i,j,k}^{n+1/2}, \tag{4.121b}
$$

$$
W_y\big|_{i,j,k} = \frac{1}{2\Delta t}\left(\psi_{yx} E_x\big|_{i,j,k}^{n} + \psi_{yy} E_y\big|_{i,j,k}^{n} + \psi_{yz} E_z\big|_{i,j,k}^{n}\right) + \left(\frac{\partial H_x}{\partial z} - \frac{\partial H_z}{\partial x}\right)\bigg|_{i,j,k}^{n+1/2}, \tag{4.121c}
$$

and $\vartheta_{uv} = 2\varepsilon_{uv} + \sigma_{uv}\Delta t$, $\psi_{uv} = 2\varepsilon_{uv} - \sigma_{uv}\Delta t$ with $u, v = x, y, z$. Furthermore in (4.120), $\det\{\mathbf{T}\}$ is the determinant of matrix $\mathbf{T} = \overline{\overline{\varepsilon}}/\Delta t + \overline{\overline{\sigma}}/2$. Despite the superficial difficulty, the preceding formulae do not heavily augment the computational burden, as their material coefficients are calculated and stored only once during the preprocessing of the simulation. Spatial discretization of (4.120) and (4.121) is conducted via the typical central differencing, whereas for the $\mathbf{E}$ or $\mathbf{H}$ vectors that reside at different locations in the cell, the average of their nearest neighbors is used. For example, in (4.120c), E_x and E_y are not provided at the (i, j, k) node with respect to E_z. For this problem, one may utilize the ensuing expressions

$$
E_x\big|_{i,j,k}^{n} = \frac{1}{4}\left(E_x\big|_{i+1,j,k+1}^{n} + E_x\big|_{i,j,k+1}^{n} + E_x\big|_{i+1,j,k-1}^{n} + E_x\big|_{i,j,k-1}^{n}\right), \tag{4.122}
$$

$$
E_y\big|_{i,j,k}^{n} = \frac{1}{4}\left(E_y\big|_{i,j+1,k+1}^{n} + E_y\big|_{i,j,k+1}^{n} + E_y\big|_{i,j+1,k-1}^{n} + E_y\big|_{i,j,k-1}^{n}\right) \tag{4.123}
$$

In a parallel framework, two of the spatial derivatives of (4.120c) are approximated by

$$\frac{\partial H_x}{\partial z}\bigg|_{i,j,k}^{n+1/2} = \frac{1}{4\Delta z}\left(H_x\big|_{i,j+1,k+1/2}^{n+1/2} + H_x\big|_{i,j-1,k+1/2}^{n+1/2} - H_x\big|_{i,j+1,k-1/2}^{n+1/2} - H_x\big|_{i,j-1,k-1/2}^{n+1/2} \right), \quad (4.124)$$

$$\frac{\partial H_z}{\partial x}\bigg|_{i,j,k}^{n+1/2} = \frac{1}{4\Delta x}\left(H_z\big|_{i+1/2,j+1/2,k+1}^{n+1/2} - H_z\big|_{i-1/2,j+1/2,k+1}^{n+1/2} + H_z\big|_{i+1/2,j-1/2,k+1}^{n+1/2} - H_z\big|_{i-1/2,j-1/2,k+1}^{n+1/2} \right.$$

$$\left. + H_z\big|_{i+1/2,j+1/2,k-1}^{n+1/2} - H_z\big|_{i-1/2,j+1/2,k-1}^{n+1/2} + H_z\big|_{i+1/2,j-1/2,k-1}^{n+1/2} - H_z\big|_{i-1/2,j-1/2,k-1}^{n+1/2} \right)$$

$$(4.125)$$

with the derivatives of the other components receiving a similar computation as well.

4.8   SURFACE IMPEDANCE BOUNDARY AND INTERFACE CONDITIONS

Surface impedance boundary conditions (SIBCs) are implemented in an attempt to discard the internal volume of conducting or lossy dielectric objects from the simulation procedure. Particularly in EMC applications, they are of critical importance because they also circumvent the use of very fine grid resolutions—attributed to the shorter wavelengths in the prior structures—throughout the solution volume. On the other hand, the objective of an interface formulation is the fulfillment of physical continuity laws at the boundary of adjacent media and the smooth evolution of propagating waves. In the case of large dissimilarities among the constitutive properties of these materials, a more systematic treatment than the simple usage of a time-domain method, is demanded. For these reasons, the present section is devoted to the description of a frequency-dependent SIBC as well as to a couple of interface algorithms for diverse time-domain approaches.

4.8.1   Generalized Frequency-Dependent SIBCs

The primary concept of a SIBC lies on Ohm's law, which establishes a relationship between the tangential electric field vector and the surface current vector at a point on the interface between free space and the conducting medium with frequency-dependent constitutive parameters ε, μ, σ [77–81]. Supposing the harmonic (exponential) dependence of all complex quantities in the frequency domain, the first-order impedance boundary condition—linking E_x and H_y components—is given by $E_x(\omega) = Z_s(\omega)\,H_y(\omega)$, where $Z_s(\omega)$ is the surface impedance of the conductor. Because the medium is dispersive, the above relation is rewritten in the form of

$$E_x(\omega) = j\omega\left[\frac{Z_s(\omega)}{j\omega}\right]H_y(\omega) \qquad (4.126)$$

Defining $\bar{Z}_s(\omega)\,H_y(\omega)\,/\,j\omega$ and plugging this expression in the frequency-domain surface impedance of a good conductor $Z_s(\omega) = (j\omega\mu\,/\,\sigma)$, one gets $\bar{Z}_s(\omega) = [\mu\,/\,(j\omega\sigma)]^{1/2}$. Via this outcome, (4.126) becomes

$$E_x(\omega) = \bar{Z}_s(\omega)\left[j\omega H_y(\omega)\right],\qquad(4.127)$$

whose time-domain counterpart is acquired through an inverse Fourier transform as

$$E_x(t) = \bar{Z}_s(t) * \left[\frac{\partial H_y(t)}{\partial t}\right],\qquad(4.128)$$

with the asterisk signifying the convolution process. The value of $\bar{Z}_s(t)$ is derived by the Laplace transform of $Z_s(\omega)$, namely, $Z_s(s) = [\mu\,/\,(\sigma s)]^{1/2}$ for $s = j\omega$, then taking the inverse operation, $Z_s(t) = [\mu\,/\,(\pi\sigma t)]^{1/2}$ when $t > 0$, and $Z_s(t) = 0$, when $t < 0$. Also, the expansion of the convolution in (4.128) gives

$$E_x(t) = \int_0^t \sqrt{\frac{\mu}{\pi\sigma\tau}}\,\frac{\partial H_y(t-\tau)}{\partial(t-\tau)}\,\mathrm{d}\tau = -\sqrt{\frac{\mu}{\pi\sigma}}\int_0^t \frac{1}{\sqrt{\tau}}\,\frac{\partial H_y(t-\tau)}{\partial\tau}\,\mathrm{d}\tau,\qquad(4.129)$$

which, by means of a discrete implementation, yields

$$E_x\big|_{i,j_b,k}^{n} \simeq \sqrt{\frac{\mu\Delta t}{\pi\sigma}}\sum_{\ell=0}^{n-1}\left\{\frac{\partial H_y\big|_{i,j_b-1/2,k}^{n-\ell}}{\partial\left[(n-\ell)\Delta t\right]}\,Z_0(\ell)\right\},\qquad(4.130)$$

where j_b is the position of the interface according to y-axis and $Z_0(\ell)$ is denoted by

$$Z_0(\ell) = \int_\ell^{\ell+1}\frac{\mathrm{d}\alpha}{\sqrt{\alpha}}\qquad(4.131)$$

Now, (4.130) is ready to be inserted in the update formula of any FDTD-based technique, as a regular field component value, replacing the corresponding term on the interface. Apparently, the direct calculation of the sum (i.e., all past H_y values should be stored for the full convolution) in the resulting time-domain formula is impractical, especially for large-scale 3-D EMC problems. To surpass this difficulty, the recursive evaluation of (4.130) will be performed in terms of the analysis, depicted in the previous section. Such a concept eliminates the requirement for the full time history of field quantities and thus achieves impressive computational savings. The main idea resides on the use of

Prony's method to expand the integrand of the convolution in an equivalent sum of exponentials and afterward apply the recursive algorithm.

Toward this direction and adopting a central-differencing scheme, (4.130) can be, firstly, approximated by

$$E_x|_{i,j_b,k}^n = \sqrt{\frac{\mu \Delta t}{\pi \sigma}} \sum_{\ell=0}^{n-1} \left[\left(H_y|_{i,j_b-1/2,k}^{n-\ell+1/2} - H_y|_{i,j_b-1/2,k}^{n-\ell-1/2} \right) Z_0(\ell) \right]$$

(4.132)

Because through Prony's method $Z_0(\ell) = \sum_{s=1}^{N} b_s e^{b_s \ell}$, the sum of (4.132) is efficiently replaced with

$$\sum_{\ell=0}^{n-1} \left[\left(H_y|_{i,j_b-1/2,k}^{n-\ell+1/2} - H_y|_{i,j_b-1/2,k}^{n-\ell-1/2} \right) Z_0(\ell) \right] = \sum_{s=1}^{N} \psi_s|_{i,j_b-1/2,k}^n ,$$

(4.133)

for N denoting the order of approximation. In (4.133), the s th recursive sum is computed via

$$\psi_s|_{i,j_b-1/2,k}^n = \left(H_y|_{i,j_b-1/2,k}^{n-1/2} - H_y|_{i,j_b-1/2,k}^{n-3/2} \right) b_s e^{b_s} + e^{b_s} \psi_s|_{i,j_b-1/2,k}^{n-1},$$

with $\psi_s|_{i,j_b-1/2,k}^1 = \psi_s|_{i,j_b-1/2,k}^0 = 0$

(4.134)

Because of its temporal locality, the aforesaid scheme combined with the SIBC abstraction is fast and the most significant: fairly affordable for an assortment of real-world applications.

4.8.2 Corrective Media Interface Formulations

Interfaces with arbitrary curvatures. The influence of staircase modeling and the incomplete fulfillment of jump conditions on either side of an arbitrary material interface—not aligned to the axes of the coordinate system—may spoil the stability and convergence of a time-domain approach. Bearing in mind the frequent appearance of such discontinuities in EMC structures, it becomes evident that a more sophisticated treatment must be pursued. This paragraph presents a competent algorithm that alters the spatial stencils near interfaces to accurately weight their contribution in the entire simulation.

Assume the curvilinear boundary of Figure 4.13, established in the secondary grid, with $\hat{\mathbf{n}} = [\hat{n}_u, \hat{n}_v, \hat{n}_w]$ as an outward normal unit vector in the (u, v, w) system of coordinates. The covariant $\mathbf{E}_{cv}^{type} = \left[e_u^{type}, e_v^{type}, e_w^{type} \right]^T$ and $\mathbf{H}_{cv}^{type} = \left[b_u^{type}, b_v^{type}, b_w^{type} \right]^T$ components in the two areas for type = 1, 2 are related by the usual physical continuity conditions. To evade the problem of the curved boundary, two coefficients, $\kappa_{i,j,k}^{(1)}$ and $\kappa_{i,j,k}^{(2)}$, are assigned to the distinct regions, as measures

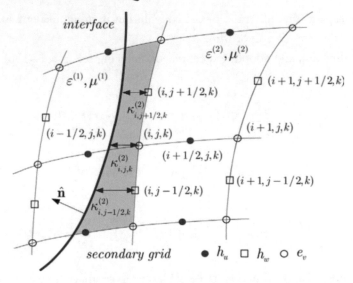

FIGURE 4.13: An arbitrary media interface that does not coincide with the mesh axes.

of the distance from the first/last cell to the location of the wall relative to cell dimensions. It is concluded that $\kappa_{i,j,k}^{\text{type}} \in [0,1/2]$ and satisfies $\kappa_{i,j,k}^{(1)} + \kappa_{i,j,k}^{(2)} = 0.5$. Thus,

$$h_u^{(1)}\Big|_{i,j,k}^{n+1/2} = \left(1 + \kappa_{i,j,k}^{(1)}\right) h_u\Big|_{i-1/2,j,k}^{n+1/2} + \kappa_{i,j,k}^{(1)} h_u\Big|_{i-3/2,j,k}^{n+1/2}, \tag{4.135}$$

is promptly evaluated, whereas $h_w^{(2)}$ is acquired by

$$\bar{h}_w\Big|_{i,j\pm1/2,k}^{n+1/2} = h_w\Big|_{i,j\pm1/2,k}^{n+1/2} + \kappa_{i,j\pm1/2,k}^{(2)} \left(h_w\Big|_{i,j\pm1/2,k}^{n+1/2} - h_w\Big|_{i+1,j\pm1/2,k}^{n+1/2}\right), \tag{4.136}$$

$$h_w^{(2)}\Big|_{i,j,k}^{n+1/2} = 0.5 \left(\bar{h}_w\Big|_{i,j+1/2,k}^{n+1/2} + \bar{h}_w\Big|_{i,j-1/2,k}^{n+1/2}\right), \tag{4.137}$$

where the bar indicates auxiliary magnetic variables. In this sense, $h_v^{(1)}$ may be extracted by four components, found via the subsequent equation along the w-axis, as

$$\bar{h}_v\Big|_{i,j\pm1/2,k\pm1/2}^{n+1/2} = h_v\Big|_{i-1/2,j\pm1/2,k\pm1/2}^{n+1/2} + \kappa_{i,j\pm1/2,k\pm1/2}^{(1)} \left(h_v\Big|_{i-1/2,j\pm1/2,k\pm1/2}^{n+1/2} - h_v\Big|_{i-3/2,j\pm1/2,k\pm1/2}^{n+1/2}\right). \tag{4.138}$$

From these extrapolated quantities, $h_v^{(1)}$ reads

$$h_v^{(1)}\Big|_{i,j,k}^{n+1/2} = 0.25 \left(\bar{h}_v\Big|_{i,j-1/2,k-1/2}^{n+1/2} + \bar{h}_v\Big|_{i,j+1/2,k-1/2}^{n+1/2} + \bar{h}_v\Big|_{i,j+1/2,k+1/2}^{n+1/2} + \bar{h}_v\Big|_{i,j-1/2,k+1/2}^{n+1/2}\right), \tag{4.139}$$

Consequently, one gets

$$b_u^{(1)}\Big|_{i,j,k}^{n+1/2} = b_u^{(1)}\Big|_{i,j,k}^{n+1/2} + \hat{n}_u \left(\mu^{(1)} - \mu^{(2)}\right) \frac{\hat{n}_u \, b_u^{(1)}\Big|_{i,j,k}^{n+1/2} + \hat{n}_v \, b_v^{(1)}\Big|_{i,j,k}^{n+1/2} + \hat{n}_w \, b_w^{(2)}\Big|_{i,j,k}^{n+1/2}}{\mu^{(2)}\hat{n}_u^2 + \mu^{(2)}\hat{n}_v^2 + \mu^{(1)}\hat{n}_w^2}$$

$$(4.140)$$

As a last step, the time-domain expression of b_u spatial derivative with respect to w is

$$\frac{\partial}{\partial w}\left[b_u\Big|_{i,j,k}^{n+1/2}\right] = \frac{2}{\left(2\kappa_{i,j,k}^{(2)} + 1\right)\Delta u}\left(b_u\Big|_{i+1/2,j,k}^{n+1/2} - b_u^{(2)}\Big|_{i,j,k}^{n+1/2}\right).$$

$$(4.141)$$

The rest of **E** components are similarly advanced, whereas the **H** ones use mesh duality. The small augmentation of algorithmic complexity is reimbursed by the highly precise management of curvilinear parts.

Boundaries with inhomogeneous dielectrics. When a higher-order FDTD technique reaches the boundary of an intermediate inhomogeneous dielectric, update formulae must be adequately tailored to eliminate potential instabilities. Let us consider a (2, 4) FDTD approach in the homogeneous regions and a locally modified scheme near its boundaries. The analysis focuses on a 2-D dielectric slab that divides the domain into three areas: two filled with air, i.e., $\varepsilon = \varepsilon_0$ for $i < S_A$ and $i > S_B$, and one with the dielectric ε for $S_A \leq i \leq S_B$. The main trait of this material is that it has a piecewise constant permittivity, whose discontinuity points are at electric nodes. Across the interface E_z and H_y quantities are continuous. So, for H_y at S_A and S_B, one can use the following fifth-order extrapolation, based on evidence from the vacuum,

$$H_y\Big|_{s,j}^{n+1/2} = \frac{1}{128}\left(325 H_y\Big|_{s\mp 1/2,j}^{n+1/2} + 35\, H_y\Big|_{s\mp 9/2,j}^{n+1/2}\right)$$

$$- \frac{1}{64}\left(210\, H_y\Big|_{s\mp 3/2,j}^{n+1/2} - 189\, H_y\Big|_{s\mp 5/2,j}^{n+1/2} + 90\, H_y\Big|_{s\mp 7/2,j}^{n+1/2}\right) \qquad (4.142)$$

with the $\pm$ signs referring to $s = S_A$ and S_B, respectively. To avoid the usual derivative approximations across the interfaces, $\partial H_y / \partial x$ at these nodes is found through

$$\frac{\partial H_y}{\partial x}\Big|_{s,j}^{n+1/2} = \mp \frac{1126}{315\Delta x}\, H_y\Big|_{s,j}^{n+1/2} \pm \frac{1}{16\Delta x}\left(\frac{315}{2}\, H_y\Big|_{s\pm 1/2,j}^{n+1/2} \mp 35\, H_y\Big|_{s\pm 3/2,j}^{n+1/2}\right.$$

$$\left. \pm \frac{189}{10}\, H_y\Big|_{s\pm 5/2,j}^{n+1/2} \mp \frac{45}{7}\, H_y\Big|_{s\pm 7/2,j}^{n+1/2} \pm \frac{35}{36}\, H_y\Big|_{s\pm 9/2,j}^{n+1/2}\right) \qquad (4.143)$$

Then, E_z at $s = S_A$ and S_B is given by

$$E_z|_{s,j}^{n+1} = E_z|_{s,j}^{n} - \frac{9\Delta t}{8\varepsilon\Delta y}\left(H_x|_{s,j+1/2}^{n+1/2} - H_x|_{s,j-1/2}^{n+1/2}\right)$$
$$+ \frac{\Delta t}{24\varepsilon\Delta y}\left(H_x|_{s,j+3/2}^{n+1/2} - H_x|_{s,j-3/2}^{n+1/2}\right) + \frac{\Delta t}{\varepsilon}\frac{\partial H_y}{\partial x}\bigg|_{s,j}^{n+1/2} \qquad (4.144)$$

If the dielectric boundary is located at $S_A + 1/2$, the role of E_z and H_y is reversed and

$$E_z|_{S_A+1/2,j}^{n} = \frac{1}{128}\left(315 E_z|_{S_A,j}^{n} + 35\,E_z|_{S_A-4,j}^{n}\right)$$
$$- \frac{1}{64}\left(210\,E_z|_{S_A-1,j}^{n} - 189\,E_z|_{S_A-2,j}^{n} + 90\,E_z|_{S_A-3,j}^{n}\right), \qquad (4.145)$$

$$\frac{\partial E_z}{\partial x}\bigg|_{S_A+1/2,j}^{n} = -\frac{1126}{315\Delta x}E_z|_{S_A+1/2,j}^{n} + \frac{1}{16\Delta x}\left(\frac{315}{2}E_z|_{S_A+1,j}^{n} - 35\,E_z|_{S_A+2,j}^{n} + \frac{189}{10}E_z|_{S_A+3,j}^{n}\right.$$
$$\left. - \frac{45}{7}E_z|_{S_A+4,j}^{n} + \frac{35}{36}E_z|_{S_A+5,j}^{n}\right)$$

$$(4.146)$$

Hence, H_y at $S_A + 1/2$ is obtained via

$$H_y|_{S_A+1/2,j}^{n+1/2} = H_y|_{S_A+1/2,j}^{n-1/2} + \frac{\Delta t}{\mu}\frac{\partial E_z}{\partial x}\bigg|_{S_A+1/2,j}^{n} \qquad (4.147)$$

It should be emphasized that the performance of (4.142)–(4.147) worsens with the increase of the contrast among the dielectric media constants. This deficiency may be accredited to the more rapid loss of smoothness at the interface and the inevitable growth of lattice reflection errors.

REFERENCES

1. A. Taflove and S. C. Hagness, *Computational Electrodynamics: The Finite-Difference Time-Domain Method*, 3rd ed. Norwood, MA: Artech House, 2005.
2. K. S. Kunz and R. J. Luebbers, *The Finite Difference Time Domain Method for Electromagnetics*. Boca Raton, FL: CRC Press, 1993.
3. A. P. Zhao and A. V. Räisänen, "Application of a simple and efficient source excitation technique to the FDTD analysis of waveguide and microstrip circuits," *IEEE Trans Microwave Theory Technol.*, vol. 44, no. 9, pp. 1535–1539, Sept. 1996.

4. T.-W. Lee and S. C. Hagness, "A compact wave source condition for the pseudospectral time-domain method," *IEEE Antennas Wireless Propagat. Lett.*, vol. 3, no. 1, pp. 253–256, 2004.

5. X. Gao, M. S. Mirotznik, and D. W. Prather, "A method for introducing soft sources in the PSTD algorithm," *IEEE Trans. Antennas Propagat.*, vol. 52, no. 7, pp. 1665–1671, July 2004, doi:10.1109/TAP.2004.831304.

6. J. S. Nielsen and W. J. R. Hoefer, "Generalized dispersion analysis and spurious modes of 2-D and 3-D TLM formulations," *IEEE Trans. Microwave Theory Technol.*, vol. 41, no. 8, pp. 1372–1384, Aug. 1993, doi:10.1109/22.241679.

7. M. Okoniewski, M. Mrozowski, and M. A. Stuchly, "Simple treatment of multi-term dispersion in FDTD," *IEEE Microwave Guided Wave Lett.*, vol. 7, no. 5, pp. 121–123, May 1997, doi:10.1109/75.569723.

8. J. W. Nehrbass, J. O. Jevti., and R. Lee, "Reducing the phase error for finite-difference methods without increasing the order," *IEEE Trans. Antennas Propagat.*, vol. 46, no. 8, pp. 1194–1201, Aug. 1998, doi:10.1109/8.718575.

9. J. B. Schneider and C. L. Wagner, "FDTD dispersion revisited: Faster-than-light propagation," *IEEE Microwave Guided Wave Lett.*, vol. 9, no. 2, pp. 54–56, Feb. 1999, doi:10.1109/75.755044.

10. J. S. Juntunen and T. D. Tsiboukis, "Reduction of numerical dispersion in FDTD method through artificial anisotropy," *IEEE Trans. Microwave Theory Technol.*, vol. 48, no. 4, pp. 582–588, Apr. 2000, doi:10.1109/22.842030.

11. K. Suzuki, T. Kashiwa, and Y. Hosoya, "Reducing the numerical dispersion in the FDTD analysis by modifying anisotropically the speed of light," *Electron. Commun. Jpn. Part II: Electron.*, vol. 85, no. 1, pp. 50–58, Jan. 2002, doi:10.1002/ecjb.1086.

12. E. A. Forgy and W. C. Chew, "A time-domain method with isotropic dispersion and increased stability on an overlapped lattice," *IEEE Trans. Antennas Propagat.*, vol. 50, no. 7, pp. 983–996, July 2002, doi:10.1109/TAP.2002.801373.

13. K. L. Shlager and J. B. Schneider, "Comparison of the dispersion properties of several low-dispersion finite-difference time-domain algorithms," *IEEE Trans. Antennas Propagat.*, vol. 51, no. 3, pp. 642–653, Mar. 2003, doi:10.1109/TAP.2003.808532.

14. S. Wang and F. L. Teixeira, "A finite-difference time-domain algorithm for arbitrary propagation angles," *IEEE Trans. Antennas Propagat.*, vol. 51, no. 9, pp. 2456–2463, Sept. 2003.

15. A. P. Zhao, "Rigorous analysis of the influence of the aspect ratio of Yee's unit cell on the numerical dispersion property of the 2-D and 3-D FDTD methods," *IEEE Trans. Antennas Propagat.*, vol. 52, no. 7, pp. 1630–1637, July 2004.

16. K. Chamberlin and D. Vidacic, "Analysis of finite-differencing errors to determine cell size when modeling ferrites and other lossy electric and magnetic materials using FDTD," *IEEE Trans. Electromagn. Compat.*, vol. 46, no. 4, pp. 617–623, Nov. 2004, doi:10.1109/TEMC.2004.837840.

17. T. Zygiridis and T. D. Tsiboukis, "Higher order finite difference schemes with reduced dispersion errors for accurate time domain electromagnetic simulations," *Int. J. Numer. Model.: Electron. Networks Devices Fields*, vol. 17, pp. 461–486, 2004, doi:10.1002/jnm.551.

18. N. V. Kantartzis, T. D. Tsiboukis, and E. E. Kriezis, "A topologically consistent class of 3-D higher-order curvilinear FDTD schemes for dispersion-optimized EMC and material modeling," *J. Mater. Process. Technol.*, vol. 161, pp. 210–217, 2005, doi:10.1016/j.jmatprotec.2004.07.027.

19. T. T. Zygiridis and T. D. Tsiboukis, "Development of higher order FDTD schemes with controllable dispersion error," *IEEE Trans. Antennas Propagat.*, vol. 53, no. 9, pp. 2938–2951, Sept. 2005.

20. N. V. Kantartzis and T. D. Tsiboukis, *Higher-Order FDTD Schemes for Waveguide and Antenna Structures*, San Rafael, CA: Morgan & Claypool Publishers, 2006, doi:10.2200/S00018ED1V 01Y200604CEM003.

21. B. Engquist and A. Majda, "Absorbing boundary conditions for the numerical simulation of waves," *Math. Comput.*, vol. 31, pp. 629–651, 1977, doi:10.2307/2005997.

22. A. Bayliss and E. Turkel, "Radiation boundary conditions for wave-like equations," *Commun. Pure Appl. Math.*, vol. 23, pp. 707–725, 1980, doi:10.1002/cpa.3160330603.

23. E. L. Lindman, "Free space boundary conditions of the time dependent wave equation," *J. Comput. Phys.*, vol. 18, no. 1, pp. 66–78, May 1975, doi:10.1016/0021-9991(75)90102-3.

24. G. Mur, "Absorbing boundary conditions for the finite-difference approximation of the time-domain electromagnetic field equations," *IEEE Trans. Electromagn. Compat.*, vol. 23, no. 4, pp. 377–382, Nov. 1981.

25. R. L. Higdon, "Absorbing boundary conditions for difference approximations to the multi-dimensional wave equation," *Math. Comput.*, vol. 47, pp. 437–459, 1986, doi:10.2307/2008166.

26. Z. P. Liao, H. L. Wong, B. P. Yang, and Y. F. Yuan, "A transmitting boundary for transient wave analyses," *Sci. Sin. Ser. A*, vol. 27, pp. 1036–1076, 1984.

27. M. J. Grote and J. B. Keller, "Nonreflecting boundary conditions for Maxwell's equations," *J. Comput. Phys.*, vol. 139, no. 2, pp. 327–342, Jan. 1998.

28. K. K. Mei and J. Fang, "Superabsorption—A method to improve absorbing boundary conditions," *IEEE Trans. Antennas Propagat.*, vol. 40, no. 9, pp. 1001–1010, Sept. 1992, doi:10.1109/8.166524.

29. O. M. Ramahi, "Complementary operators: A method to annihilate artificial reflections arising from the truncation of the computational domain in the solution of partial differential equations," *IEEE Trans. Antennas Propagat.*, vol. 43, no. 7, pp. 697–704, July 1995, doi:10.1109/8.391141.

30. J.-P. Bérenger, "A perfectly matched layer for the absorption of electromagnetic waves," *J. Comput. Phys.*, vol. 114, no. 2, pp. 185–200, Oct. 1994, doi:10.1006/jcph.1994.1159.

31. J.-P. Bérenger, "Making use of the PML absorbing boundary condition in coupling and scattering FDTD computer codes," *IEEE Trans. Electromagn. Compat.*, vol. 45, no. 2, pp. 189–197, May 2004.

32. J.-P. Bérenger, *Perfectly Matched Layer (PML) for Computational Electromagnetics*. San Rafael, CA: Morgan & Claypool Publishers, 2007, doi:10.2200/S00030ED1V01Y200605CEM008.

33. W. C. Chew and W. H. Weedon, "A 3D perfectly matched medium from modified Maxwell's equations with stretched coordinates," *Microwave Opt. Technol. Lett.*, vol. 7, no. 13, pp. 599–604, 1994, doi:10.1002/mop.4650071304.

34. Z. S. Sacks, D. M. Kingsland, R. Lee, and J.-F. Lee, "A perfectly matched anisotropic absorber for use as an absorbing boundary condition," *IEEE Trans. Antennas Propagat.*, vol. 44, no. 12, pp. 1630–1639, Dec. 1995.

35. S. D. Gedney, "An anisotropic perfectly matched layer-absorbing medium for the truncation of FDTD lattices," *IEEE Trans. Antennas Propagat.*, vol. 44, no. 12, pp. 1630–1639, Dec. 1996, doi:10.1109/8.546249.

36. L. Zhao and A. C. Cangellaris, "GT-PML: Generalized theory of perfectly matched layers and its application to the reflectionless truncation of finite-difference time-domain grids," *IEEE Trans. Microwave Theory Technol.*, vol. 44, no. 12, pp. 2555–2563, Dec. 1996, doi:10.1109/22.554601.

37. F. L. Teixeira and W. C. Chew, "PML-FDTD in cylindrical and spherical grids," *IEEE Microwave Guided Wave Lett.*, vol. 7, no. 9, pp. 285–287, Sept. 1997, doi:10.1109/75.622542.

38. F. Collino and P. Monk, "The perfectly matched layer in curvilinear coordinates," *SIAM J. Sci. Comp.*, vol. 19, no. 6, pp. 2061–2090, 1998, doi:10.1137/S1064827596301406.

39. P. G. Petropoulos, "Reflectionless sponge layers as absorbing boundary conditions for the numerical solution of Maxwell's equations in rectangular, cylindrical and spherical coordinates," *SIAM J. Appl. Math.*, vol. 60, no. 3, pp. 1037–1058, 2000.

40. N. V. Kantartzis, "A generalised higher-order FDTD-PML algorithm for the enhanced analysis of 3-D waveguiding EMC structures in curvilinear coordinates," *IEEE Proc. Microwave Antennas Propagat.*, vol. 150, no. 5, pp. 351–359, Oct. 2003, doi:10.1049/ip-map:20030269.

41. K. Sankaran, C. Fumeaux, and R. Vahldieck, "Cell-centered finite-volume-based perfectly matched layer for the time-domain Maxwell system," *IEEE Microwave Theory Technol.*, vol. 54, no. 3, pp. 1269–1276, Mar. 2006, doi:10.1103/PhysRevE.55.7696.

42. R. W. Ziolkowski, "Time-derivative Lorentz materials and their utilization as electromagnetic absorbers," *Phys. Rev. E*, vol. 55, pp. 7696–7703, 1997.

43. E. M. Tentzeris, R. L. Robertson, J. F. Harvey, and L. P. B. Katehi, "PML absorbing boundary conditions for the characterization of open microwave circuit components using multiresolution time-domain techniques (MRTD)," *IEEE Trans. Antennas Propagat.*, vol. 47, no. 11, pp. 1709–1715, Nov. 1999, doi:10.1109/8.814951.

44. M. Kuzuoglou and R. Mittra, "Frequency dependence of the constitutive parameters of causal perfectly matched anisotropic absorbers," *IEEE Trans. Microwave Guided Wave Lett.*, vol. 6, no. 12, pp. 447–449, Dec. 1996, doi:10.1109/75.544545.

45. J. A. Roden and S. D. Gedney, "Convolutional PML (CPML): An efficient FDTD implementation of the CFS-PML for arbitrary media," *Microwave Opt. Technol. Lett.*, vol. 27, no. 5, pp. 334–339, Dec. 2000, doi:10.1002/1098-2760(20001205)27:5<334::AID-MOP14>3.3.CO;2-1.

46. D. Jiao, J.-M. Jin, E. Michielssen, and D. J. Riley, "Time-domain finite element simulation of three-dimensional scattering and radiation problems using perfectly matched layers," *IEEE Trans. Antennas Propagat.*, vol. 51, no. 2, pp. 296–305, Feb. 2003, doi:10.1109/TAP.2003.809096.

47. T. Rylander and J.-M. Jin, "Perfectly matched layer in three dimensions for the time-domain finite element method applied to radiation problems," *IEEE Trans. Antennas Propagat.*, vol. 53, no. 4, pp. 1489–1499, Apr. 2005.

48. S. Wang, R. Lee, and F. L. Teixeira, "Anisotropic-medium PML for vector FETD with modified basis functions," *IEEE Trans. Antennas Propagat.*, vol. 54, no. 1, pp. 20–27, Jan. 2006, doi:10.1109/TAP.2005.861523.

49. Q. H. Liu, "PML and PSTD algorithm for arbitrary lossy anisotropic media," *IEEE Microwave Guided Wave Lett.*, vol. 9, pp. 48–50, 1999.

50. F. L. Teixeira and W. C. Chew, "Lattice electromagnetic theory from a topological viewpoint," *J. Math. Phys.*, vol. 40, no. 1, pp. 169–187, Jan. 1999, doi:10.1063/1.532767.

51. M. Fusco, M. Smith, and L. Gordon, "A three-dimensional FDTD algorithm in curvilinear coordinates," *IEEE Trans. Antennas Propagat.*, vol. 39, no. 10, pp. 1463–1471, Oct. 1991, doi:10.1109/8.97377.

52. Q. H. Liu and J. Q. He, "An efficient PSTD algorithm for cylindrical coordinates," *IEEE Trans. Antennas Propagat.*, vol. 49, no. 9, pp. 1349–1351, Sept. 2001.

53. R. Holland, "Finite-difference solution of Maxwell's equations in generalized nonorthogonal coordinates," *IEEE Trans. Nuclear Sci.*, vol. NS-30, no. 6, pp. 4589–4591, Dec. 1983.

54. J.-F. Lee, R. Palendech, and R. Mittra, "Modeling three-dimensional discontinuities in waveguides using the nonorthogonal FDTD algorithm," *IEEE Trans. Microwave Theory Technol.*, vol. 40, no. 2, pp. 346–352, Feb. 1992.

55. S. L. Ray, "Numerical dispersion and stability characteristics of time-domain methods on nonorthogonal meshes," *IEEE Trans. Antennas Propagat.*, vol. 41, no. 2, pp. 233–235, Feb. 1993, doi:10.1109/8.214617.

56. N. Madsen, "Divergence preserving discrete surface integral methods for Maxwell's equations using nonorthogonal unstructured grids," *J. Comput. Phys.*, vol. 119, no. 1, pp. 34–45, June 1995, doi:10.1006/jcph.1995.1114.

57. S. Gedney, F. Lansing, and D. Rascoe, "A generalized Yee-algorithm for the analysis of MMIC devices," *IEEE Microwave Theory Techol.*, vol. 44, pp. 1393–1400, 1996, doi:10.1109/22.536021.

58. R. Schuhmann and T. Weiland, "A stable interpolation technique for FDTD on non-orthogonal grids," *Int. J. Numer. Model.*, vol. 11, no. 6, pp. 299–306, Nov.–Dec. 1998, doi:10.1002/(SICI)1099-1204(199811/12)11:6<299::AID-JNM314>3.0.CO;2-A.

59. S. D. Gedney and J. A. Roden, "Numerical stability of nonorthogonal FDTD methods," *IEEE Trans. Antennas Propagat.*, vol. 48, no. 2, pp. 231–239, Feb. 2000, doi:10.1109/8.833072.

60. N. V. Kantartzis, T. I. Kosmanis, T. V. Yioultsis, and T. D. Tsiboukis, "A nonorthogonal higher-order wavelet-oriented FDTD technique for 3-D waveguide structures on generalised curvilinear grids," *IEEE Trans. Magn.*, vol. 37, no 5, pp. 3264–3268, Sept. 2001, doi:10.1109/20.952591.

61. R. Nilavalan, I. J. Craddock, and C. J. Railton, "Quantifying numerical dispersion in nonorthogonal FDTD meshes," *IEEE Proc. Microwave, Antennas Propagat.*, vol. 149, no. 1, pp. 23–27, Feb. 2002, doi:10.1049/ip-map:20020144.

62. T. I. Kosmanis and T. D. Tsiboukis, "A systematic topologically stable conformal FDTD algorithm for modeling curved dielectric interfaces in 3-D," *IEEE Trans. Microwave Theory Technol.*, vol. 51, no. 3, pp. 839–847, Mar. 2003, doi:10.1109/TMTT.2003.808617.

63. D. Poljak and V. Doric, "Time-domain modeling of electromagnetic field coupling to finite-length wires embedded in a dielectric half-space," *IEEE Trans. Electromagn. Compat.*, vol. 47, no. 2, pp. 247–253, May 2005, doi:10.1109/TEMC.2005.847374.

64. F. Edelvik and T. Weiland, "Stable modeling of arbitrarily oriented thin slots in the FDTD method," *IEEE Trans. Electromagn. Compat.*, vol. 47, no. 3, pp. 440–446, Aug. 2005, doi:10.1109/TEMC.2005.853160.

65. C. Yang and V. Jandhyala, "Combined circuit-electromagnetic simulation using multiregion time domain integral equation scheme," *IEEE Trans. Electromagn. Compat.*, vol. 48, no. 1, pp. 2–9, Feb. 2006, doi:10.1109/TEMC.2006.870708.

66. D. F. Kelley and R. J. Luebbers, "Piecewise linear recursive convolution for dispersive media using FDTD," *IEEE Trans. Antennas Propagat.*, vol. 44, no. 6, pp. 792–797, June 1996, doi:10.1109/8.509882.

67. D. M. Sullivan, "Frequency-dependent FDTD methods using Z transforms," *IEEE Trans. Antennas Propagat.*, vol. 40, no. 10, pp. 1223–1230, Oct. 1992, doi:10.1109/8.182455.

68. J. L. Young, "Propagation in linear dispersive media: Finite difference time-domain methodologies," *IEEE Trans. Antennas Propagat.*, vol. 43, pp. 422–426, 1995, doi:10.1109/8.376042.

69. K. P. Prokopidis, E. P. Kosmidou, and T. D. Tsiboukis, "An FDTD algorithm for wave propagation in dispersive media using higher-order schemes," *J. Electromagn. Waves Appl.*, vol. 18, no. 9, pp. 1171–1194, 2004, doi:10.1163/1569393042955306.

70. J. Choi, M. Swaminathan, N. Do, and R. Master, "Modeling of power supply noise in large chips using the circuit-based finite-difference time-domain method," *IEEE Trans. Electromagn. Compat.*, vol. 47, no. 3, pp. 424–429, Aug. 2005, doi:10.1109/TEMC.2005.851719.

71. G. Orjubin, F. Petit, E. Richalot, S. Mengué, and O. Picon, "Cavity loses modeling using lossless FDTD method," *IEEE Trans. Electromagn. Compat.*, vol. 48, no. 2, pp. 429–430, May 2006.

72. A. P. Duffy, A. J. M. Martin, A. Orlandi, G. Antonini, T. M. Benson, and M. S. Woolfson, "Feature selective validation (FSV) for validation of computational electromagnetics (CEM). Part I—The FSV method," *IEEE Trans. Electromagn. Compat.*, vol. 48, no. 3, pp. 449–459, Aug. 2006, doi:10.1109/TEMC.2006.879358.

73. A. Orlandi, A. P. Duffy, B. Archambeault, G. Antonini, D. E. Coleby, and S. Connor, "Feature selective validation (FSV) for validation of computational electromagnetics (CEM). Part II—Assessment of FSV performance," *IEEE Trans. Electromagn. Compat.*, vol. 48, no. 3, pp. 460–467, Aug. 2006, doi:10.1109/TEMC.2006.879360.

74. Q. H. Liu and G.-X. Fan, "A frequency-dependent PSTD algorithm for general dispersive media," *IEEE Microwave Guided Wave Lett.*, vol. 9, no. 2, pp. 51–53, Feb. 1999.

75. J. D. Paul, C. Christopoulos, and D. W. P. Thomas, "Generalised material models in TLM— Part 1: Materials with frequency dependent properties," *IEEE Trans. Antennas Propagat.*, vol. 47, pp. 1528–1534, 1999.

76. J. D. Paul, C. Christopoulos, and D. W. P. Thomas, "Generalised material models in TLM— Part 1: Materials with anisotropic properties," *IEEE Trans. Antennas Propagat.*, vol. 47, pp. 1535–1542, 1999.

77. J. G. Maloney and G. Smith, "The use of surface impedance concepts in the finite-difference time-domain method," *IEEE Trans. Antennas Propagat.*, vol. 40, no. 1, pp. 38–48, Jan. 1992.

78. J. A. Roden and S. D. Gedney, "The efficient implementation of the surface impedance boundary condition in general curvilinear coordinates," *IEEE Trans. Microwave Theory Technol.*, vol. 47, pp. 1954–1963, 1999.

79. H. Wang, M. Xu, C. Wang, and T. Hubing, "Impedance boundary conditions in a hybrid FEM/MoM formulation," *IEEE Trans. Electromagn. Compat.*, vol. 45, no. 2, pp. 198–206, May 2003.

80. S. Kellali, B. Jecko, and A. Reineix, "Implementation of a surface impedance formalism at oblique incidence in FDTD method," *IEEE Trans. Electromagn. Compat.*, vol. 35, no. 3, pp. 347–356, Aug. 2003.

81. M. K. Kärkkäinen, "FDTD surface impedance model for coated conductors," *IEEE Trans. Electromagn. Compat.*, vol. 46, no. 2, pp. 222–233, May 2004, doi:10.1109/TEMC.2004.826891.

Authors Biographies

Nikolaos V. Kantartzis received the Diploma degree and Ph.D. degree in electrical and computer engineering from the Aristotle University of Thessaloniki (AUTH), Thessaloniki, Greece, in 1994 and 1999, respectively.

In 1999, he joined the Applied and Computational Electromagnetic Laboratory, Department of Electrical and Computer Engineering, AUTH, as a Postdoctoral Research Fellow. He coauthored *Higher-Order FDTD Schemes for Waveguide and Antenna Structures* (Morgan & Claypool Publishers, 2006). He has authored or coauthored several refereed journal papers in the area of electromagnetic compatibility, computational electromagnetics and especially higher order finite-difference time-domain methods, PMLs, and vector finite elements. His main research interests include electromagnetic compatibility modeling, time- and frequency-domain algorithms, double-negative metamaterials, waveguides, and antenna structures.

Theodoros D. Tsiboukis received his diploma degree in electrical and mechanical engineering from the National Technical University of Athens, Athens, Greece, in 1971, and his doctor of philosophy degree from the Aristotle University of Thessaloniki (AUTH), Thessaloniki, Greece, in 1981.

From 1981 to 1982, he was with the Electrical Engineering Department, University of Southampton, Southampton, UK, as a senior research fellow. Since 1982, he has been with the Department of Electrical and Computer Engineering (DECE), AUTH, where he is currently a professor. He has served in numerous administrative positions including director of the Division of Telecommunications, DECE (1993–1997), and chairman of DECE (1997–2001). He is also the head of the Advanced and Computational Electromagnetics Laboratory, DECE. He has authored or coauthored eight books and textbooks including *Higher-Order FDTD Schemes for Waveguide and Antenna Structures* (Morgan & Claypool Publishers, 2006). He has authored or coauthored over 135 refereed journal papers and more than 100 international conference papers. He was the guest editor of a special issue of the International Journal of Theoretical Electrotechnics (1996). His main research interests include electromagnetic-field analysis by energy methods, computational electromagnetics (finite-element method, boundary-element method, vector finite elements, method of moments, finite-difference time-domain method, alternating-direction implicit finite-difference time-domain method, integral equations, and absorbing boundary conditions), metamaterials, photonic crystals, and electromagnetic compatibility problems.

Prof. Tsiboukis is a member of various societies, associations, chambers, and institutions. He was the chairman of the local organizing committee of the 8th International Symposium on Theoretical Electrical Engineering (1995). He has been the recipient of several awards and distinctions.

Printed in the United States
by Baker & Taylor Publisher Services